Geochemical Methods of Prospecting for Non-Metallic Minerals

T0174108

GEOCHEMICAL METHODS OF PROSPECTING FOR NON-METALLIC MINERALS
(NEW AND EXPANDED EDITION)

*I.L. Komov, A.N. Lukashev
and A.V. Koplus*

CRC Press
Taylor & Francis Group
Boca Raton London New York

CRC Press is an imprint of the
Taylor & Francis Group, an **informa** business

First published 1994 by VSP

Published 2019 by CRC Press
Taylor & Francis Group
6000 Broken Sound Parkway NW, Suite 300
Boca Raton, FL 33487-2742

© 1994 by Taylor & Francis Group, LLC
CRC Press is an imprint of Taylor & Francis Group, an Informa business

First issued in paperback 2019

No claim to original U.S. Government works

ISBN-13: 978-0-367-44959-9 (pbk)
ISBN-13: 978-90-6764-179-1 (hbk)

Visit the Taylor & Francis Web site at
http://www.taylorandfrancis.com

and the CRC Press Web site at
http://www.crcpress.com

CIP-DATA KONINKLIJKE BIBLIOTHEEK, DEN HAAG

Komov, I.L.

Geochemical methods of prospecting for nonmetallic minerals / I.L. Komov, A.N. Lukashev and A.V. Koplus ; transl. from the Russian by V.I. Sverdlov. - Utrecht : VSP
Orig. publ.: Utrecht : VNU Science Press. 1987. - With ref.
ISBN 90-6764-179-0 bound
NUGI 817
Subject headings: non-metallic minerals ; geochemical prospecting

Typeset in Lithuania by TEV, Vilnius.

Contents

Preface

Geochemical methods of prospecting for and evaluation of minerals are applied widely today at all the stages of geological exploration. It so happened historically that geochemical methods were developed in Russia and Ukraine, primarily to prospect for oil and ore minerals. The indicators of the latter were elements constituting the main component of the ore proper, such as copper, zinc, tin, tungsten and molybdenum. Yet geochemical methods of prospecting for many classes of non-metallic minerals have not been elaborated. It was chiefly because of the insufficient sensitivity of analytical methods that the occurrence of exogenic and endogenic haloes around such types of deposit as rock crystal, Iceland spar, asbestos, sulphur, gem and semi-precious stones was questioned. The specific feature of many non-metallic minerals is that both the main components of mineralization and their accessories (admixture elements) are used when prospecting for the latter. Admixture elements occur around the productive bodies and deposits, the main elements being chlorine, fluorine, the alkaline elements and mercury.

Cases are known of the successful application of geochemical methods when prospecting for and exploring certain types of non-metallic minerals (such as salts, fluorite, boron, apatite and rock crystal) that contributed to the identification of a number of deposits of this type of raw material. However, when exploring non-metallic minerals, geochemical methods are used on quite a limited scale and frequently are not included into the overall complex of logical exploration. Geochemical methods are used very little in prospecting for deposits of talc, barite, magnesite, graphite, Iceland spar, asbestos, and precious and semi-precious stones. The small range of geochemical exploration in reconnaissance and prospecting surveys for non-metallic minerals is explained by the poor treatment of these problems in the literature. The basic data on geochemical methods of prospecting for non-metallic minerals are discussed mainly in articles published in journals, and there are no review works. Thus, there is a need for a generalized overview of the accumulated factual material on the geochemical methods of prospecting for non-metals, demonstrating certain specific features of applying these methods in the search for different types of deposits. This task seems imperative in view of the anticipated comprehensive increase of consumption and utilization of non-metallic minerals in the next few years.

Information on the types of raw materials is presented in the following sequence:

(1) general data (genetic types, conditions of formation, geological prospecting indications);

(2) indicator minerals and elements;

(3) geochemical methods of prospecting along dispersion trains and haloes, plus hydrogeochemical and geobotanical methods;

(4) primary endogenic haloes;

(5) vertical geochemical zonality; and

(6) methods, stages, and sequence of work.

Data on the investigations carried out by the authors (diamonds, coloured stones, fluorite, rock crystal, Iceland spar) and data from other researchers are discussed in the book. The section on fluorite has been written by A. V. Koplus, the sections on phlogopite and Iceland spar by A. N. Lukashev, the other chapters by I. L. Komov.

CHAPTER 1

Types of non-metallic minerals and general data on geochemical methods of prospecting

The group of non-metals, which is widespread amongst the variety of minerals, is of great economic significance. Non-metals are minerals which, as a rule, do not serve as a raw material for the extraction of metal. The division of solid minerals into ore and non-metallic minerals is quite conditional. Non-metallic minerals are used in the industry to extract perfect minerals (piezo-optic, abrasive, precious), mineral phases of a certain chemical composition (mining chemical raw material), or are used in total as a rock or aggregate. These deposits are characterized by various conditions of formation, and variety of composition. The industry currently processes new types of raw materials that were not considered useful in the past. Large-scale utilization of non-metallic minerals in industry is the reason for the comprehensive increase in their extraction and the sharp rise of their cost. Intensive work is being carried out in Russia and abroad, prospecting for and exploring non-metals. Considerable changes have occurred in the structure of their cost and extraction over the last 25–40 years. For example, building stone (dimension, monumental, crushed stone) is the most expensive non-metallic mineral in the USA (going by cost of raw material), while the raw material for cement is second, and sand, gravel and lime are third [19, 93]. A relative rise is observed in the cost of the extracted chemical raw material. The extraction of natural soda, vermiculite and perlite is increasing.

The deficit of metals and the necessity of their replacement by other materials, the sharp increase in the requirements for mineral fertilizers in agriculture, the expansion of construction, including highway and hydraulic engineering, and the demand for heat-insulating materials have advanced non-metallic raw materials to the front ranks in the mine industry. Scientific fundamentals of the laws governing the formation of deposits of non-metallic minerals, and methods for prospecting have been studied and developed. The works of P. M. Tatarinov, V. P. Petrov, I. I. Bock, A. A. Yakzhin, A. S. Sokolov, V. S. Sobolev, D. S. Korzhinsky, B. Ya. Merenkov, A. A. Ivanov and V. M. Borzunov are of particular interest.

There are various methods of classification of deposits of non-metallic minerals. Classification can be (a) genetic, (b) by the commercial element or group of elements, or (c) by the composition of the enclosing rocks.

Non-metallic minerals can be divided into magmatogene (endogenic), exogenic, and metamorphogenetic groups, based on conditions of formation. The first group includes the magmatic minerals proper (diamond, apatite, graphite, precious stones), pegmatitic minerals (muscovite, apatite, feldspar, rock crystal, quartz, fluorite, precious stones: topaz, beryl, tourmaline, garnet, amethyst, corundum, sapphire, ruby); carbonatitic minerals (apatite, phlogopite, chrysolite); skarn (boron, graphite, phlogopite, wollastonite, chrysotile asbestos, talc, witherite, cordierite, cyanite, andalusite, precious stones: spinel, clinohumite, forsterite, corundum, ruby, sapphire, lazurite); albitite–greisen (beryl, chrysoberyl, emerald, topaz, fluorite); and hydrothermal minerals (fluorite, chrysotile asbestos, Iceland spar, rock crystal, phlogopite, graphite, barite). The second group includes deposits that have formed as a result of weathering (apatite, vermiculite, kaolin, bentonite, barite, phosphorite, magnesite, talc, salts, chrysoprase, malachite, azurite, sulphur, gypsum, mineral pigments, quartz sand); placer (diamond, garnet, corundum, sapphire, spinel, barite); and sedimentary and volcanic-sedimentary minerals (boron, barite, phosphorite, salts, gravel, sand, clay, shale, limestone, chalk, dolomite, marl, cement raw material, gypsum, jasper, rhodonite, marls, diatomite, tripoli, gaize). The group of metamorphogenetic deposits includes metamorphic minerals proper (amphibole asbestos, graphite, corundum, emery, garnet, apatite, mica, marble, quartzite, roofing slates), and metamorphosed minerals (apatite, rhodonite).

The data indicate that the majority of non-metallic minerals are of polygenic origin. The classification of non-metallic minerals advanced by V. P. Petrov is the most convenient for rational use of geochemical methods [93].

The first group in the classification includes deposits whose use is conditioned by the perfection of the crystals, and comprises of carbon, silicon, calcium, magnesium, aluminium, potassium and beryllium. With the exception of beryllium, all these elements are widespread in the earth's crust and it is difficult to use them as indicators — the amounts of the elements in productive bodies differ little from their levels in the enclosing rocks. Accessory elements, primarily chlorine, fluorine, mercury and the alkali elements may be used as indicators in geochemical exploration. Evaluation of raw material quality (the presence of pure crystals without flaws, or fibres of a certain size) is also of major significance in reconnaissance prospecting work for non-metals.

The second group in Petrov's classification includes minerals where use is made of formations of a particular chemical composition, containing such elements as phosphorus, fluorine, boron, potassium, iodine and sulphur (mining chemical raw material). Elements constituting the major component of the ore are effective indicator elements for this type of raw material in the case of geochemical prospecting methods. This makes this group here like metalliferous minerals.

The third group includes aggregate minerals, whose use is conditioned by both chemical composition and physical properties, or only by the latter. In addition to monomineral formations (for example, magnesite, wollastonite, etc.), this group also includes polymineral formations (nephrite, igneous and metamorphic rocks). The majority of deposits in this group are characterized by being spread over a wide area and are prospected using conventional geological methods. Fluorine (vermiculite, wollastonite), caesium, rubidium (volcanic glass, perlite), and chlorine (nephrite) are indicators when applying geochemical methods of prospecting for deposits of this group. The chemical composition (absence of harmful admixtures) and physical properties (strength, occurrence of fissures) are of significance when evaluating the raw material. It is advisable to combine geophysical and geochemical methods when prospecting for aggregate raw materials.

Geochemical methods of prospecting for non-metallic minerals represent a part of the complex geological prospecting work applied for locating and evaluating the prospects of certain regions. They are based on investigating spatial laws governing the arrangement of concentrations of chemical elements. The theoretical fundamentals of geochemical methods of prospecting are embodied in the teaching of V. I. Vernadsky and A. E. Fersman on the migration, dispersion and concentration of chemical elements in the earth's crust. Other comprehensive contributions to the elaboration and application of geochemical methods have been made by A. P. Vinogradov, N. I. Safronov, E. A. Sergeyev, S. D. Miller, A. P. Solovov, A. A. Saukov, D. P. Malyuga, N. N. Sochevanov, V. I. Krasnikov, I. I. Ginsburg, S. V. Grigorian, E. M. Kvyatkovsky, V. V. Polikarpochkin, L. V. Tauson, L. N. Ovchinnikov, A. A. Beus, A. N. Yeremeyev, A. I. Perelman and M. G. Valyashko.

Geochemical laws are established on the basis of the study of the distribution of chemical elements in various types of rocks, soils, natural waters, and plants. The average contents of chemical elements, characterizing the composition of various rocks in the earth's crust (lithosphere), as well as the composition of other geospheres (hydrosphere, atmosphere), have been called clarkes, as suggested by A. E. Fersman. The specific features of distribution of chemical elements in rocks, which have not been subjected to the effect of any processes, when spatial changes in the quantitative characteristics are of systematic character, condition the local geochemical background. As a result of geological processes associated with magmatic, hydrothermal and other activities, elements are redistributed in local zones of rocks that were subjected to the effect of solutions accompanied by the introduction and removal of chemical elements. Endogenic geochemical anomalies form in local zones with a distinctly different content of elements as compared to the geochemical background. The quantitative indices of the content of indicator elements, in this case, markedly differ from the background ones. The areas of loose formations and soils with an anomalous distribution of chemical elements as compared to the geochemical background form exogenic geochemical anomalies. The characteristic of the substance composition of the deposits, i.e. totality of elements that are present in the productive zone, constitutes its geochemical spectrum.

Geochemical methods of prospecting are based on detecting anomalies that are associated with the primary concentration or secondary dispersion of chemical elements by special sampling of natural formations. The following problems are solved in this case:

(1) identification of geological formations, within which the discovery of deposits is possible;

(2) prospecting for concealed productive bodies whose geochemical anomalies show on the erosional surface and are identified in workings or boreholes.

Lithogeochemical, hydrogeochemical, biogeochemical and atmogeochemical methods of prospecting are used, depending on the type of sampled material and the objective of the exploration. The first ones are based on identifying secondary haloes and dispersion trains, as well as primary haloes of deposits. Secondary dispersion haloes (hypergene lithochemical anomalies) are formed in surface formation (soils, eluvium and deluvium) during hypergene alteration of the deposits. Dispersion trains represent an area of increased content of chemical elements in the sediments of a temporary or permanent drainage system, which has formed as a result of hypergene destruction of deposits and their primary haloes. Secondary dispersion haloes are divided into mechanical and salt haloes by their phase status. In the first case, the mineral phases are present in a solid state in the form of stable primary or secondary minerals in the zone of hypergenesis that migrate mechanically. The elements in the salt halo are in the form of dissolved compounds. There are residual dispersion haloes (i.e. haloes that form in the zone of hypergenesis of a previously existing ore body or its primary halo) and superimposed haloes, wherein primary mineralization did not exist before the development of secondary process of dispersion [46]. Exposed (outcropping to the surface) and closed or buried dispersion haloes (at a certain depth from the surface in the representative horizon) are differentiated depending on the potentialities of sampling. Buried haloes are detected by drilling shallow holes to the level of the representative horizon. The following classification of secondary dispersion haloes has been advanced [46]:

(1) Residual exposed, fan-shaped haloes, forming in intensively denuded areas in modern eluvial–deluvial formations.

(2) Superimposed, exposed, diffusion haloes, that are characteristic of conditions in accumulation–denudation plains in arid and moderately humid areas.

(3) Superimposed, exposed, accumulation haloes, associated with the processes of salt accumulation of elements.

(4) Close to type III, but forming in the absence of an allochthonous sheet and displaced toward the presently existing run-off.

(5) Leached and extremely depleted haloes, originating in eluvial–deluvial deposits in humid areas.

(6) Buried residual haloes, characteristic of closed areas of two-stage structure.

(7) Buried superimposed haloes.

Dispersion trains are differentiated into mechanical and salt trains, based on the phase status of the migrating components. They are also differentiated into exposed and buried trains by their potential for detection. Most elements form mechanical dispersion trains in the open drainage system of actively denuded areas of deposits. Salt dispersion trains are characteristic of the areas with leached surface secondary haloes and concealed productive bodies. The productivity index, which is directly proportional to the content of the element and the halo dimensions, is used to characterize haloes and dispersion trains.

The primary geochemical halo of an element is a circumore space, which is enriched or impoverished in this element as a result of its redistribution during the formation of deposits [11]. The multicomponent composition of the chemical elements in the primary haloes is conditioned by the character of the enclosing rocks, the types of non-metal and the conditions of their formation. Isometric (gently pitching lodes) and symmetrical (steeply dipping zones) haloes develop around the deposits. The major specific features of primary haloes are their dimensions, exceeding considerably those of the productive deposits around which they have developed, and the zonality of the structure. Geochemical haloes of maximum dimensions, composition and intensity correspond to large deposits. This regularity makes it possible to predict the scale of mineralization. Geochemical zonality of haloes is displayed most clearly in the change of the ratio of the sum (or products) of the elemental content in the supra-ore parts of the halo to the content of the sub-ore parts. It is most advisable to use the zonality index of indicator element, i.e. the ratio of the productivity of a halo formed by the given element to the sum of the productivities of the haloes of all indicator elements, with the aim of comparative assessment of the zonality of haloes.

Hydrogeochemical methods of prospecting are applied to detect and contour anomalies in subsurface and surface waters. The essence of biogeochemical methods of prospecting consists in exploring the distribution of chemical elements in herbaceous and arboreal vegetation, and peat. The complexity of sampling and the dependence of the element content on the season and species of the plants restrain the application of this method. The atmospheric geochemical (gas) method of prospecting is applied to explore and identify anomalies in the subsurface air and in the near-surface atmosphere, associated with productive deposits. Haloes of helium (a stable product of radioactive decay), mercury (in overlapping deposits), and radon are detected when prospecting for minerals.

Dispersion haloes of solid mineral phases or their associations, contained in non-metallic minerals, are detected in loose deposits. The heavy concentrate and heavy concentrate-decreptophonic (steaming halo) methods of prospecting are applied to detect and contour haloes and dispersion trains of small mineral fractions in alluvial and proluvial deposits. Steaming haloes, represented by secondary gaseous–liquid inclusions, accompany various non-metallic minerals (mica in pegmatites, rock-crystal-bearing veins, fluorite). Steaming haloes are detected by the heavy concentrate, which is washed out of the alluvial and deluvial deposits, or from pounded rocks. The light fraction of the heavy concentrate is subjected to decreptophonic investigation, consisting in counting the total number of ruptures (pulses) in the gaseous–liquid inclusions when samples are heated. Comparative evaluation of detected decrepit anomalies is performed by calculating the intensity of steaming (in conditional units), related to the given cross-section or area. A decrepit activity of samples that exceeds the background a dozen times indicates the proximity of ore deposits.

Heavy concentrate geochemical sampling of alluvial rocks and bedrocks with the aim of identifying secondary and primary haloes consists in exploring the distribution of minerals and chemical elements in heavy concentrates. Analysis of the mineral composition of washed heavy concentrates and their specific geochemical features assists in grading anomalies, determining the character of the main source of removal, which also enhances the contrast of the haloes. How the problem of deciding what fraction should be analysed is solved depends on the type of the deposit and its potential mineral composition. Heavy concentrate surveying has great potential because it explores a concrete mineral form of elements, and enhances the representativeness of sampling. The results of geochemical analysis of heavy concentrates are presented in the form of heavy concentrate geochemical maps, plans and horizon cross-sections.

1.1. GENERAL PRINCIPLES OF GEOCHEMICAL MIGRATION

The geochemical field is understood as a geological space with close levels and dispersions of concentrations of chemical elements in mineral associations. The geochemical fields of the earth can be divided into fields of dispersion and concentration. Geochemical concentration fields differ in the conditions governing the formation of those mineral associations within which they develop and also in concentration levels.

The degree and character of the concentration of matter in geochemical fields is determined by concentration coefficients (CC) and the dispersion of the material conditions. According to CC values geochemical fields have bodies of high (commercial) (100–1000), medium (10–20) and low (1–10) concentration.

The elements are not evenly distributed in the rocks, and since the residual soil concentration reflects that of the parent rock, the latter has a controlling influence over the soil content. This influence decreases with increasing maturity of the soil. Trace element contents in juvenile (young) soils closely reflect the content of the parent rock.

In a residual soil, geochemical anomalies are characteristically formed directly over mineralization (although later modification by downslope creep may occur) during the normal processes of soil formation. That is, as the bedrock weathers and the soil forms from the weathering bedrock, the element from the mineralization, like the other components, is incorporated in the soil horizons. This anomaly is consequently derived *in situ* by normal weathering processes involving both mechanical and chemical modification. Such anomalies are characterized by relatively strong metal-bonding in contrast to hydromorphic anomalies. In a residual environment, the anomaly may be further modified by down-slope creep. This is a purely mechanical function and does not chemically alter metal-bonding.

Chemical weathering involves a process of solution which is influenced by such factors as ionic potential, hydrogen ion concentration (pH), redox potential (Eh), colloidal conditions, and microorganic and biological activity.

Originally, the size of the anomaly may be roughly the dimensions of the mineralization, with lateral spreading due to natural causes such as slumping and spreading during the normal process of rock weathering and soil compaction acting on the anomaly until it may be several times larger than the bedrock expression.

In order to keep the idealized models as simple as possible, the same legend is used for all models. Space does not allow the inclusion of this legend in each diagram; hence, it is only given in full in the caption of Fig. (1.1).

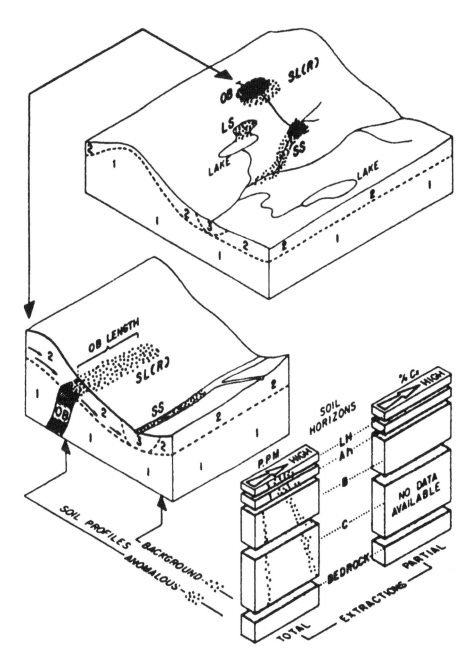

Figure 1.1. Model A1 (Cordillera). Idealized models for geochemical dispersion of mobile elements in well-drained residual soil. [14] *Anomaly types*: SL(R) = residual soil anomaly, SL(M) = mechanically smeared soil anomaly (by glacial action), SS = stream sediment anomaly, LS = lake sediment anomaly, SP = seepage anomaly, BG = bog anomaly. *Overburden types*: 1 = bedrock, 2 = residual soil, 3 = recent alluvium, 4 = till, 5 = over-burden of remote origin. *Others*: OB = orebody, ::: = the density of dots indicates anomaly strength.

1.2. ORIENTATION STUDIES

The geochemical factors which influence the dispersion of the elements vary considerably both in time and space. Each field area must be considered a new problem and an orientation study made to find out the nature of the local factors which have contributed to dispersion in the area.

The orientation programme should be carried out over an area known to contain mineralization of a type similar to that sought, which has not had the natural geochemical patterns disturbed or contaminated. It should include an appraisal of the geology, mineralization, geomorphology and climate, in relation to the existing geochemical patterns.

The geochemical samples collected at the earth's surface are all strongly affected by the processes of weathering. These processes also control secondary dispersion in drainage systems, soils, and transported material. Consequently, factors affecting weathering should be carefully studied during an orientation survey. These factors include the resistance of the primary rock-forming minerals, and the physical features of rainfall, temperature, relief, and drainage.

Table (1.1) gives the relative stabilities of the more common rock-forming minerals in the zone of weathering. In general, the order of resistance to weathering of the rock-forming mineral group is: oxides > silicates > carbonates and sulphides. Thus, knowledge of the geochemical behavior of the elements present is very important.

Geological factors, such as structure and rock type, are very important in an exploration survey.

Stream sediment sampling is the principal method of regional geochemical exploration. Depending on a specific environment it could be combined with sampling of loose deposits, hydrogeochemistry and biogeochemistry. Combination with heavy mineral concentrate techniques is recommended in some special cases (exploration for diamond). The use of remote sensing and airborne geophysics information is very advantageous. Medium-scale systematic geochemical exploration requires a combination of stream sediment and soil sampling techniques with the above-mentioned auxiliary methods if necessary.

Soil sampling is a leading technique during detailed geochemical exploration. Close combination with ground geophysics is essential. Biogeochemistry could be used as an auxiliary technique in some special environments. Petting, trenching and preliminary drilling could be necessary in order to study revealed mineralization.

Lithochemical methods of prospecting based on elements with high dispersion capacity and capable of forming extensive aureoles, have been proved to be very promising in the location of deep-seated deposits. These methods are based on the wide participation of these elements in endogenic formation, as well as their great mobility in endogenic conditions. These elements usually occur on the outer zones of geochemical aureoles of deposits.

The typification of the natural conditions of prospecting work is based on the study of the particular formation and classification of secondary haloes in different lithogenetic types of rocks and systems of geochemical barriers and relics.

Table 1.1.

Chemical stability of minerals in the zone of weathering [2]. (Rock-forming minerals followed by common accessories columnised to the left; ore minerals to the right)

	Mineral	Major elements	Trace elements	Ore minerals	Major elements	Trace elements
Very stable	Quartz	Si	Ge, (Li, Al, Fe, Mg, Alk).	Gold	Au	Te, Ag, Cu, Fe, Pt, Hg.
	Zircon	Zr, Si.	Nb, Ta, Hf, Y, U, Th, Pb.	Platinum	Pt Metals.	Au, Ag, Cu, Fe, Ni, Cr, Sb, Te, As, Hg.
	Corundum	Al	Mg, Fe, Ti, Cr, (Ni, V).	Rutile	Ti	Nb, Ta, Fe, (Sn, Cr, V).
	Spinel series	Mg, Fe, (Mn, Zn), Al.	Ti, Cr, (Sn).	Chromite series	Cr, (Mg, Fe), Al.	Ti, Mn, Zn, Pb, Ca.
	Topaz	Al, F, Si.	Fe			
	Tourmaline	Na, Mg, Fe, Al, Si,B.	Li, F, Ca, Cr, Ga, Sc, Sn, Cu.			
Stable	Muscovite	K, Al, Si.	Cr, F, Rb, Sr, Ba, Ga, V, Cs, Li, Te, Na, Ca, Mn.	Columbite-tantalite	Nb, Ta, (Fe, Mn).	RE, Ti, Zr, Sb, Bi.
	Alkali feldspar	K, Al, Si, (Na, Ca).	Ba, Rb, Sr, Ti, (Fe, Mn, Mg), Cu, Ga, Li, Tl, Cs.	Galena	Pb, S.	Bi, Sb, As, Ag, Se, Te, Tl, (Pt, Pd, Mo, Ni, Hg).
	Sodic plagioclase	Na, Al, Si, (Ca).	Ba, Sr, Rb, Ti, Ga, Cu, Fe, Mn.	Barites	Ba, (Sr), S.	Pb, Ca, (Hg, Co).
	Garnets	Ca, Mg, Fe, Al, Si.	Mn, Cr, Ga, Y, Ti.	Ilmenite	Fe, Ti, (Mg, Mn).	V, Cr, Zn, Ni, Co.
	Andalusite	Al, Si.	— (Na, K).	Cassiterite	Sn	Ta, Nb, Fe, Ti, Mn, In, Ag, Pt, (Sc).
	Kyanite	Al, Si.	— (Cr, Na).	Monazite	Th, La, Ce, P.	RE, U, Al, Fe, Ca, Si.
	Sillimanite	Al, Si.	— (Ca, Na, K).	Magnetite	Fe, (Mg, Ti).	Cr, Ni, Co, V, Mn, Zn.

Major elements in brackets are subsidiary; trace elements in brackets are rarely present; RE = Rare earths. Alk = Alkali elements.

Table 1.1.
(Continued)

	Mineral	Major elements	Trace elements	Ore minerals	Major elements	Trace elements
Fairly stable	Amphiboles	Ca, Mg, Fe, Al, Si.	Cr, Mn, Ti, Li, Zn, Sr, Ba, F.	Wolframite	W, Fe, Mn.	RE, Nb, Ta, Sc, Zr, Hf, In, Sn.
	Pyroxenes	Ca, Mg, Fe, Si; (Na, Li, Al).	Mn, Ni, Cr, Ti, V, (Co, Pb, Zr, Ce, Sc, Ba).	Haematite	Fe	Mn, Ti.
	Apatite	Ca, P, F, Cl.	RE, Sr, Mn, U, Th, (Pb).	Scheelite	Ca, W.	RE, Nb.
	Sphene	Ca, Ti, Si, F.	RE, Sr, Ba, Th, Sn, Nb, Ta, V.			
	Staurolite	Fe, Mg, Al, Si.	— (Zn, Co, Ni).			
	Chloritoid	Fe, Mg, Mn, Al, Si.	Ti, Y.			
	Epidote group	Ca, Fe, Mn, Al, Si.	RE, Y, Th, U, Nb, V, Sn, Zn, (Cr, Pb, Sr).			
Unstable (easily weathered)	Calcic Plagioclase	Ca, (Na), Al, Si.	Sr, Mg, Ba, Ga, Cu, Ti, Mn.	Pyrrhotite-pentlandite	Fe, Ni, S.	As, Sb, Co, Mn, (Cu).
	Feldspathoids	Na, K, Al, Si, (Ca).	Mg, Mn, Ti, F, S, (Mo, Be).	Sphalerite	Zn, S, (Fe).	Cd, Mn, As, Ga, In, Ge, Co, Hg.
	Alkalic Amphiboles	Na, Mg, Fe, Al, Si, (Ca).	(Ti, Mn, Li, Zn, F).	Chalcopyrite	Cu, Fe, S.	Au, Ag, Co, Ni, Mn, Pt, Se, In.

Major elements in brackets are subsidiary; trace elements in brackets are rarely present; RE = Rare earths. Alk = Alkali elements.

Table 1.1.
(Continued)

Mineral	Major elements	Trace elements	Ore minerals	Major elements	Trace elements
Augite	Ca, Mg, Al, Si, (Ti).	Cr, Ni, Co, V, Zn, Mn, Pb, Cu, Zr, Sc, Ga, Li, Sr, RE.	Arsenopyrite	As, Fe, S.	Au
Biotite	K, Mg, Fe, Al, Si.	Rb, Sr, Ba, Li, Cs, F, Na, K, Co, Ni, Sc, V, Zn, Cd, Cu, Ga, Mn.	Pyrite	Fe, S.	Ni, Co, (Au, Ag, Tl, V, Tl).
Hornblende Series	Na, Fe, Mg, Ca, Al, Si.	Mn, Li, Ga, Sc, Ni, Co, V, Sr, Rb, Cu, Cr, Zr, Zn, Pb, Ba, RE.	Molybdenite	Mo, S.	Ag, Au, Re, Pt.
Chlorite	Mg, Fe, Al, Si.	Cr, Ti, Mn, Ni.	Fluorite	Ca, F.	RE, Sr.
Olivine	Mg, Fe, Si.	Cr, Ni, Co, Zn, Ti, Cu, Mo, Mn, Li, Ge.			
Glauconite	K, (Na, Ca), Fe, Mg, Al, Si.	Tl			
Calcite	Ca, C, (Mg, Mn).	Fe, Ba, Sr, Co, Zn.			
Dolomite	Ca, Mg, C, (Mn).	Fe, Pb, Zn, Co, Ba.			
Gypsum	Ca, S.	Mg, Si, Fe, Ti, Sr, B, Cu, Al, Mn, (Li, Be, Y, Zr, Bi, Ba, Au, Ca).			
Halite	Na, Cl.	Br, Cu.			

Major elements in brackets are subsidiary; trace elements in brackets are rarely present; RE = Rare earths. Alk = Alkali elements.

Geochemical exploration by secondary aureoles — 80 mech fraction — has been applied in humid tropical regions. When planning prospecting by secondary aureoles and dispersion trains and experimental work for the choice of sampling fraction it is necessary to take into consideration [15]:

(1) targets of prospecting in close relation to the scale of operations (detailed, regional, etc.);

(2) genetic types of deposits expected to be discovered and the possible mineral forms of existence of ore-forming chemical elements and indicator-elements of the mineralization;

(3) the character of the hypergene migration of chemical elements and of the creation of secondary aureoles and dispersion trains in specific landscape-geochemical conditions, including the character of the leaching and dilution of appropriate anomalies in paleo- and recent conditions;

(4) the type of geochemical sample analyses scheduled to be used in the process of the operation.

In non-metallic deposits the location and study of mineralization in desert conditions can be carried out successfully using geochemical methods of prospecting along with other methods.

The outcome of lithochemical shots by secondary dispersion haloes depends on the choice of fractions and the depth of material selected for testing. The secondary haloes are 1.5–5 times bigger than the primary haloes. In fractions of $-1 + 0.5$ mm, characterizing mainly aeolian sand, the minimum concentration of elements is noted in all cases. The representative fraction for F, Cl, K, Na, Hg and As is 1–8 mm with an optimal depth of 0.25–0.3 mm. On plots with a good exposure of rock and mineralization zones, a fraction with less than 1 mm is representative and this makes shots by primary geochemical haloes advisable. The maximum development of haloes is noted in the crossing nodes of structures and in magnetoactive zones.

Hydrogeochemical reconnaissance studies on the river and stream drainage system are carried out to obtain the ore characteristics of large territories, provinces and regions. Low density sampling of river waters and sediments is used (1–2 sites per 100 km^2). Usually streams 1–50 km long are studied.

The most reasonable approach is multi-element analysis giving an adequate metallogenetic picture of the territory and outlining areas containing various types of mineralization. In water, steadily migrating elements such as F, B, Cl, S, As and Mo are analysed; in sediments such elements as Pb, Zn and Sn are determined, anomalous contents of which are found mainly in alluvium. F, B, S and As form large anomalous areas of tens and even hundreds of thousands of square kilometres, probably resulting from their enrichment in a huge volume of rock.

The hydrogeochemical method is useful for the systematic sampling of surface waters. The existence of a hidden deposit must be fixed by variation in different chemical indices (pH, mineralization level, water type, SO_4^{-2} concentrations, etc.) compared with the background and by the increase in concentrations of ore elements and their associated minerals.

In terms of hydro- and biogeochemical studies the contrasting dispersion aureoles of F, P and Cu have been revealed on deposits and in promising areas.

A degree of coincidence is observed between the contours of primary and secondary aureoles and the hydrogeochemical contour. In this case the coincidence of the contours is in a direct correlation with the relief forms: with in the limits of a relatively smoothed relief form these is a full coincidence, and on steep slopes there is a displacement of up to several hundreds of metres from the centre of the primary aureoles.

Lakes may provide some information about the metallogeny of the drainage system. Investigations carried out in areas covered with loose permafrost deposits have shown that lake waters and deposits overlying ore bodies contain anomalous concentrations of chemical elements. This is explained by migration, caused by electrochemical processes taking place in interstitial waters.

The efficiency of the application of hydrochemical prospecting to hydrothermal deposits in folded mountain regions is determined, first and foremost, by the possibility of confining the exploration areas to such geological and structural features as are favourable for the formation of hydrochemical anomalies.

As crust weathering and biological landscape productivity are decisive in the formation of hydrogeochemical anomalies in folded mountainous regions, so there are optimal conditions for the effective use of hydrochemical prospecting in the taiga landscapes with a medium mountain structure, where the weathering crust has sufficiently high permeability and moderate evaporation restrains the quality of false hydrochemical anomalies forming on account of the evaporation concentration of ore elements.

A technique of hydrogeochemical prospecting and surveying at scales of $1:50\,000$ with sampling from springs and small streams has been carried out within a region of known ore manifestation and outside it. The sampling network is of $2.5-4$ points per 1 km^2. A broad range of chemical elements in water including macrocomponents and microcomponents, gases, pH, Eh, live and dissolved organic matter have been analysed.

The method was inferred to be effective. The type and areal distribution of hydrogeological anomalies depend on the geochemical environment, the intensity of water exchange and the position of the ore body or veins in the area or in the section.

Hydrogeochemical exploration on closed areas can be used to discover hidden ore bodies of fluorite at depths ranging from tens of metres to $200-300$ m. The most indicative in this respect is the exploration of fluorite deposits based on the fluorine content in natural waters.

In regional hydrogeochemical investigations, besides the main task of estimating the prospects for detecting concealed endogenic mineralization, the question of mapping abyssal fault tectonics has also been solved.

The distinctive diagnostic characteristics of tectonically activated zones are the general presence of a high mercury content in subsurface waters and the relatively frequent increased content of rare alkali elements. Some differences have been found in the macrocomponent composition of waters.

The mobility of elements in water depends upon the properties of the elements and the particulars of the geochemical environment, the parameters of which are determined by such factors as gas composition, the biological productivity of the landscape, the trend in the transformations of organic substances, etc.

Each type of non-metallic deposit is characterized by a specific chemical composition of the ground waters reflecting the paragenetic association of the mineral and forming elements. The degree of manifestation of this association in ground waters is defined by the actual geological and hydrogeological conditions and the migration ability of the chemical elements (microelements).

Data available in Russia and Canada show the high effectiveness of hydrogeochemical prospecting in the permafrost areas. The method is of special value here in view of such peculiarities as the vast forest cover, the great thickness of loose deposits and abundant swamps and rough slipped rocks.

The most favourable areas for hydrogeochemical prospecting are areas with island and discontinuous permafrost where deep hydrogeochemical investigations are possible. Chemical investigations of ice layers, especially those formed by subpermafrost waters, make it possible to obtain geochemical information on deep levels of the Earth's crust and to evaluate their ore-bearing potential.

Hydrochemical data are commonly presented as the element content in water relative to its total salinity (dissolved salts); this makes it possible to compare the data obtained in different geological and geographical conditions.

The problem of the geochemical study of anthropogenic dispersion flows in surface water streams and their comparison with dispersion flows of deposits has only appeared recently.

Ore and anthropogenic sources of dispersion flows differ according to the means of supply and the forms of the chemical elements. For ore flows the main supply is related to the activity of slope processes and ground water discharge and depends on the meteohydrological environment. Apart from surface run-off the year round, the direct discharge of contaminated substances in to the water flow plays a considerable part in the formation of anthropogenic flows. It is believed that the soluble forms of elements are essential for ore flows and their supply is relatively even in time.

The associations of chemical elements in ore dispersion flows are well correlated with the mineralogical and geochemical features of deposits. The qualitative composition of the association in the solid and water phase ingredients of flows is stable over considerable areas and corresponds to the metallogenic features of the latter. The accumulation of a very wide range of chemical element complexes is typical for the stream sediments of anthropogenic dispersion flows. Elements not present together or associated in ore flows are encountered very rarely in these complexes. Associations of non-ferrous, rare and trace elements, among which Ag, Hg and Cd are concentrated most intensively and Pb, Ni, Cu, Zn, Sn, Sr, Bi and P are present almost constantly, are often found in dispersion flows in industrial areas. The rate of their accumulation is $(n \times 10)-(n \times 100)$, which is considerably higher than in ore flows. In dispersion flows in agricultural areas Hg, Cu, P and a large group of metals (Y, Yb, Sc, Be, Zr, Nb), sometimes Ag, Co and Mo are commonly associated.

Azonal types of waters with high mineralization (nitrate, sulphate-calcium, chloride, etc. with high contents of phosphorus, ammonia compounds, reduced form of iron) which often have no natural analogues, are found in the water-soluble parts of flows. The zonal or almost zonal hydrochemical type of water is characteristic for ore flows. Azonal waters (sulphate and others) are only encountered near ore deposits. A contrasting change in water type in area is a characteristic feature of an anthropogenic dispersion flow.

The distribution of chemical elements transported in soluble form in ore flows is usually regular and decreases comparatively smoothly down stream from the source of the chemical elements.

Anthropogenic flow anomalies of soluble forms are unstable in time, space and according to the level of concentration. Anthropogenic flows in stream sediments have a sharply varying distribution downstream. Ore flows in stream sediments are characterized by a smoother distribution of elements. Data on the comparison of ore and anthropogenic dispersion flows are necessary in order to interpret the anomalies outlined by prospecting for mineral deposits.

This technique appears to be the most effective when applied in biogeochemistry. The content of elements within plants directly depends upon the amount of these elements in the water flowing through fractures in the ground.

Biogeochemical prospecting was carried out on aeolian sediment areas. Barium ores were discovered by the biogeochemical method.

Geobotanical methods used in reconnaissance work involve the recognition of indicator species. These are plants and trees which are specific for a certain element, that is, an indicator species is only found where the element concerned occurs in anomalous concentrations. The technique is economical since observations covering large areas can be made quickly, and no samples or analyses are required. However, a field man skilled in the recognition of the indicator species is required.

Abnormal concentrations of elements in the soils can cause poisoning in vegetation, resulting in stunted or abnormal growth. This feature has also been used as a guide to ore.

Biogeochemical methods are used more for detailed surveys, and are used in conjunction with or separately from geobotanical techniques.

The concentration of elements in plants is dependent on a variety of physical, chemical, and organic factors which control the intake of nutritional solutions. Consequently, biogeochemical anomalies are much more heterogeneous than soil anomalies. However, geobotanical surveys require no sample collection at all. The two combined are a very useful tool, especially in areas where soil is absent or unobtainable due to snow cover or permafrost conditions.

Care should be taken to collect all biogeochemical samples during the same season, since seasonal variations in concentration occur within plant tissues. This is in response to the nutritional requirements of the plant, which will be high in the growing season and low at other times. The intake of elements for different species from the same area will vary considerably, so that it is essential to sample only individuals of the same species, or even sub-species.

Table (1.2) illustrates the differential and selective absorption of elements by and between plant species, compared with the concentration in the bedrock and derived (residual) soil horizons. Note the selective concentration of Ba, P and Zn and the lack of concentration of Cr and Ti in the plant tissues. The low juniper absorbs Ni and Cr, whereas the species do not. Elements are concentrated differentially in the tissues of the various organs of a plant, so that the same part of each individual must be sampled. Note also the element differentiation between soil horizons and the relation to the bedrock content.

Plants accumulate the volatile forms of chemical elements with the most intensity. They accumulate them from the liquid phase and from the water-soluble forms of the solid phase approximately 100 times less intensively and from the solid phase of soils and rocks in the root-inhabited zone approximately 300 000 times less intensively. The majority of chemical elements are accumulated by the thin absorbing roots of the lower parts of the root system. Directly proportional dependences of the contents of the accessible forms of chemical elements in the native rocks and in the plants occur for a small quantity (5 per cent on the average) of non-barrier plant bioobjects [63]. The contents of some volatile elements and metals in such bioobjects are a good reflection of their quantity in the native rocks including the magmatic systems. The real possibility of biogeochemical sampling from planned squares of up to 1 hectare and more stipulates the acquisition of averaged geochemical characteristics of rocks which are not exposed on the surface.

It is most advisable to use biogeochemistry for mapping and investigating the heterogeneities in magmatic and ore-magmatic systems: contacts of rocks, xenoliths, dykes and zones of tectonic strikes which are distinguished by the content or accessibility for plants of even one of the 30–70 determined chemical elements. It

Table 1.2.

Selective absorption of elements by plants compared with their concentration in soils and bedrock [2]. Samples from Eagle Bluff, Alaska

Element	Soils (ppm dry wt)			Rock (ppm)		Vascular plants (ppm ash)			
	Horizon			(C–D) Basaltic greenstone	Aspen stems	White birch stems	White spruce stems and leaves	Low juniper stems and leaves	
	A	B	C						
Ba	1 000	2 000	1 000	200	2 000	10 000	7 000	1 000	
Cr	70	100	70	150	< 10	< 10	< 10	50	
Cu	40	30	40	150	100	150	30	50	
Mn	500	500	150	1 000	1 000	3 000	700	150	
Ni	75	50	50	100	10	50	10	150	
P	60	80	80	80	18 000	24 000	24 000	12 000	
Sr	150	500	100	100	1 000	1 000	1 500	1 000	
Ti	3 000	3 000	3 000	7 000	50	1 000	150	1 000	
V	100	100	100	300	15	15	15	30	
Zn	50	50	25	50	1 200	6 000	1 000	20	

Data taken from Shacklette 1966, p. F13. [156].

is especially advisable to use biogeochemistry to reveal, contour and map separate ore-bearing stockwerks, zones and ore bodies and also to study the complicated inner structure of ore fields and ore-bearing structures, using the majority of metals and also the volatile forms of mercury and fluorine.

Biogeochemical anomalies 1–10 m wide are usually more local than biogeophysical anomalies 20–100 m wide; geophysical anomalies in the electrical resistance of rocks have the greatest width of 100–500 m. Biogeophysical ore-resonance anomalies usually correspond to the thickness of some zonal biogeochemical anomalies.

The factual data obtained and also the results of mapping old excavations (archeological method) make it possible to recommend an original complex of exploration and prospecting operations in covered areas. The selection of locations is made on the basis of geological, geophysical, soil-geochemical, hydrogeochemical and archeological investigations carried out earlier. Biogeophysical anomalies of the probable species and geochemical types of mineral deposits are revealed and mapped by automobile and then pedestrian surveys with an ore resonator. More detailed biogeochemical exploration and prospecting using non-barrier or practically non-barrier bio-objects of widespread plants are conducted in the most promising biogeophysical anomalies. In these operations it is recommended that a crossed system of profiles with a sampling interval of 2–10 m should be used when searching for veined ore bodies, and an interval of 5–30 m when searching for stockwork ore bodies. In profile intervals with concentrations of old excavations this interval is diminished to 1–3 m. Supposed ore biogeochemical anomalies are investigated by excavations, trenches and drill cores.

1.3. THE COMBINATION OF BIOGEOCHEMISTRY AND BIOGEOPHYSICS IN EXPLORATION AND PROSPECTING FOR HIDDEN AND BURIED ORE BODIES AND DEPOSITS

The biogeochemical or biolocational method named dowsing was used in many countries in antiquity and the Middle Ages. In our day it is used in many countries in exploration for hidden deposits of non-metallic minerals.

Our investigations in 1990–1993 in covered areas revealed that the majority of productive resonance biogeophysical anomalies with π-form metallic frames often coincide with biogeochemical anomalies of F, Hg and Cl in the corresponding non-barrier bio-objects of plants.

The development of geochemical prospecting methods by buried dispersion haloes has had special significance in recent times under the conditions of the exhaustion of easily detected outcropping deposits and for the economic and technical possibilities of increasing the exploration of deep-seated metalliferous deposits.

Alluvial dispersion haloes are characterized by the forms of the distribution of terrigenous, chemogenic and saline elements. The high content of P, Cl, F and U in carbonate and organic form is conditioned by calcium-sodium and calcium-magnesium types of water, a weak alkaline medium and probably by secondary mineralization.

Investigations have shown that the secondary epigenetic haloes, distinctly displayed in soils, are formed over deposits overlapped by allochthonous sediments of different lithological geochemical types and with a different hydrodynamic regime of the ground water (from flooded coarse clastic sediments to carbonate and gypsum-bearing clays).

These haloes are regular geochemical evidence of overlapped and concealed-overlapped deposits. They are observed with any thickness of the covering sedimentary mantle developed in the regions under investigation (up to 150 m).

There are two types of epigenetic secondary haloes [98]:

(1) a non-infiltration electrodiffusional type developing over concealed mineralization and displayed in soils with in the limits of the vertical projection of the productive zone on the surface (undisplaced halo);

(2) the infiltrational type, developing in the aquiferous horizon and displayed in soils at their exit areas (displaced halo).

In several regions ore-enclosing rocks are often overlapped by transported (allochthonous) young sediments — Quarternary glacial, Meso-Cenozoic alluvial lacustrine and lagoon-marine.

For the present, geochemical prospecting methods are used little in these conditions and predominantly as a type of prospecting for deep-seated mineralization (by investigation of drill cores of haloes, drilling down to the basement rocks).

Epigenetic haloes are characterized predominantly by a mobile mode of occurrence of the indicator elements (saline, sorptive and metalloorganic ones). The form of occurrence which makes it possible to reveal soil haloes depends upon the landscape-geochemical conditions.

The utilization of epigenetic haloes makes it possible to localize promising areas in covered regions, to organize drilling, to detect mineralization or primary haloes and to carry out geochemical interpretation and assessment.

During prospecting it is necessary to consider the landscape-geochemical features of the area, determining: 1) the effective mode of occurrence of the indicator elements; 2) the halo morphology in plan; 3) the possibility of detecting infiltration haloes in the soil.

1.4. GEOCHEMICAL PROSPECTING OF DEPOSITS IN THE MORAINE COVER

The complex and essentially allochthonous character of the cover creates a number of difficulties in carrying out lithochemical prospecting for deposits by secondary haloes and dispersion flows and makes it necessary to use special prospecting methods.

A technique of lithochemical surveying at scales of 1 : 50 000 with sampling from a depth of 0.5–0.8 m from the loose formations is proposed. Under the conditions of not too rough a terrain and marsh-ridden territory, it is recommended that, instead of the generally accepted sampling of bottom deposits (dispersion flows), sampling should be carried out along the contours of swamps and lakes [13, 14].

A lithochemical survey is accompanied by the sampling of heavy concentrates. The fine fraction of samples is subjected to spectral and special analysis. Detailed lithochemical surveys on a scale of 1 : 10 000 within promising fields are carried out with sampling based on a regular net work from the same representative horizons.

The shift in lithochemical anomalies from the basic sources is estimated at all stages of work using geological and geophysical data and by means of exploring the roundness of fine clastic material in the working sections.

The described technique is effective when the thickness of the allochthonous cover is comparatively thin ($\sim$ 10 m).

1.5. PRIMARY GEOCHEMICAL HALOES

The elaboration of geochemical criteria for the location and estimation of mineralization by primary aureoles is based on the identification of the indicator elements of various non-metallic types of deposits, the estimation of the productivity of the ore-bearing rocks, and the determination of their zonality indices.

The use of primary geochemical haloes is determined by important properties:

(1) primary haloes and ore mineralization have an identical genetic nature;

(2) primary haloes have a broader contour than ore bodies;

(3) chemical elements in ore bodies and in corresponding primary haloes are distributed zonally.

The investigation of hundreds of different deposits has shown that primary haloes extend up to 200–300 m beyond the geological and structural features of ore mineralization.

Stratiform deposits of hydrothermal origin are characterized by the concordant development of primary haloes. The latter are considerably bigger than the corresponding ore bodies and in the nature of their zonality are complete analogues of the primary geochemical haloes developed around deposits with transverse ore bodies.

Haloes, like ore deposits, develop exclusively within definite productive beds of ore-bearing rocks.

The anomalous area in every type of mineralization is characterized by the regular combination of primary haloes with zones of evacuated elements, which indicates their redistribution (under the ore formation of the element dispersed in the ore-bearing rock).

Primary geochemical haloes in deposits of sedimentary origin differ from endogenous ones by a more even distribution of the indicator elements within the productive beds. The regular distribution peculiarities of indicator elements in the near-ore space of stratiform deposits can be used to locate areas with highly concentrated ore elements (ore bodies) and thus provide a rational position for exploratory boreholes.

Primary haloes of leaching have great theoretical significance. With their help it is possible to determine the level of erosional shear, the prospects for mineralization and the location of ore bodies and deposits.

Bedrock sampling based on drill cores is the basic element of conclusive geochemical exploration directly aimed at the discovery and evaluation of ore bodies using the properties of primary geochemical haloes. The use of all available geophysical information, if required, combined with geophysical methods is very important for successful exploration.

The major objective of geochemical prospecting is to locate anomalies, associated with ore bodies in the heterogeneous geochemical field, representing a form of spatial distribution of elements. A productive geochemical anomaly is confined to favourable geological structures, unlike false anomalies, that are conditioned by random variations of the field's regional component. These structures are characterized by correlation of increased or decreased values are compared to the background ones at several (at least three) points on the profile or in adjacent prospecting profiles (sections). Anomalies resulting from the introduction or removal of elements are observed around ore bodies of non-metallic deposits. Anomalies resulting from the removal of elements are small in size, but their combination (commissure) with anomalies resulting from the introduction of elements, as a rule indicates the epicenters of ore bodies.

When classifying anomalies, it is necessary to consider their character, contrast, and composite nature. The anomaly contrast characterizes the ratio of the maximum or minimal content of chemical elements within the given anomaly to the background level. The mineralization ratio is the relation of the number of samples with anomalous content to the total number of samples taken in the sample interval. Additive (arithmetic mean summation of the content of elements in each sample that are normalized in the units of mean background content) and multiplicative (multiplication of contents) indices are used to enhance weak haloes and detect the epicentre of maximum mineralization. The quantitative characteristic of the primary haloes consists also in evaluating the clarkes, coefficients of concentration, and summary contents of elements in the geochemical cross-sections. The abundance ratio is the ratio of an element content in a concrete geological body to its clarke content. The coefficient of anomalousness K_a is the ratio of the content of a chemical element within a certain cross-section of the halo C_a to its background content: $K_a = C_a/C_b$. The coefficient of concentration K_c, which is the ratio of the element content in the circumore rocks (or ore bodies) to the lower anomalous value $K_c = C_x/C_a$, is used to compare the results of analysing samples taken within the ore zone. The summary content of metal M and the productivity of the halo are determined to calculate the relative introduction (or removal) of the elements in the circumore rocks. The summary content of the metal in a given cross-section of the halo (linear productivity) is determined as the sum of products of sampling steps $(l_1, l_2, \ldots, l_n)$ by the anomalous concentrations of the element $(C_1, C_2, \ldots, C_n)$ in the given profile on deducting the value of the 'local' background (in per cent) [46]:

$$M = l_1(C_1 - C_b) + l_2(C_2 - C_b) + \cdots + l_n(C_n - C_b)$$

or when sampling at uniform intervals

$$M = \Delta x \left(\sum_{i=1}^{n} C_i - nC_b \right),$$

where Δx is the sampling step, n is the number of sampling points, and C_b is the background content of the element. The dimension of the parameter is a metre-geobackground (mgb) if the content of the element is expressed in units of the geochemical background (coefficient of anomalousness).

These parameters make it possible to gain a general estimation of the introduction of ore- and halo-forming elements, and they can be used for evaluating the anomalies.

Geochemical zonality is one of the most outstanding features of the primary geochemical haloes of non-metallic deposits in general and hydrothermal and metamorphic ones in particular. The oxial geochemical zonality of the primary haloes of deposits is universal, that is to say there is an ordered distribution of element haloes up the dip of ore bodies from Ni at the bottom to Hg and Ba at the top. The established universal geochemical zonality of the primary haloes is expressed by the following sequential series of the most widespread indicator elements of the non-metallic deposits (up the dip of the ore bodies): Ni−Co−Cr−Cu−Ti −W−Sn−Mo−Bi−Zn−Pb−Ag−Sb−Ba−Hg−F−Cl−Br−I.

The universal geochemical zonality of the primary haloes of deposits is of great practical significance. It makes it possible to answer successfully the following very important questions in the assessment of geochemical anomalies:

(1) to distinguish geochemical anomalies represented by non-commercial zones of disseminated mineralization;

(2) to assess the level of erosional shear of geochemical anomalies;

(3) to evaluate the approximate reserves of blind and little-eroded mineralization forecast by geochemical data;

(4) to interpret the conditions governing the formation of the deposit.

Contemporary ideas about the migration of elements permit the existence of a wide range of complex compounds. The formation and stability of these combinations are determined by thermodynamic conditions and the composition of the solutions. The position of elements in haloes is correlated with the level of accumulation of the elements F, Cl, B and S.

Reliable geochemical criteria for zonation, vertical amplitude and the level of the erosional cut-off are the ratios K/Na, Li/Rb, Hg/Cu; F/Cl — in the liquid phase and CO_2/CH_4, and H/C/N/O in the gaseous phase of the ores and near-ore altered rocks. O, N, C is dominant on near surface deposits.

The complex coefficient $H = (V \times CO_2)/CH_4$ is most informative, where: V is the gas saturation of the rocks Sm^3/kg. On the upper horizons of the bodies of deposits the values of H range from 10 000 on the middle horizons to 1000 on the low horizons and 100 on the underlying horizons [5].

The primary zonation of the distribution of chemical elements is the most important criterion that enables us in geochemical prospecting to identify the level of the erosional shear of mineralization, to predict its depth and to distinguish supra-ore from sub-ore aureoles when prospecting for blind deposits. The problem of studying zonation includes two aspects. The first is a qualitative one, concerned with establishing the series in the succession of element deposition. The second

aspect is a quantitative one related to finding the geochemical indices of zonation
(ν) monotonously changing with depth, or in another direction, the dimensionless
relations between the average content of (productivities) elements.

Deposits display a great number of common indices of monotonous zonation
of the type $(Cl \times F \times Mg \times Ba)/(Zn \times Mo \times V \times Ni)$ exceeding the number of
occasional ones.

The geochemical indices of zonation ν established separately for each type of
deposit studied, provide an unequivocal evaluation of the erosional shear level.

Geochemical studies show that the dispersion patterns of selected elements in
soils are areally extensive and spatially related to the mineral deposits. Vertical
and lateral zoning is also evident among the elements. Depending on the depth
of non-metallic deposits the mineral zones are outlined by anomalous silver. The
depth of a deposit can be roughly estimated by studying the relative concentration
and distribution of silver. Computer-generated perspectives can economically and
rapidly portray a large mass of geochemical data.

The stability constant of the elements strictly determines its place in the zonal
series. This dependence of the precipitation sequence is so strong that the solution,
concurrent with supply, extracts unbalanced elements from the country rock along
the column and transports them up to their place in the series.

The zonality of haloes is determined by the different stability of the chloride
complexes, in the form of which metals are transported. The precipitant, sulphur
(reacting as hydrogen sulphide) is supplied in homogeneous deposits by solution
as sulphur-dioxide.

1.6. APPLICATION OF GEOCHEMICAL METHODS DEPENDING ON THE SCOPE AND OBJECTIVES OF EXPLORATION

The following stages of geochemical exploration dependent on the scales of geological surveys can be singled out:

(1) Regional geochemical exploration (scales $1:200\,000$).

Its main objective is to reveal the existence of geochemical anomalies in various geological substances (such as stream sediments, water, bedrock, etc.) and to
delineate the promising areas for the following detailed exploration).

Regions characterized by a particular type of mineralization occur throughout the world, and are termed Metallogenic Provinces. They vary in size from
relatively small areas to large regions of subcontinental size, and have usually developed under structural controls. An appraisal of the geology and mineralization
on a regional scale will help to delimit any such provinces which may heighten interest in the area. These provinces furnish target areas for reconnaissance surveys,
suggest the type of mineralization to be anticipated, and guide the organization of
an exploration programme.

(2) Medium-scale geochemical exploration (on a scale of $1:50\,000$) has the
main aim of adding detail to geochemical anomalies revealed during reconnaissance in order to make a more conclusive evaluation of their possible potential.

(3) Detailed geochemical exploration (scale 1 : 10 000 or larger). Its objectives include the detailed delineation and evaluation of revealed geochemical anomalies in order to justify follow-up drilling or other serious prospecting activity.

(4) Geochemical evaluation of exploratory drilling results. The main objectives of this conclusive stage are: a) to evaluate the potential of the flanks and deep-seated horizons of known deposits and b) to direct the detailed prospecting of primary geochemical haloes related to blind bodies.

1.7. ELEMENT ASSOCIATIONS

Certain elements tend to group together into mineral associations in a specific environment due to similar migrational capacities (relative mobilities). This capacity is dependent mainly on the valency and ionic radii of the atoms concerned, and to a lesser extent on other physical properties such as energy, positive potential, polarization, radioactivity, specific gravity, melting and boiling points, and mechanical hardness. Groupings of the commoner elements according to their valency and ionic radii are given in Fig. (1.2).

Elements with similar radii and valency tend to associate together in nature, as substitutions in the crystal structure of minerals are easiest between elements of these groupings.

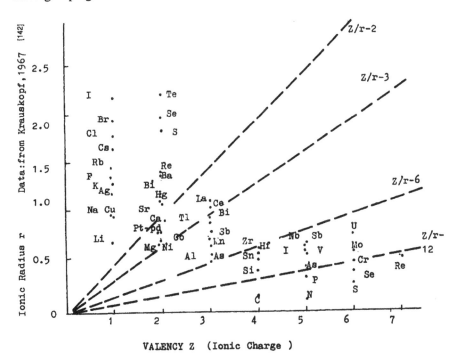

Figure 1.2. Groupings of the elements according to their valency and ionic radius [2].

Element associations are mainly based on similar relative mobilities under the conditions of formation. Consequently, they are susceptible to changes in the environment.

However, some elements retain their associations throughout a wide range of geological conditions. Thus, many elements remain in association throughout the hypogene cycle (igneous activity and metamorphism). However, such groups frequently become disassociated during the supergene cycle (weathering and sedimentation). Some geochemical associations are specific for a given set of environmental conditions. These include some sedimentary ores, apatite ores, and specific ores. The more common geochemical associations of the trace elements are given in Table (1.3).

Table 1.3.

Common geochemical associations of the trace elements [2]

Rock type	Association
1. Plutonic associations	
Ultramafic–alkaline rocks	Nb–Ta–Re.
Felsic–alkaline rocks	Ti–Nb–Ta–Re–Zr–U–Th–Mo.
Ultramafic rocks	Cr–Co–Ni–Cu.
Mafic rocks	Ti–V–Sc.
Granitic rocks	Ba–Li–Be–W–Mo–Sn–Nb–Ta– Sc–U–Hf–Zr–Ti–Sr.
Pegmatitic rocks	Be–Li–B–La–Rare earths–Nb–Ta– Th–U–Cs–Sc–Zr–Rb–Hf. Sn–Mo–Nb–W.
2. Hydrothermal sulfide ores	
General associations	Cu–Pb–Zn–Mo–Au–Ag–As–Hg– Sb–Se–Co–Ni–U–V–Bi–Cd.
Some specific association	Cu–Re–Mo. Hg–As–Sb–Se–Ag–Zn–Cd–Pb. Bi–Sb–As. Pb–Zn–Cd–Ba. Au–Ag–Cu–Co–As. Ni–Cu–Pt–Co. Au–Ag–Te–Hg.
3. Contact metamorphic rocks	W–Sn–Mo, (B–Li–Be).
4. Some sedimentary associations	
Black shales	Cu–Pb–Zn–Cd–Ag–Au–V–Mo–Ni–Cr– Nb–Ta–V–U–Rare earths–Sb–As–Co–Sb–Pt.
Phosphorite	V–Mo–Ni–Ag–Au–Nb–Rare earths–Zn–As–Pb.
Evaporites	Li–Ba–Cs–Sr–Rb–B–I.
Bauxite	Nb–Ga–Ti–Be.
Laterite	Ni–Cr–V.
Manganese oxides	Co–Ni–Mo–Zn.
Placers and sands	Sn–Au–Pt–Nb–Ta–Hf–Ti–Zr–Th–Rare earths.
Red beds, continental	U–V–Se–As–Mo–Pb–Cu–Pb.
Red beds, of volcanic origin	Cu–Pb–Zn–Ag–V–Se.

Compiled from Ginzburg 1967 [137], Krauskopf 1955 [141] and 1967 [142], Mason 1958 [146], Hawkes & Webb 1962 [138].

1.8. GEOCHEMICAL EXPLORATION OF NON-METALLIC DEPOSITS BY SUPERIMPOSED HALOES OF HALOGENS

Halogens can serve as direct indicators in the location of non-metallic minerals if they are part of veins and minerals, or indirect ones, when they take part in the ore-forming or secondary process and are not part of the ore minerals.

It was previously considered almost an axiom that the geochemical anomalies connected with blind deep-seated deposits overlapped by thick series of sedimentary rocks cannot be formed on the surface.

The geochemical method of prospecting by the superimposed haloes of halogens (fluorine, iodine) makes it possible to discover some non-metallic deposits (fluorite, rock crystal), and to locate tectonic fault zones.

The superimposed formation of fluorine haloes — the chemical element which is the most active and has the most reactional power — has been thought especially impossible.

On the basis of a study of the distribution of halogens in natural including local fluorine haloes, objects it has been found that superimposed halogen haloes, are formed above blind deep-seated veins, above the areas of secondary transformations of rocks and zones of wall-rock alteration, and also above zones of tectonic activation, including the areas of the relaxation of deep-seated subsurface waters.

The determination of mobile forms of halogen content in superimposed aureoles is carried out by the method of desorption taking into account the kinetics and dynamics of the sorption-desorption process. This makes it possible to establish the most rational time of desorption during which the content of halogen and its forms have the greatest contrast and the most locally related source — the ore mineralization.

One or more of the elements: Cl, F, B, CO_2, S, Si, Br and I are universal in practically all types of mineral deposits, yet we have not employed them to full advantage in geochemical prospecting methods. Cl and F are universal indicators of practically all types of epigenetic deposits.

Fluorine is an indicator of fluorite and fluorine-apatite mineralization. The values of the hydrochemical background obtained within an area with no fluorite mineralization (80 000 km^2) do not exceed 0.2 per cent. Fluorine concentration in rivers and streams draining ore belts, provinces and regions exceeds the background value by a factor of 2–10.

Progress in the geochemical exploration of deposits is due to the application of the newest methods of fluorine detection in rocks. It would be advisable to make more extensive use of the neutron-activation methods of fluorine determination. These methods essentially increase the effectiveness of exploration at all stages of work. At the stage of prospecting on a scale of 1 : 50 000 it is highly effective to perform geochemical work using an automobile neutron-activation survey of fluorine. The effectiveness of detailed prospecting (scale 1 : 5 000) increases, when applying a pedestrian neutron-activation survey of fluorine.

Endogenic non-metallic deposits are accompanied by primary aureoles of iodine and bromine; they considerably exceed the sizes of primary aureoles in the other elements. The participation of iodine and bromine in the processes of endogenic ore formation proceeds from the hypothesis of halogenic metal transportation in postmagmatic solutions. It is confirmed by the presence of these elements in modern ore hydrotherms (Cheleken, the Red Sea, New Zealand, etc.) and in gas-liquid inclusions of hydrothermal minerals. The low energy capacities of iodine and bromine ions do not allow them to form proper minerals, and high values of the ion radius prevent their isomorphic entry into the crystal lattices of other minerals. Being unbound, (on the basis of temperature, concentration and pressure gradients in the centre of ore-formation), iodine and bromine ions in the form of iodides and bromides in the pore and pellicular waters are dispersed in the host rocks, forming primary aureoles.

In the universal zonality of primary aureoles, iodine and bromine take a marginal (extreme) position far from the ore bodies. This determines the significance of primary aureoles of iodine and bromine in the search for deep-bedded, closed deposits.

Iodine has been found to be the best geochemical path-finder. The most characteristic feature of aureoles of iodine is their widespread occurrence exclusively in the space overlying the deposits, irrespective of their composition or the country rocks.

The penetration depth of lithochemical methods based on highly mobile elements is no less than 500 m. In salt deposits aureoles of iodine have been recognized in the products of weathering. In boron deposits aureoles of iodine were detected through a loess layer of 120 m. There is a decided prevalence of iodine in the vertical zonation of endogenic aureoles of deposits and it is of primary importance in the search for hidden mineralization.

1.9. MERCURY AUREOLES AND EXPLORATION

Mercury aureoles have been located around mineralization of many types and are associated with many non-metallic deposits. Among the deposits sought using mercury dispersion aureoles are boron, barium, apatite and rock crystal.

There has been considerable recent interest in the use of mercury vapour as a tracer for various classes of mineral deposits.

In the past the accurate determination of the vapour by atomic absorption was notoriously unreliable due mainly to the many possible interferences. A dual-beam dual-wavelength spectrometer has been developed by using the 2536-A resonance line. Differential measurements of absorption during the periods of 'field' and no 'field' yield a direct value of mercury vapour concentration [40].

A more advanced technique makes use of the different planes of polarization of the centre and split-emission components. With an electro-optical modulator and analyser the two components may be sequentially routed through the spectrometer. The extension of the technique to other elements and its application to the chemical analysis of airborne particulates are discussed, and mention is made of the possibility of isotopic measurements. The mercury spectrometer has been extensively field tested, both in geological and pollution applications. Data from atmospheric and soil-gas tests are presented, and the applicability of the instrument to geochemical exploration is considered.

A gaseous-mercury survey is conducted by field portable target mercury photometers; the sensitivity of the instrument is 1×10^{-8} mg/l. Mercury content in soil, basic rocks and water is determined by a mercury-atomic-absorption photometer with a sensitivity of 1×10^{-7} and 2×10^{-8} g/l.

Mercury is an important indicator element for geochemical prospecting for non-metallic deposits and unlike other indicator elements forms not only lithogeochemical and hydrochemical, but also atmochemical haloes, which can be reliably discovered by modern technical methods.

The mercurometric method of prospecting for deposits uses primary haloes of mercury dispersion; the most important advantage of the method is the large sizes of the mercury haloes in comparison with the haloes of other elements. Primary haloes of mercury dispersion have been discovered in hydrothermal deposits of various genetic types and different compositions.

High mercury concentrations in the range of $(n \times 10^{-3})-(n \times 10^{-5}\%)$ are found in deposits. These concentrations are often of no industrial significance but are important for prospecting. Of all metals, mercury has the highest volatility. As a result, primary haloes of mercury are of a large size and are expressed especially intensively in the supra-ore part of deposits. Mercury is a stable supra-ore element in the sequence of vertical zonation.

Several temperature forms of mercury are usually present in ore bodies. Samples from ore bodies and enclosing rocks are subjected to differential thermal analysis using the atomic fluorescence apparatus 'Flur-1'. The following intervals of the mercury-bearing components of thermal decomposition are clearly distinguished in thermograms: 1 — 190–230 °C, 2 — 260–280 °C, 3 — 320–340 °C, 4 — 400–440 °C, 5 — 550–620 °C, which can testify to the presence of mercury in mineralized rocks in various forms, mercury-bearing minerals and isomorphic admixtures included.

The quality of the forms of mercury in primary haloes decreases considerably away from the ore bodies and deposits, and the thermal spectrum becomes narrower and narrower. There is a greater number of forms of mercury and its thermal spectrum is wider in the supra-ore part of non-metallic deposits in comparison with the sub-ore part. Temperature form 1 sharply predominates in near-ore aureoles of dissemination, and forms 2 and 3 are found in supra-ore levels and can be used as indicators of blind ore bodies.

There exists a constant process of mercury transition from any forms in ores and primary haloes to vapour mercury. This leads to the formation of a mercury atmosphere around ore bodies in overlapping rocks and the near-surface air.

The above peculiarities of mercury distribution are generally of the same type for all the non-metallic deposits studied. The results of the investigation make it possible to conclude that besides determining total mercury content, it is also necessary to study the forms of its presence in ore bodies and near-vein aureoles. The additional information thus obtained will make it possible to conduct prospecting and exploration work more rationally.

Deposits, unlike any other rocks, represent an anomalous source of mercury vapour. Thus, dispersion haloes of free occluded mercury vapours and sorptive mercury in rocks, subsoils, soils and waters are formed above non-metallic deposits due to constant mercury respiration. From soils mercury vapours are transferred to the atmosphere, leading to the formation of unstable mercury gas dispersion haloes in the near-surface atmosphere. At present the methods and instruments for the determination of ultra minor concentrations of mercury vapour in soil air and the near-surface atmosphere are available. Mercury respiration is also characteristic of non-metallic and ore deposits, petroleum, gas and coal deposits as well as the zones of deep faults.

Mercury vapour concentrations of up to $(3000-5000) \times 10^{-8}$ with the background being 1×10^{-8} mg/l have been established over some rock crystal deposits. The intensity of gas-mercury haloes is caused not so much by concentrations of mercury in veins and primary and secondary haloes as by the presence of easily sublimated forms of mercury, including native mercury.

Gas mercury haloes in soil air with an intensity of $(3-40) \times 10^{-8}$ and a stable background of $(0.5-1) \times 10^{-8}$ mg/l were discovered above blind, buried and concealed-buried ore deposits. The depth penetration of the gas-mercury method can be estimated as 100 m for non-metallic deposits. The processes of regional and local metamorphism (corresponding to high-grade green-schist and amphibolic facies of metamorphism) lead to the depletion of mercury in the ores and haloes and to size limitations of the haloes of primary dispersions of mercury, which reflects on its gas haloes. This must be taken into account when using the mercurometric method of exploration within shield and other widely developed regions.

Mercury haloes themselves are universal hidden ore indicators for practically all kinds of mineral resources. These haloes can be used successfully for the solution of different genetic problems and for the purpose of geological mapping (detecting faults, dome structures).

1.10. CONCENTRATIONS OF RADIOELEMENTS

Each group of non-metallic deposits has its own radiogeochemical characteristics according to the content of radioactive elements and the degree of their dissemination. The dissemination parameters (standard deviation and variability coefficient) are less mobile, because their value is determined by secondary processes. In deposits of apatites and phosphorites geological formations have increased (as compared to the background) values of radioactivity and high variability coefficients. Data on quantitative analyses of the parameters of the distribution of radioactive elements can be applied when evaluating the potential ore mineralization of deposits.

Concentrations of uranium, thorium, potassium and their relations are rather sensitive to the development of various postmagmatic processes. Radiogeochemical investigations of non-metallic deposits have shown that many ore fields, especially of such elements as P, F, B and others, are separated from the ore-free host formations according to the concentrations of the elements and their relations. Fluorine-beryllium mineralization, as well as the intrusions of a subalkaline composition accompanying it, is distinguished by a high thorium content with the subordinate role of uranium and potassium.

Thermographic and thermoluminescence methods are effective in prospecting for non-metallic minerals. The former is applied to determine the ignition losses, and the latter to establish the intensity of rock thermal luminescence at heating. Usually, when approaching productive deposits, both the intensity of thermal luminescence (owing to an increase in the number of mineral-activators) and the loss of mass (owing to the presence of water in the minerals) increase. Methods of γ-irradiation (quartz, Iceland spar, and fluorite), infra-red spectroscopy, and electron paramagnetic resonance (EPR) (diamond, quartz) are highly effective together with geochemical methods. These methods make it possible to evaluate the quality of the raw material and the degree of erosional truncation of the productive bodies. The interaction of the fast particles of the solid quanta with the minerals represents a complex cascade process, associated with ionization and structural changes in the crystalline lattice of the crystals. Owing to the fact that we do not deal under natural conditions with the processes proper but with their results, the structure of the minerals as determined by irradiation makes it possible to gain major geochemical information on the changes in the composition of the solutions and on their physico-chemical parameters. The methods of EPR and infra-red spectroscopy make it possible to determine the relative content of indicator elements and the nature of hydrogen bearing flaws, their concentration, and, therefore, to judge the geochemical conditions of deposit formation and the raw material quality.

A combination of various methods makes it possible to search for deposits, to determine the prospects of deep horizons and flanks of deposits, and to evaluate the quality of minerals.

CHAPTER 2

Deposits of the first group

Geochemical methods of prospecting for 14 genetic types of non-metallic minerals, whose deposits are characterized by independent mineral phases are discussed in this section. Deposits of this group include gems (diamond, topaz, beryl, ruby, sapphire, emerald, garnet, chrysolite, chrysoprase and turquoise); abrasives (diamond, garnet and corundum); piezooptic minerals (rock crystal and Iceland spar), anti-abrasives (agate and chalcedony), as well as mica, asbestos and graphite. The ore of the latter is rock containing minerals with a certain correlation of components (for example, the iron content in fluorphlogopite should not exceed 14 per cent). With the exception of malachite, chrysoprase, and turquoise, the entire group of these deposits relates to endogenic formations. Diamond, garnet, ruby, sapphire, rock crystal, and agate are concentrated also in placers. The majority of these useful minerals are used by industry without violation of their initial integrity. Only asbestos and graphite are subjected to mechanical treatment. Despite the variety of these raw materials, they are characterized by a series of common features:

(1) Useful minerals contain mainly petrogenic elements, that are widespread in the earth's crust: C, Si, Al, Ca, Mg and K (Be is an exception).

(2) Deposits are evaluated by the quantity and quality of the commercial mineral, physical properties of the minerals (minimal size of crystals and mineral grains, homogeneity of internal structure, and absence of growth defects).

(3) Many mineral species (quartz, calcite, muscovite and phlogopite) are found in the composition of widespread rocks.

(4) Preservation of the crystalline structure integrity in minerals is observed in the case of their formation within the boundaries of stable zones, wherein active structure-forming and metamorphic processes have been completed simultaneously with the formation of ore.

The geochemical methods of prospecting for the genetic types of non-metals being considered have been elaborated to a various extent, which is due to unequal investigation of the deposits (as a rule, of their geological structure) and extremely non-uniform distribution of high-grade raw materials. Fragmentary data on deposits of gems are conditioned by the character of the explorations (secondary purpose of prospecting). A certain non-uniformity of the mineralogical–geochemical criteria of evaluating deposits is conditioned by the specific character of the research carried out by the authors and the volume of material available for different genetic types of minerals.

2.1. DIAMONDS

Diamond is a scarce mineral raw material relating to one of the modifications of free carbon. Primary and placer deposits of diamonds are distinguished. The former are associated with kimberlites filling diatremes. They are grouped usually in isolated regions, belts and zones. Kimberlites contain a considerable amount of xenoliths of sedimentary rock, deep-seated rocks of the earth's crust and upper mantle. Contact metamorphism of the xenoliths and rocks, enclosing diatremes, is manifested insignificantly. To our mind, diamonds in kimberlites are of xenogenic origin. Kimberlitic magma transported diamond and other minerals of the deep-seated rocks (products of mantle substratum disintegration) to the earth's surface.

The xenogenic character of minerals in the diamond-pyrope facies and in diamonds has been confirmed by radiological and mineralogical data, and by the presence of crystalline inclusions with octahedral cut in the diamonds (negative crystals), whose formation occurred in the course of a long geological period (billions of years and more). The self-cutting of the inclusions occurred as a result of long residence of diamonds in pyrope ultrabasites and eclogites, forming the mantle substratum at a certain depth. This is evidenced by the occurrence of diamonds in xenoliths of various composition in deep-seated rocks. Kimberlite pipes in the internal parts of platforms are characterized by:

(1) a high degree of diamond content;

(2) small depth of erosional truncation (down to 200 m);

(3) abundance of pyrope peridotite inclusions; and

(4) explosion bodies (pipes) of large dimensions, absence of dykes and veins.

However, the marginal parts of kimberlite pipes are notable for the absence or low content of diamonds, wide development of dykes and veins, absence of deep-seated inclusions of pyrope peridotites, spatial association with carbonatites, and wide development of xenoliths in sedimentary rocks.

A relatively large amount of octahedral diamond crystals are found in the central parts of the provinces, the crystals being confined to large pipes. Dodecahedral individuals predominate at the periphery of kimberlite provinces, wherein the kimberlite bodies are usually small and rapidly wedging out. The specific features of diamonds from kimberlite provinces (size, colour, shape and grade), from the viewpoint of their xenogenic character, are conditioned by the regional specific features of the mantle substratum, which is heterogeneous by composition and, probably, by age. The degree of the diamond content in the kimberlites within the provinces depends on the diamond content of the zone, within which the hearth of the kimberlite magma developed, as well as on the degree of the mantle substratum disintegration. Analysis of the magnetic fields of the Siberian province has made it possible to establish that the specific features of the substance composition of the kimberlites are not associated with the composition of the crystalline basement, being conditioned by deeply occurring (mantle) processes.

Natural diamonds crystallize at high pressures and temperatures characteristic of the upper mantle. The internal stress, existing near syngenetic minerals, which are prisoners of diamonds, reaches 2500 MPa in forsterites, 1900 MPa in chromite, and 400 MPa in pyropes. The crystallization and formation of diamonds are estimated to occur at 1200–1500 °C, and at a depth of 200 km. These estimates are based on natural geothermometers, character of dislocational structure of the crystals, and the form of entry of nitrogen, and other elements. Diamond in kimberlites is a polygenic mineral genetically associated with ultrabasic and basic (eclogites) mantle rocks. Besides kimberlite pipes, Precambrian deep-seated garnet–diopside rocks, as well as eclogites may be sources of diamond placers. The finds of diamonds in ultrabasites, eclogites, and basalts are only of mineralogical interest currently. Diamond placers are confined usually to deep clastic formations in ancient river valleys and deltas, proluvial–deluvial trains, and thick crusts of weathering. Precambrian diamond-bearing provinces are known (Brazilian, Australian, etc.) wherein primary sources of diamonds have not yet been established. The common prospecting indications of diamond-bearing deposits are as follows:

(1) Presence of kimberlites within the range of activated parts of platforms with the ancient Precambrian basement.

(2) Development of fractured zones directly adjoining large sinking structures (aulacogenes, syneclises) and fan-shaped dislocations at an angle to the main faults.

(3) Development of rocks of the trappean formation, which, nevertheless, may be spatially separated from the kimberlites.

N. V. Sobolev, A. P. Bobriyevich, A. D. Kharkiv, B. I. Prokopchuk, I. P. Ilupin, E. G. Sochneva, N. N. Sarsadskikh, V. A. Milashev and others contributed greatly to the development and improvement of mineralogical and geochemical methods of prospecting for diamond deposits.

Mineralogy of diamond-bearing formations

Minerals in kimberlites are divided into kimberlitic (orthomagmatic), formed during crystallization of the kimberlite magma proper, and extra-kimberlitic (xenogenic), represented by the products of disintegration of the mantle rocks of the diamond-pyrope facies, entrained by the kimberlitic magma. The latter include diamond, pyrope, chrome-diopside, chromite and ilmenite. Orthomagmatic minerals are those of intratelluric impregnation (olivine, phlogopite and apatite) and of the solid phase of kimberlite mesostasis (olivine, phlogopite, apatite, calcite, picroilmenite, perovskite and magnetite). The secondary minerals of kimberlites are lizardite, brucite and carbonates. Crystalline inclusions of the main minerals, forming kimberlites (phlogopite, apatite, calcite, perovskite and magnetite), do not occur in diamonds proper, in pyrope, or in chrome-diopside. This correlates quite well with the xenogenic character of diamonds. Staurolite, cyanite, ilmenite, tourmaline, rutile, garnet, zircon, corundum, and, in some cases, gold, anatase, beryl, magnetite, haematite, sphene, chromite, platinum, osmiridium, hamlinite, gorceixite, goyazite, florencite, rare earth phosphates, aluminum (florencite) and monazite occur in placers of the Precambrian diamond-bearing provinces. Florencite occurs in loose placers, as well as in quartz–mica schists (Brazil), where it is associated with tourmaline, monozite, haematite, zircon and yellow pyrope. It is likely that many of the listed minerals are accessories of Precambrian diamond-bearing occurrences, whose primary sources have not yet been established.

It has been evident for some years that kimberlite and lamproite are merely transporting agents carrying diamond from the upper mantle to the crust [76]. The macrodiamonds in economic deposits are derived from sometimes highly diamondiferous peridotitic and eclogitic rock that formed in the lithospheric upper mantle. Both peridotitic and eclogitic diamonds occur in every known diamond deposit worldwide. Since some of these minerals (notably garnets and chromites) have very distinctive compositions, their presence can be detected and used as a strong indicator of eclogitic and harzburgitic diamond populations.

The scheme was derived empirically, and the localities constituting the orientation survey did not include lamproites. It has been apparent for some time that it may not be directly applicable to olivine lamproites in Australia. Although lamproitic diamonds have a similar origin to those from kimberlite, the key indicator of the mineral contents of the host rocks seems lower and less informative with respect to diamond content.

Mineral inclusions in diamond

The large, clear, well-formed crystals of diamond that occur as an accessory mineral in many kimberlites have been interpreted as products of stable growth. The occurrence of minerals inside diamond is relatively common. The inclusions are usually small, being on average about 100 μm in size. A significant characteristic of virtually all inclusions is their monominerallic nature; polymineral-lic inclusions rarely occur and include: olivine, pyroxene, garnet, ilmenite and chromite.

Meyer (1977) found higher concentrations of Cr_2O_3 and Al_2O_3 in the olivines (South West Africa) in xenoliths (Table (2.1)), but he believed that olivines from South West Africa were more fayalitic than those from Kimberley and Lesotho. In contrast, olivine inclusions in diamond from localities worldwide have a very restricted composition (Fo 93 per cent) and have Cr_2O_3 contents in excess of most terrestrial olivines. The orthopyroxenes in the ultramafic xenoliths are magnesian-rich, ranging in composition from about 83 to 95 per cent En [100 Mg/(Mg + Fe)] (Table (2.2)). Orthopyroxene inclusions in diamond are not uncommon and also have a restricted range of compositions, between En 92 and 96 per cent. Clinopyroxenes from petrographic types of rock (granular, sheared, discrete and eclogite) show marked differences in chemistry (Table (2.3)).

The chemistry of the garnets from the individual textural types of xenoliths is sufficiently distinct to characterize the various samples (Table (2.4)). Compared to the garnets from the granular and sheared xenoliths, those that occur as discrete xenocrysts have a much larger range in Mg/(Mg + Fe) but a fairly constant

Table 2.1.
Representative analyses of olivines from xenoliths in kimberlite and diamond inclusions [77]

	1	2	3	4	5
SiO_2	40.92	40.53	40.8	40.1	41.3
TiO_2	0.00	0.02	< 0.03	0.04	< 0.01
Al_2O_3	0.02	0.05	0.04	< 0.03	< 0.01
Cr_2O_3	0.01	0.04	< 0.05	< 0.03	0.05
FeO	6.87	9.09	7.25	13.2	7.98
MnO	0.11	0.16	0.11	0.14	0.11
MgO	51.16	50.29	52.4	47.6	50.1
CaO	0.01	0.07	< 0.03	0.06	0.03
NiO	0.49	0.38	0.38	0.25	0.40
	100.08	101.01	101.0	101.4	100.0
100 Mg/(Mg + Fe)	93.0	90.8	92.8	86.5	91.8

1: Granular lherzolite, Lesotho, Southern Africa (Nixon and Boyd, 1973) [153].
2: Sheared lherzolite, Lesotho, Southern Africa (Nixon and Boyd, 1973) [153].
3: Lherzolite, Montana, USA (Hearn and Boyd, 1975) [139].
4: Megacryst, Frank Smith Mine, South Africa (Boyd, 1974) [131].
5: Inclusion in Brazilian diamond (Meyer and Svisero, 1975) [149].

Table 2.2.
Representative analyses of orthopyroxenes from ultramafic xenoliths and diamonds [77]

	1	2	3	4
SiO_2	57.04	56.07	56.7	57.4
TiO_2	0.00	0.26	0.23	0.02
Al_2O_3	0.93	1.00	0.82	0.77
Cr_2O_3	0.32	0.40	0.02	0.39
FeO	4.32	5.23	8.09	4.79
MnO	0.12	0.14	0.16	–
MgO	36.41	34.4	33.2	34.7
CaO	0.31	1.26	0.93	1.38
Na_2O	0.07	0.36	0.16	0.17
	99.52	99.8	100.3	99.6
100 Mg/(Mg + Fe)	93.8	92.1	88.0	92.8

1: Granular garnet lherzolite, Lesotho (Nixon and Boyd, 1973) [153].
2: Sheared lherzolite, Premier Mine, South Africa (Danchin and Boyd, 1976) [135].
3: Enstatite xenocryst, Frank Smith Mine, South Africa (Tsai and Meyer, 1976) [160].
4: Inclusion in diamond, Premier Mine, South Africa (Meyer and Tsai, unpubl.).

Ca/(Ca + Mg + Fe) value (Fig. (2.1)). Garnet is one of the commonest minor minerals in kimberlite; wine-red-purple pyrope is regarded as one of the most diagnostic tracer minerals used in prospecting for kimberlite. The mineral inclusions (including garnets) within diamonds appear to have chemical affinities with the phases occurring within either the garnet peridotite suite rocks or the eclogite-suite rocks.

Table 2.3.

Representative analyses of clinopyroxenes from ultramafic and eclogitic xenoliths and inclusions in diamond [77]

	1	2	3	4	5	6	7	8	9
SiO_2	54.8	54.8	53.7	55.4	55.2	55.5	55.0	55.6	55.0
TiO_2	< 0.03	0.34	< 0.03	0.32	0.33	0.04	0.15	0.23	0.26
Al_2O_3	2.12	2.60	1.66	2.37	2.53	0.96	3.23	17.1	5.78
Cr_2O_3	1.60	0.45	1.59	1.04	0.30	0.33	6.15	–	0.06
FeO	1.57	5.22	1.49	2.52	5.58	3.02	1.76	2.36	5.88
MnO	0.08	0.14	0.09	0.09	0.15	–	0.07	0.03	0.02
MgO	17.2	21.0	17.4	18.6	21.0	21.2	12.9	6.55	12.3
CaO	21.5	13.3	22.5	18.7	13.4	17.9	14.7	10.1	16.6
Na_2O	1.38	1.47	1.32	1.52	1.53	0.53	4.93	7.35	3.93
K_2O	–	–	–	0.02	–	0.02	–	–	0.00
	100.3	99.3	99.8	100.5	100.0	99.5	98.9	99.3	99.8
$Ca/(Ca+Mg)$	0.473	0.313	0.482	0.420	0.314	0.379	0.450	0.527	0.492

1: Granular garnet lherzolite, Lesotho (Boyd and Finger, 1975) [133].
2: Sheared garnet lherzolite, Lesotho (Boyd and Finger, 1975) [133].
3: Garnet lherzolite, Montana (Hearn and Boyd, 1975) [139].
4: Sheared garnet lherzolite, Somerset Island, Canada (Mitchell, 1977) [152].
5: Discrete clinopyroxene xenocryst, Lesotho (Nixon and Boyd, 1973) [154].
6: Inclusion in diamond, Premier Mine, South Africa (Meyer and Tsai, 1976) [151].
7: Associated with diamond, Mir Pipe, Yakutia (N. V. Sobolev et al., 1971) [158].
8: Cumulate eclogite, Roberts Victor Pipe, South Africa (Lappin and Dawson, 1975) [143].
9: Inclusion in diamond, Urals (N. V. Sobolev et al., 1971) [159].

Table 2.4.

Analyses of representative garnets from ultramafic and eclogite xenoliths in kimberlite, and inclusions in diamond [77]

	1	2	3	4	5	6	7	8
SiO_2	42.1	42.0	41.9	41.6	41.5	40.7	41.0	40.8
TiO_2	0.83	0.88	0.04	< 0.03	1.22	0.33	0.02	0.34
Al_2O_3	20.3	19.5	18.8	21.7	20.6	21.3	14.4	21.8
Cr_2O_3	2.15	4.43	6.80	2.30	1.33	0.50	12.6	0.08
FeO	8.62	6.35	6.69	8.28	9.19	16.3	6.11	18.1
MnO	0.27	0.29	0.46	0.44	0.29	0.38	0.28	0.35
MgO	20.5	22.0	19.0	20.2	20.9	13.1	23.6	12.7
CaO	4.37	5.09	7.18	4.80	4.82	7.94	1.74	5.23
Na_2O	0.09	0.08	0.01	< 0.02	0.10	–	–	0.09
	99.2	100.6	101.0	99.3	100.0	100.1	99.8	99.5

1: Sheared garnet lherzolite, Lesotho (Nixon and Boyd, 1973) [153].
2: Sheared garnet lherzolite, Frank Smith Mine, South Africa (Boyd, 1974) [132].
3: Granular garnet lherzolite, Lesotho (Nixon and Boyd, 1973) [153].
4: Granular garnet lherzolite, Frank Smith Mine, South Africa (Boyd, 1974) [132].
5: Discrete garnet megacryst, Frank Smith Mine, South Africa (Boyd, 1974) [132].
6: Eclogite garnet, Stockdale kimberlite, North America (Meyer and Brookins, 1971) [148].
7: Cr-pyrope inclusion in diamond, Ghana (Meyer and Boyd, 1972) [147].
8: Almandine-pyrope inclusion in diamond, Siberia (N. V. Sobolev *et al.*, 1971) [158].

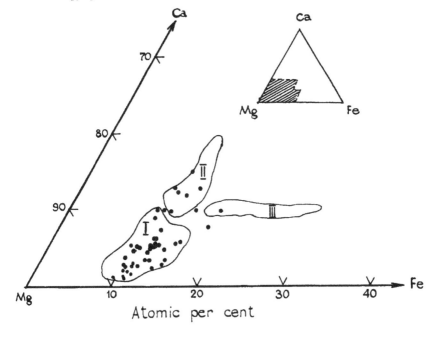

Figure 2.1. Garnets from xenoliths, megacrysts and 'ultramafic suite' inclusions in diamond. The fields of lherzolite and megacryst garnets are predominantly after [134, 77]. I — Ultramafic garnet inclusion in diamond; II — granular and sheared lherzolites; III — discrete megacrysts.

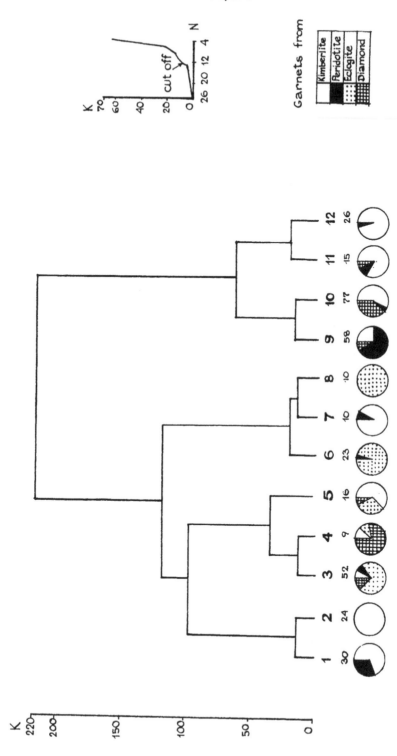

Figure 2.2. Dendrogram for the 12 clusters. Pie diagrams indicate the general source of garnets comprising each cluster. The figures above the pie diagrams refer to the number of samples in each cluster. The inset shows the behavior of the fusion coefficient justifying 'cut-off' at the 12-cluster level [20]. K — fusion coefficient, N — number of olueters.

The results of cluster analysis can be conveniently summarized in a dendrogram (Fig. (2.2)) in which the ordinate represents the measure of dissimilarity between the clusters [20]. The method requires the Euclidean distance coefficients for all possible paired combinations of samples. This distance coefficient is a measure of the dissimilarity between two samples on the basis of all measured variables, and is defined as the sum of all variables of the squared difference between the two samples, divided by the number of variables. The ordinate (Fig. 2.2) is a value corresponding to a twofold increase in the total error caused by fusion, and is known as the fusion coefficient. Clusters 1 and 2 merge to form a new cluster dominated by titanium-rich pyrope megacrysts from kimberlite; clusters 3 and 4 merge to give 'normal' eclogites (i.e. not iron-rich or containing kyanite); clusters 6, 7 and 8 all contain high-calcium garnets; clusters 9 and 10 contain high-magnesium, low-calcium, and moderate-chromium garnets; and clusters 11 and 12 contain moderate-magnesium, moderate-calcium, high-chromium garnets, mainly from kimberlites. The FeO, MgO and CaO content of the garnets in cluster 12 is very similar to that of cluster 11; by comparison, however, TiO_2 is low (0.18 per cent) and Cr_2O_3 is very high (15.9 per cent). CaO shows a positive correlation with Cr_2O_3 and TiO_2.

Searching for minerals indicative of diamond source rocks, identified by their distinctive compositions, has proved a successful prospecting aid. Used in an integrated assessment, garnets, chromites and ilmenites can predict diamond potential in a semi-quantitative way with considerable success. Exceptions can occur, most notably where sub-calcic peridotitic garnets are found in non-diamond-bearing kimberlites.

Ilmenite compositions can be useful in assessing diamond potential, presumably because they are relevant to diamond preservation.

The geochemistry of indicator minerals

Dawson and Stephens [20] have classified kimberlite garnets into 12 groups by cluster analysis, in an attempt to classify the source rocks for garnet macrocrysts. The compositions of garnet and lherzolite are not diamond diagnostic in an exploration programme. The method described by Gurney [38] has repeatedly proven to be useful in exploration programme on several continents.

The minerals closely associated with diamond have a well defined range of compositions: in diamond harzburgites the garnets are high in MgO and Cr_2O_3 and low in CaO [38]. The component of diamond derived from garnet harzburgite is assessed empirically by considering both the number of sub-calcic garnets found in a diatreme and their degree of calcium depletion (Fig. (2.3)). In respect of eclogitic sources of diamonds, the most diagnostic features of the eclogite garnets associated with diamond are the trace amounts of Na_2O in garnet ($Na_2O > 0.07$ wt per cent) and the elevated levels of TiO_2 [38]. The composition of eclogite garnets associated with diamonds worldwide with respect to these two key oxides is presented in Fig. (2.4).

The use of heavy minerals in diamond exploration has been defined to permit an assessment of the diamond potential of that source. This is routinely applied

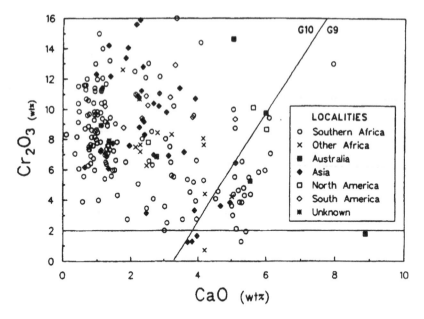

Figure 2.3. A plot of Cr_2O_3 against CaO for peridotitic diamond inclusion garnets from worldwide localities. The diagonal line distinguishing sub-calcic 'G10' garnets from calcium saturated 'G9' garnets was defined by Gurney (1984) on the basis that 85% of peridotitic garnets associated with diamond plot in the G10 field. The horizontal line drawn at 2 wt% Cr_2O_3 is used as an arbitrary division between eclogitic and megacrystic garnets (< 2 wt% Cr_2O_3) and peridotitic ($>$ wt% Cr_2O_3). Localities where peridotitic garnets plot to the left (sub-calcic) side of the 85% line have diamond potential. Localities without such garnet compositions are barren [38].

in various major diamond exploration programmes and relies on the interpretation of the composition of garnets, chromites and ilmenites mentioned above.

These minerals are sought in streams and soils for the proximity of a source rock. The criteria used vary widely and can depend on the region. The identification of garnets and chromites with specific compositions has indeed turned out to be a useful indicator.

Chromites associated with diamond show a high average Cr_2O_3 62.5 wt per cent [38]. Chromite is used in a similar manner to garnet to provide an indication of the amount of diamond in the diatreme derived from disaggregated chromite harzburgite. The useful chromite compositions are defined in Fig. (2.5).

Ilmenites with low Fe^{3+}/Fe^{2+} ratios are associated with higher diamond contents than those with higher inferred Fe^{3+} contents. High Cr^{3+} can be found in either association but is only a positive factor when it occurs with high Mg.

At each locality the contribution to the overall diamond population from each of the garnet harzburgite and eclogite formations is assessed by establishing the abundance of the garnets and chromite derived from the disaggregation of the mantle host rock. It should be clearly realized that these diamond sources are additive and that a really good contribution from any one of them would be sufficient to provide an economic grade in a diatreme.

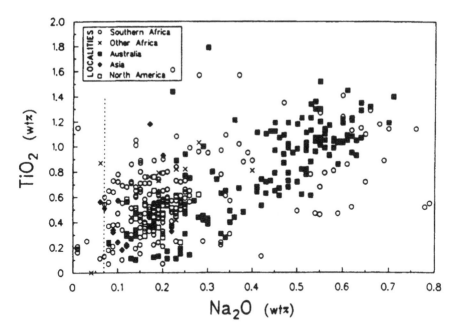

Figure 2.4. A plot of TiO_2 versus Na_2O for eclogitic diamond inclusion garnets from worldwide localities. Note that the elevated levels of both these two elements is characteristic of eclogitic garnets associated with diamonds. In exploration applications, garnets with $Na_2 \geqslant 0.07$ wt% are considered significant, but eclogitic garnets must be distinguished from megacryst garnets to assess diamond potential [38].

The system is a major aid to exploration programmes. In Botswana, where it was applied to several dozen kimberlites discovered under the Kalahari cover by Falcon Bridge Exploration (1980), heavy mineral analyses correctly identified all the barren kimberlites and all the diamond-bearing kimberlites and flagged the best ore-bearing body. This was found immediately after the first batch of heavy minerals from the source had passed in front of the microprobe. In this environment of hidden ore bodies it was an unqualified success.

In a stream sampling programme in Venezuela (1975), the presence of garnets from a nearby diamond source was picked up in the Guaniamo region where the primary source of some of the alluvial diamonds has now been found [38]. The fact that the [38] diamonds are being traced by their association with fragments of the mantle rocks from which they were originally released by disaggregation is so fundamental that whatever geological vehicle has been used to convey them to the surface there is a chance that semi-quantitative relations may be maintained.

Any assessment of the economic potential of a kimberlite based on indicator mineral compositions must view the evidence as an integrated package. The peridotitic and eclogitic diamond potential are additive and the ilmenite-based preservation index is valid only if diamonds were present in the first place [109].

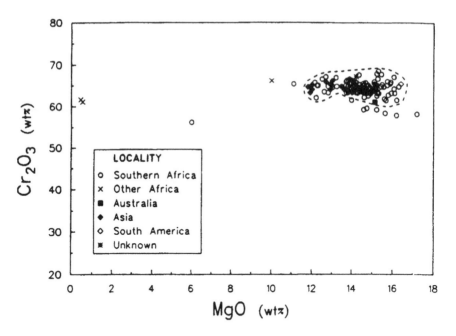

Figure 2.5. A plot of MgO versus Cr_2O_3 for chromite diamond inclusions from worldwide localities. Note the highly restricted chrome-rich character of the inclusions. The preferred compositional field for exploration applications which includes > 90% of the data points is indicated. Localities where chromite compositions plot within this field have diamond potential [38].

Heavy mineral macrocrysts from a wide variety of kimberlites have been examined and the results interpreted in the light of the above consideration. It has been found that [38]:

(1) The relationship described in garnets and chromites can be used to predict the presence of diamond with a high degree of certainty.

(2) In some regions of the world, including the Kalahari craton, the primary diamond content of a kimberlite may have been substantially modified by resorption due to the oxidation process.

(3) Peridotitic garnets and chromites, and eclogitic garnets and ilmenites, which survive well in the secondary environment and are frequently the first tangible minerals from the programme, can give an early signal about the diamond potential of the rocks containing them, providing their compositions.

Fundamentals of lithogeochemical, hydrogeochemical, and biogeochemical methods of prospecting for kimberlites

An advantageous prerequisite for the application of geochemical methods of prospecting for kimberlites is a substantial difference in chemical composition between the kimberlites and the enclosing rocks. Kimberlites contain an increased content of TiO_2 (1.5 per cent), Cr_2O_3 (0.01–0.2 per cent), NiO (0.01–0.7 per cent),

P_2O_5 (0.55 per cent), Cl (0.25 per cent), F (0.4 per cent), and B (0.02 per cent). The predominance of potassium over sodium in kimberlites makes it possible to differentiate kimberlites from similar ultrabasic rocks. The specific feature of kimberlites is their enrichment in strontium and rare earth elements, whose content in the pipes is much higher (by one or two orders of magnitude) than in the enclosing rocks. Kimberlites, as compared to enclosing rocks, traps, and other basic rocks, are characterized by an increased content of uranium (2.5×10^{-4} per cent) (in traps 1.5×10^{-4}, in limestones 1.0×10^{-4}). When kimberlites undergo weathering and denudation, indicator elements (nickel, cobalt, chromium, zinc, copper, etc.) are concentrated in the loose alluvial and deluvial formations and condition the formation of secondary haloes and dispersion trains, whose dimensions exceed greatly the volume of the kimberlite pipes and spread over an appreciable distance therefrom. Lithogeochemical prospecting by secondary haloes and dispersion trains is applied on a wide scale when prospecting for diamond-bearing deposits. This type of prospecting is applied mainly at the stage of medium- and large-scale geological mapping (1 : 200 000 and 1 : 50 000) in combination with heavy concentrate sampling. Analysis of available data demonstrates that methods of lithogeochemical prospecting by secondary dispersion haloes are quite effective if it is possible to apply them in the given area. This is confirmed by data of lithogeochemical prospecting by secondary dispersion haloes in mountain taiga and tropical conditions in Siberia, India, Sudan, Congo and Sierra Leone. The configuration of dispersion haloes of elements around the kimberlite bodies is illustrated in Fig. (2.6). When kimberlites are localized in carbonate rocks, positive geochemical haloes of Ni, Cr and Co form around them. These elements are concentrated mainly in the highly changed basic mass of kimberlites and are liberated more easily in the process of weathering than the admixtures contained in the minerals [54].

The dimensions of the haloes exceed the parameters of the pipes by 1.5–2 times. The haloes of cobalt and nickel are detected in close proximity to kimberlite bodies; the displacement of the epicenters of the anomalies from the central part of the diatremes on slopes with a steepness of several degrees is only 50–150 m. The downslope displacement of haloes reaches 150 m in zones affected by solifluction processes. The content of elements in secondary dispersion haloes surrounding kimberlite bodies in Siberia is as follows: nickel 0.02–0.3 per cent (contrast range 3–7), chromium 0.05–0.08 per cent (contrast range 3–4), and titanium 1.0–3.0 per cent (contrast range 3–5). The contrast of anomalies diminishes with the distance from the kimberlite bodies. The series of geochemical mobility is as follows: Ni–Co–Cr–V–Zr–Cu [54]. Negative haloes of lead are found over kimberlite bodies occurring in carbonate rocks.

Kimberlite bodies located amongst traps generate secondary haloes of nickel, chromium and cobalt. Increased concentrations of vanadium and gallium are characteristic of the substratum rocks. The dimensions and shape of the haloes over these formations depend on the thickness of the overlying eluvial–deluvial formations and the intensity of the solifluction processes. In loose formations, faults, controlling the arrangement of kimberlite bodies, are detected by the haloes of mercury and radioactive elements. As a rule, the anomalies are oriented linearly. Mineralogical methods can be used to prospect for primary and placer deposits

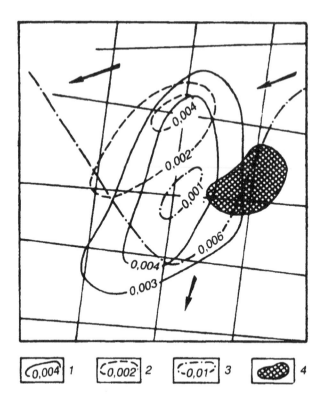

Figure 2.6. Secondary dispersion haloes of nickel, cobalt and chromium in eluvial–deluvial deposits over a kimberlite body [54]. 1–3 — Isolines of contents (1 — nickel, 2 — cobalt, 3 — chromium); 4 — contour of a kimberlite body. Arrows show direction of slope.

of diamonds. Associations of minerals, occurring in the diamonds proper, are of major significance. Investigation of the diamonds and their accessories, as well as investigation of the changes in concentration, grain size and degree of grain erosion allow the discovery of basic bodies. Diamonds are transported over considerable distances (up to 15 km) from the kimberlite pipes. They are not eroded, but crack when they are transported by river streams. The number of fragments and damaged crystals increases with the distance of diamond transportation by rivers. If the diamonds have passed the intermediate traps, the traces of erosion are expressed in smoothed edges, face junctures, the appearance of percussion figures, and crescent-shaped cracks [74]. Owing to the low content of diamonds (hundred thousandths and millionths of a per cent) in kimberlite pipes, prospecting is carried out for accessory minerals, which, possessing higher stability and higher concentration as compared to diamonds, occur more frequently in alluvial formations. Owing to their high density (3.2–4.7), paragenetic accessories of diamond (ilmenite, pyrope, chrome-spinellid, chrome-diopside and zircon) are concentrated in the heavy fraction of loose formations and are easily diagnosed under field conditions.

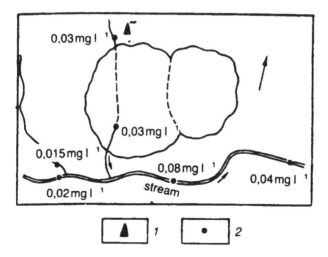

0,03 mg l [1]

0,03 mg l [1]

0,015 mg l [1]

0,08 mg l [1]

0,04 mg l [1]

0,02 mg l [1]

stream

▲ 1   • 2

Figure 2.7. Zinc content in hydrochemical samples around a kimberlite pipe [62]. 1 — a kimberlite pipe, 2 — hydrochemical anomalies.

Experimental investigations have demonstrated the possibility of using hydrogeochemical and biogeochemical methods in prospecting for kimberlite bodies. The mineral phases of the kimberlites and the products of their weathering are dissolved in water, conditioning the formation of hydrochemical anomalies. Zinc is an indicator element of kimberlites and its aqueous haloes occur in river valleys (downstream) up to 5 km from the pipes (concentrations of this element content are 10–100 times greater than background levels). Compounds of zinc (bicarbonate salts) relate to the group of easily soluble formations, distinguished by substantial mobility. The specific features of zinc distribution in hydrochemical samples taken around a diamond-bearing pipe are shown in Fig. (2.7). When approaching the pipe, the zinc content in the haloes reaches 0.05 mgl^{-1} compared with a background value of 0.005 mgl^{-1}. A sharp increase of zinc concentration in water samples is observed near the contact with the kimberlites. In addition to Zn, Ni, Co, Cr, V, Ga and Cu are also used as indicators. Water with an overall increased content of the above-listed metals usually control the zone of propagation of kimberlite bodies. Roots of many plants, penetrating to a depth from 2 to 12 m, concentrate elements of buried haloes in loose formations. Increases in Ni, Cr and Ti content have been found in the ashes of plants collected directly adjacent to kimberlite bodies. The concentration of these elements in the vegetation is up to five times less at some distance from the kimberlites. Thus, the vegetative cover may serve as an indicator of kimberlite bodies. In mountain-taiga areas, sodded and weakly manifested in the topography, kimberlite pipes are covered with loose formations that preserve moisture non-uniformly, as are the surrounding sedimentary rocks. Drifts above kimberlites have a higher moisture content and an increased concentration of phosphorus, chlorine, and potassium, and these conditions encourage the growth of relatively rich vegetation on the overlying sites, which are detected in colour infra-red aerial photographs taken at a low altitude. The vegetative cover on the majority of pipes is characterized by the presence of alder undergrowth, which is distinguished by its considerable height and close standing as compared to the undergrowth on the surrounding rocks. The height of the stand over kimberlites is 9–11 m at an average trunk diameter of 0.25 m. Deciduous thin forest on enclosing limestone is distinguished by coppice and extremely rare stand. The locations of kimberlite bodies in aerial photographs are marked by a darker photographic hue, that is conditioned by closer standing of the woody plants and shrubs. The distinction of the kimberlite bodies from the enclosing rocks by magnetic susceptibility and content of radioactive elements is an advantageous application of geophysical methods (airborne magnetic, gravimetric and γ-spectrometric surveys). The common pattern is an increased magnetic susceptibility in kimberlites, and a high value of the uranium–thorium ratio (20 and more) against 2–5 units in the deluvium of traps.

Practical recommendations for lithogeochemical and mineralogical methods of prospecting for kimberlites by secondary haloes and dispersion trains

Lithogeochemical exploration by secondary haloes and dispersion trains is carried out in three stages: a reconnaissance survey, prospecting, and detailed work.

Lithogeochemical exploration by dispersion trains is carried out at the first stage in combination with other techniques, heavy concentrate and hydrochemical survey being carried out at the same time as geological surveying at a scale of 1 : 200 000. It is necessary to perform reconnaissance lithogeochemical exploration by the secondary dispersion haloes in weakly dissected open regions of the arid zone, owing to which the common geochemical and mineralogical characteristics of the region being explored are determined. These operations are done when passing routes across the predominant strike of the rocks and faults with consideration of crossing the developed typhons in the area. Dispersion haloes and dispersion trains of kimberlite bodies may be discovered in this case.

Prospecting lithogeochemical exploration by secondary haloes and dispersion trains at a scale of 1 : 50 000 is carried out in regions where positive results have been obtained after reconnaissance mineralogical and geochemical exploration or other work. Prospecting by dispersion trains should be carried out simultaneously with hydrochemical sampling. The applicability of either method in particular areas depends on the development of the drainage system and the potential influx of minerals and rocks from deposits into river valleys. Combined application of heavy concentrate sampling and prospecting by river dispersion trains is most effective.

Prospecting is performed by the lithogeochemical method in areas with weakly manifested paths of modern run-off. The discovery of kimberlite bodies is the purpose of prospecting lithogeochemical exploration by secondary haloes and dispersion trains.

Detailed lithogeochemical exploration by secondary dispersion haloes is carried out in localities with diamondiferous indications. The purpose of this work is to detect, contour, and evaluate kimberlite diatremes.

Lithogeochemical prospecting by dispersion trains at a scale of 1 : 200 000 to 1 : 50 000 is performed by sampling sandy-argillaceous fractions of river bed sediments by the grid of routes. On reconnaissance surveys at a scale of 1 : 200 000 the distance between sampling points in the river bed is 250–500 m (two to three samples per 1 km^2). When prospecting at a scale of 1 : 50 000, sampling intervals are every 100 m and 20 samples are taken per km^2. The routes follow the beds of rivers, streams, and ravines. Heavy concentrate sampling and measurements of the magnetic susceptibility of small fractions by means of the cappameter EKP-3 are performed simultaneously with lithogeochemical sampling of the alluvium. Prospecting and detailed lithogeochemical surveys should be done on a grid of profiles, corresponding to one of the commonly accepted scales, in order to detect dispersion haloes (Table (2.5)) [46].

Considering the isometric shapes of the kimberlite bodies and the potential minimal dimensions of the secondary haloes, it is advisable to perform lithogeochemical sampling when prospecting at scales of 1 : 50 000, 1 : 25 000, or 1 : 10 000, on grids of 150 × 150, 100 × 100, or 50 × 50 m, respectively. The grid can be changed, depending on the assumed dimensions of the pipes and the conditions of applying lithogeochemical prospecting. Areas can be differentiated: mountain-taiga and tropical landscapes. These in their turn, are divided into territories covered by products of local rock weathering and those covered by formations brought in from remote areas. Lithogeochemical prospecting in new regions should start by sampling various horizons of deluvial formations. Analysis and generalization of the results of geochemical work are carried out in the process of this exploratory work on geologically studied plots with a very dense grid of sampling along the profile (every 5–10 m) at depths of 0.2–0.5 m. The most representative fraction and the initial mass of the samples are collected. Silty clay or sandy fractions of alluvium are sampled at the surface or from 15–20 cm holes within the dry floodplain part of the river bed during lithogeochemical prospecting by dispersion trains.

Table 2.5.

Sampling grid at lithogeochemical surveys by secondary dispersion haloes

Scale	Distance, m		Density of samples per 1 cm^2 of map	Number of samples per 1 km^2 area
	Between profiles	Between sampling points		
Reconnaissance surveys				
1 : 200 000	2000	100	20	5
1 : 100 000	1000	100–50	10–20	10–20
Prospecting surveys				
1 : 50 000	500	50	10	40
1 : 25 000	250	50–25	5–10	80–160
Detailed surveys				
1 : 10 000	100	20–25	5–4	400–500
1 : 5 000	50	20–10	2.5–5	1000–2000
1 : 2 000	25–20	10	1.6–2	4000–5000

When prospecting by dispersion haloes, a sandy clay fraction of eluvial–deluvial formations is sampled. The samples are taken from beneath the humus horizon of soil at a depth of 20–30 cm.

When the predominating secondary dispersion haloes in the region are weakened at the surface, or are buried, the samples should be taken from the representative horizon, which coincides with the upper humus horizon of the soil profile (exposed haloes), or with the alluvial horizon at a depth of 0.5–0.8 m from the surface (buried haloes) under conditions of retarded denudation [46].

Deep lithogeochemical prospecting by secondary dispersion haloes in prospective regions is performed by sampling holes at the depth of the representative horizon. The mass of the sample should be 50–150 g, and grinding should be performed up to 200 mesh. Samples are analysed for the following chemical elements: nickel, copper, chromium, manganese, vanadium, gallium, niobium, zinc, lead, zircon and mercury. It is advisable to make analysis for nickel under field conditions. To this end, a fraction < 80 mesh is treated with a mixture of $HCl + HNO_3$ (1 : 3), or HF, and the nickel is detected in the extract. Disentanglement of geochemical anomalies with a weak contrast, and evaluation of the local geochemical background are performed by variation-statistical processing of the data. The parameters of element distribution in the samples are determined with consideration of the lithological specific features of the enclosing rocks. High and average anomalous haloes are differentiated by the content of elements of the ferric group. Multiplicative haloes of ($Cr \times Ni \times Co$) are constructed to enhance the contrast of the lithogeochemical haloes. It is advisable to apply mineralogical and geochemical methods jointly with the development of combined heavy concentrate and geochemical maps. In some cases, when it is difficult to diagnose the minerals, the heavy concentrate samples (every second one) are subjected to spectral analysis, whereupon the contrast of the geochemical anomalies enhances comprehensively.

Data are plotted on the maps that relate to the qualitative and quantitative composition of the minerals, the geochemical anomalies, and sites, prospective for diamonds, are contoured. The potential directions are determined for the washdown products of the kimberlites. The geochemical indications of known (standard) pipes and trap formations are considered in this case.

Alignment of the zones of development of accessory minerals with secondary haloes and geophysical anomalies makes it possible to predict kimberlite bodies more objectively.

Mineralogical methods are applied for sampling alluvial and deluvial formations in order to: (a) detect primary and placer deposits; (b) determine potential sources of removal; (c) establish the distance of mineral transportation; and (d) detect small dykes of traps that trace faults. Special attention is given to sampling from small, sometimes drying up tributaries of large river valleys and ravines. It is advisable to make single holes in the mouths of dry rivers in order to wash in heavy fractions. Water streams circumjacent to magnetic anomalies located at a distance from the outcrops of traps are sampled most thoroughly. The initial mass of the sample is determined on the spot, and the optimal one is a volume of 20 l. When locating the sampling plots, it is necessary to consider the fact that the garnet concentrate usually accumulates downstream of the spit head. Considering the comparatively low density of garnet ($3.64-3.74$ g cm^{-3}), it is necessary, when carrying out mineralogical explorations, to preserve 25–30 per cent of the light fraction in the sample (grey heavy concentrate). Small-scale (1 m^3) sampling of deposits for diamonds is performed on plots with an appreciable accumulation of diamond accessories. Specific regions for detailed work are outlined in the case of positive results. Heavy concentrate is sieved through screens with 0.5–1 mm meshes, and the presence of pyrope and its approximate content in the large fraction (+1.0 mm) is determined visually. The borders of potential spread of the kimberlite bodies are contoured on the basis of heavy concentrate sampling. Sites of potential occurrence of pipes are marked roughly by the comprehensive increase of pyrope, picroilmenite, and other elements in the samples. The genetic type of the minerals and their size, the predominance of picroilmenite in the heavy concentrate as compared to pyrope, the presence of a kelyphitic coating in the pyrope grains, preserving at a distance of 0.2–1.0 km, as well as a leucoxene and perovskite jacket on picroilmenite, which is stable at a distance of 3 km from the mother source, are considered when generalizing the results of the investigations. Experience shows that pyrope and picroilmenite are transported to a distance of 15 km, magnetite up to 8 km, chrome-diopside up to 5 km, and olivine and perovskite up to 2 km. The content of grains with a diameter more than 0.5 mm at a distance of 300 m from the diatremes is: pyrope 75%, picroilmenite 91%, and at a distance of 2.2 km pyrope 30%, picroilmenite 65% [58].

Heavy concentrate sampling of deluvium allows one to determine the contours of potential basic diamondiferous areas, and to prospect for individual kimberlite bodies near the explored sites. An indication of new diatremes is the presence of diamonds in deluvial sediments, differing by their morphological and physical specific features from the explored bodies (diamonds with a certain set of indications are usually present in each specific pipe). Chips in accessory minerals of diamonds near kimberlite bodies are characterized by a bright lustre, and the samples of deluvium contain minerals of the light fraction (phlogopite, chlorite, serpentine). Heavy concentrate sampling of deluvial formations, when prospecting at a scale of 1 : 50 000, is performed at 100 m intervals, and the lines of the route are laid along the slopes of river valleys above the boundary of river bed alluvium. Deluvial samples are elutriated from clay and washed on trays. Sometimes additional sampling (up to four samples) is performed up the slope at a distance of 100–150 m from the basic prospecting lines in order to contour the local area.

By paragenous associations of minerals it is possible to establish roughly the dykes of traps, crossing the zones of dislocations with a break in continuity, as well as the occurrence of carbonatites that can be associated with kimberlite bodies. Diopside, augite, ilmenite and magnetite are indicator minerals of small bodies of trap formation, and pyrochlore, columbite, thorite and zircon are those indicator carbonatites.

The results of sampling are classified, and typical mineral parageneses are established. The per cent/weight mineral ratio is given for 1 m^3. The change (percentage) in the yield of the heavy concentrate heavy fraction along the river beds is shown on the maps in the logarithmic scale by bands of various width, and the percentage of the minerals is shown within the bands by stripes. Outlines are drawn of definite prospective zones, and the results are given of mineralogical analysis of deluvial formations and pounded bedrock. The percentage (by mass) of the pyrope and ilmenite in the heavy fraction is given.

When prospecting for kimberlite bodies, it is advisable to apply a set of methods, including mineralogical sampling and lithogeochemical surveying.

Geophysical methods are applied in diamondiferous areas in the process of regional explorations, detailed prospecting, and evaluation work. They involve airborne magnetic and gravimetric surveys at a scale of 1 : 200 000. Magnetic survey is most effective. When prospecting for kimberlite pipes, airborne magnetic surveys should be accompanied by ground magnetic exploration at a scale of 1 : 10 000–1 : 5 000. It is difficult in some cases to unambiguously identify the anomalies produced by kimberlites. Airborne magnetic survey with simultaneous recording of the magnetic field at several levels by means of two magnetometers is performed to sort out false anomalies (trap and tuff outliers, accumulations of dolerites, and magnetite-bearing sandstones in the deluvium) and to ensure trustworthy interpretation of true anomalies. This method makes it possible to separate those anomalies caused by pipe-shaped bodies extending to a considerable depth from the other anomalies detected.

Diamond prospecting is the most recent field to which the use of atomic instruments has been applied.

South Africa, as one of the leading diamond-producing areas in the world, has contributed much to the advancement of new techniques for diamond exploration, aerial radioactivity surveys being the latest development.

It is known that diamonds originate in 'kimberlite pipes', vertical, funnel-shaped bodies of rock that were once deep within the earth's crust. The diameter of the pipes can be as wide as 800 m. Geologists believe these pipes were formed when a gaseous explosion inside the earth produced a funnel-shaped crater which was later filled by molten kimberlite oozing up from the depths of the earth.

It is now possible to locate these pipes by carrying out scintillometric aerial surveys. The scintillometer is one hundred times more efficient than a standard Geiger counter. It is an extremely sensitive γ-ray detector, γ-rays being the most penetrating of the radioactive rays. This radioactivity black-out serves to guide geologists to the areas where diamonds may be found.

Geophysicists have discovered that different types of rock show different degree of radioactivity. Granite has a high degree of radioactivity, while basalt kimberlite belongs to the basalt family and exhibits very little radioactivity. Kimberlite pipes are usually surrounded by rock of granitic origin. An airborne scintillometric survey would show the presence of these kimberlite pipes by the rise and fall in radioactivity.

A plane equipped with an airborne scintillometer would fly a grid pattern 15–70 m above the ground. The γ-radiation picked up by the scintillometer would be continuously recorded on a chart. If the chart shows a sudden drop in the radioactive level and then returns to normal, further pedestrian investigation of the area should definitely be made with a Geiger or scintillation counter. This would pinpoint the area of low radioactivity. By taking readings at 15–20 m intervals, the outline of the kimberlite pipe could be mapped out. The evidence gained from such a survey is correlated with the results of a magnetometric survey. If the evidence of both surveys is corroborative, the chances are that a kimberlite pipe will be found in that area.

When kimberlites occur in granites, it is possible to use radiometric methods of prospecting. The pipes are found by negative anomalies of the gamma field and enhanced anomalies of the magnetic field. The γ-spectrometric method (increased content of thorium and uranium) allows detection of pipes filled with kimberlites of the mica and tuff types and overlain with drifts up to 15 m in thickness.

Heavy concentrate sampling is an effective method of prospecting for placer deposits of diamonds. When prospecting for placer deposits, it is necessary to consider the intensity of denudation processes and the development of crusts of weathering, contributing to the formation of rich diamondiferous deposits that also contain garnets with cone-shaped, pyramidal and imbricate sculptures.

The heavy concentrate method makes it possible to locate the areas of development of ancient terrigenous formations as the most probable collectors of diamonds. The main criterion of the productivity of placers is the presence of diamonds. When sampling alluvial formations, it is necessary to take into consideration that accumulation of diamondiferous material occurs in negative forms of the relief: karst depressions, lake basins, etc. Special attention should be given to river valleys wherein the steps of terraces are clearly displayed on the slopes, indicating repeated washout and redeposition of elastic products with an outwash of fine particles and concentration of the useful components. Samples should be taken at the bend of the river bed longitudinal profile from steep to gently sloping sections, as well as at the stretches of narrow, deeply downcut valleys with a thinning of the alluvium.

When prospecting for placers, holes are drilled, test pits are sunk (in winter) along the transverse profiles of river valleys, and heavy concentrate samples are taken. The distance between the workings is determined in each case, depending on the specific features of the placer structure. Considering the jet-like distribution of diamonds in alluvial formations and their confinement to the subterrane, it is preferable to sink test pits to a depth of 20 m, reaching the bedrock. Sampling is trustworthy and representative in this case. Attention is paid to the potentiality of detecting kimberlites in the basic occurrence. Primary sedimentary collectors indicate the closeness of kimberlite bodies. The kimberlite pipes are the socle from which diamonds are supplied directly. When evaluating placers, the most prospective are sites in the marginal parts of zones of maximum washout, the bottom and upper parts of rifts, and the central parts of deep ravines [74]. The rudaceous (oligomict or monomict) composition of the sediments, located in the carbonate rocks of the substratum, is a favourable factor to be considered in prospecting for placers.

When prospecting for placers, geochemical sampling of loose formations is performed within the borders of mature river valleys where it is difficult to detect mineralization by the heavy concentrate method.

Specific features of prospecting for kimberlite pipes under conditions
unfavourable for the formation of exposed secondary haloes

In the presence of buried and semi-buried haloes, surveys of dispersion trains are combined with geophysical exploration, and taking samples from boreholes and workings.

Fan-shaped false (mixed) haloes that are remote from the primary sources form due to the development of solifluction processes; dislocation and screening of the haloes are observed, depending on the exposure of the slopes, their steepness, and the composition of loose formations. Before starting lithogeochemical surveys in these regions, an analysis is made of the maps of the Quaternary deposits and aerial photographs are interpreted, making it possible to perform zoning of the territory by the conditions of exploration. Experimental work is carried out to determine the optimal sampling depth, which should exceed the thickness of the solifluction formations. Spot-medallions are used for sampling; these make it possible to prospect at the level of the representative horizon at a minimal depth; the density of the spread of forms of the frozen microrelief usually satisfies a prospecting grid at a scale of $1 : 1\,000$. The presence of an ochreous sub-horizon in the soil profile and an increase in the thickness of the active layer are indications of the closeness of basic outcrops of kimberlite bodies.

It is difficult to prospect for kimberlites by conventional mineralogical methods in regions with a developed crust of weathering, owing to intensive alteration of the minerals (magnetite, pyrope). Under these conditions, it is necessary to study the clay composition (content of elements of the ferric group, as well as boron, flourine, and mercury). When prospecting for kimberlites filled in the upper part

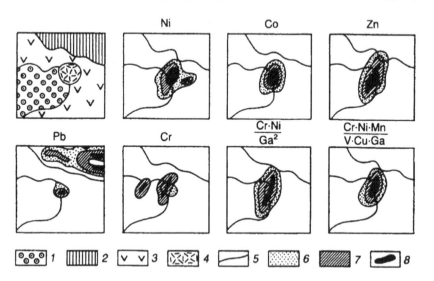

Figure 2.8. Secondary dispersion haloes of indicator elements around a kimberlite pipe [116]. 1 — terrigenous formations; 2 — carbonate-terrigenous rocks; 3 — traps; 4 — kimberlite rocks; 5 — geological boundaries; 6 — geochemical anomalies of low contrast; 7 — geochemical anomalies of medium contrast; 8 — geochemical anomalies of high contrast.

with tufogenic-sedimentary formations, it is necessary to perform geochemical sampling of the borehole core, to detect indicator elements characteristic of kimberlites, and to carry out a gravimetric survey. Lithochemical surveys are done differently when kimberlites are localized in traps and limestones. Under these conditions (Fig. (2.8)), multiplicative indices $(Cr \times Ni)/Ga^2$ in the zones of development of carbonate rocks and $(Cr \times Ni \times Mn)/(V \times Cu \times Ga)$ near the rocks of the trappean formation are indicators of kimberlite bodies. The content of Cr, Ni and Nb in kimberlites is 1.5–3 times greater than in traps and terrigenous formations. Superposition of magnetic and geochemical anomalies makes it possible to recommend them with certainty for checking by drilling operations. The use of multiplicative characteristics makes it possible to evaluate the kimberlite nature of lithochemical anomalies under conditions where pipes are adjacent to sheet intrusions, or one even partially overlapped by the latter. In addition haloes associated with kimberlites differ from anomalies associated with the basic rocks by increased concentrations of Nb, Zr, rare earth elements and the Ni/Co value, and by a reduced number of indicator elements and the absence of radiometric anomalies in the case of alkaline or ultrabasic formations. The following indications of anomalies formed by the outcrops of kimberlites are considered: high-contrast, spatial superposition of indicator elements, and increase in their content with depth. When kimberlite pipes are overlain with traps, and Upper Paleozoic and Mesozoic sedimentary formations, it is necessary to perform heavy concentrate sampling, sinking test pits and drilling boreholes first in river valleys along the outcrops of these rocks every 500 m, and then in the zones of development of these deposits [58]. Basal horizons in the Upper Paleozoic formations and the crust of weathering of Lower Paleozoic rocks are sampled. It is advisable to sink test pits with a cross-section of 4 m² to a depth of 30 m to ensure reliable prospecting, thereby extracting and analysing a sufficient quantity of accessory minerals of diamonds. Boreholes are drilled under traps if there are prerequisites for detecting kimberlite bodies [58].

Biogeochemical methods of prospecting are applied on territories covered with glacial or alluvial deposits brought from afar, in boggy areas, and in regions with buried defluxion haloes. They may supplement geochemical methods, particularly in winter, when it is impossible to perform lithochemical surveys. Samples of vegetation should be taken on a square grid 40×40 m. When kimberlite bodies are localized at a depth, but within the zone of aeration and groundwater flow, it is necessary to apply the hydrochemical method of prospecting for primary deposits. It may play the leading role in the areas covered with re-deposited loose formations, and for detecting kimberlite bodies in zones of development of trappean intrusions, wherein it is difficult to apply geophysical methods.

Sites are outlined for metallometric survey as a result of geochemical exploration and detection of anomalies by dispersion trains. Sampling by secondary haloes at the stage of small-scale reconnaissance prospecting is necessary if the drainage system prevents the use of the above method. Special attention is given in hydrochemical prospecting to sampling in the mouths of shallow brooks. The water is analysed under field conditions. If anomalies of zinc, nickel, copper and cobalt are found in the waters of one of the tributaries, prospecting is continued in that area. Grading of hydrochemical anomalies is performed with the application of normalized estimates of the contents of components.

For many years prospecting for kimberlite has been based on heavy-mineral surveys, aerial photographic investigation, geophysical surveys on the ground, and geochemical exploration methods.

Anomalies of Cr, Ni and Co have been discovered together with Mn, Ti, Zn, Pb, Ba, Sr, La, Zr, Rb, etc. in sediments downstream from kimberlite bodies or in the overburden above kimberlite bodies.

An experiment into the effectiveness of the stream-sediment reconnaissance survey for kimberlite was carried out by Liv Jimin, and Shao Yue in the Zhennan Region. Table (2.6) shows the anomalous dispersion train of Cr, Ni, Nb, La, Zn, Co, Ba and Pb in the stream-sediment samples downstream of a known kimberlite pipe. The dispersion distance of multiplicative anomalies of these elements may reach over 1.6 km. Figure (2.9) indicates that all the index values of Guizhou Province are greater than those of Shandong Province, and the values of Shandong are greater than those of Liaoning. The LRE (light rare earth elements) and TRE (total rare earth elements) products of Liaoning are of the order of 10^8 and 10^{12}, respectively. Enrichments with the LRE and TRE products of Liaoning are observed in the upper part of kimberlite pipe No. 51 (J90, J93). The LRE and TRE

Table 2.6.

The anomalous dispersion train in the stream-sediment samples downstream of a known kimberlite pipe in Fu Xian, Liaoning Province [103]

Sample No.	Distance (m)	Cr ppm	Ni ppm	Nb ppm	La ppm	Y ppm	Zn ppm	Co ppm	Ba ppm	Pb ppm
D 82N 144	0	400	500	35	25	35	300	30	600	60
146	70	700	400	30	40	25	200	30	400	20
147	140	1000	400	35	35	15	350	35	700	40
148	220	450	350	30	30	12	150	20	600	30
149	310	350	350	30	30	12	200	40	400	30
150	410	350	350	50	40	15	200	20	500	30
162	590	150	20	15	18	10	30	10	200	15
163	720	150	70	20	18	10	60	15	200	18
168	860	100	40	15	30	20	35	10	500	25
169	1000	200	30	10	15	10	30	5	300	10
170	1150	150	35	10	15	10	20	5	200	10
204	1600	200	30	12	25	15	30	7	200	15
205	1750	150	20	T	T	10	T	3	100	10
206	1950	170	20	8	20	10	/	10	200	10
208	2120	120	35	10	20	12	20	14	200	13
209	2540	100	20	10	20	12	T	10	200	13
210	2890	120	20	8	/	12	T	5	200	10
211	3240	150	20	8	T	10	T	4	200	10
212	3600	130	10	8	20	10	/	5	200	10
213	4020	150	15	15	/	T	20	10	/	10
214	4200	150	20	10	/	10	T	6	350	10

T = trace content; / = not detected.

	LRE product	HRE product	TRE product
GUIZHOU	5.6×10^{12}	7.9×10^{5}	9.6×10^{18}
SHANDONG	5.0×10^{9}	3.0×10^{5}	1.2×10^{15}
LIAONING	6.2×10^{8}	3.0×10^{4}	1.7×10^{12}

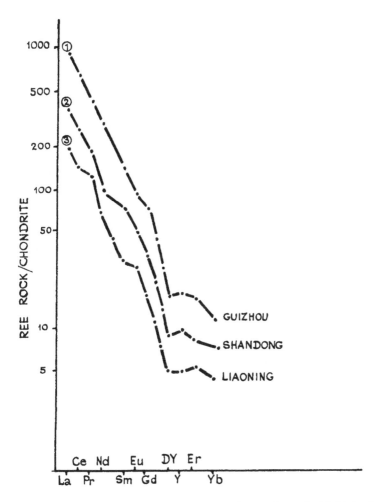

Figure 2.9. The distribution curves of the rare earth elements in porphyritic kimberlite in Guizhou, Shandong and Liaoning Provinces [103].

products of Shandong are of the order of 10^9 and 10^{15}, respectively, which means they are also roughly equivalent to the LRE and TRE products in the middle part of kimberlite pipe No. 51 (J94, J87). The LRE and TRE products of Guizhou Province are of the order of 10^{12} and 10^{18}, respectively, which had no equivalent in pipe No. 51.

The distribution curves of the rare earth elements of the kimberlite in the three Chinese provinces are parallel, with the Guizhou curve in the upper, the Shandong curve in the middle, and the Liaoning curve in the lower position (Fig. (2.9)). The Shandong and Liaoning curves are strikingly similar to curves of J87 and J90, respectively (Fig. (2.10)).

In Liaoning, most of the kimberlites discovered are blind ores, rich in diamond of the gem types. In Shandong, only diamonds for industrial use are found in the kimberlite. In Guizhou, no diamond has been found in the kimberlite discovered to date.

Vertical zoning of the REE (rare earth element) distribution along kimberlite pipes, with the LRE and TRE products increasing from the uppermost part of the kimberlite pipe to its root is observed in the deposits of China. The high value of the LRE and TRE products suggests that most samples from Guizhou were taken from the root of the pipe, where no diamond has been found.

Differentiation between kimberlite with basic rocks and kimberlite with ultra-basic rocks has been possible by analysing Nb, cold extractable Zn and Mg and other characteristic elements [40].

A reconnaissance soil survey was made in and around the mine area; mica-kimberlite and mica-peridotite veins were delineated by Ni and Cr anomalies.

The analysis of aerial photographs of kimberlite bodies as well as observations on the ground have shown that vegetation plays a very important part in this search. It has been found that the vegetative cover is more intensely developed on the pipes than on the surrounding rocks (greater height of trees and shrubs, closer grass and scrub cover complexity of mosses and lichens). The vegetation on kimberlite outcrops is not homogeneous. This heterogeneity depends not only on the influence of the kimberlites but also on the combination of all the factors of the physical and geographical surroundings. With a definite correlation of the physical and geographical surroundings the vegetation of kimberlites acquires specific features, it differs sharply from the vegetation on the rocks enclosing the kimberlites and shows up well on aerial photos. The isolation of the vegetation on kimberlites and its ability to be deciphered can be used as an auxiliary means in searching for beds of diamonds using the aeromethod (taking into account the variation in the concrete features of vegetation in different physical and geographical surroundings).

It is possible to evaluate a productive breccia associated with kimberlite by employing geochemical methods. The samples were collected from the mine area in China and were divided into three groups [103]: 1) various kinds of porphyric kimberlite; 2) productive breccia containing kimberlite material; 3) non-productive breccia and magmatic rock. The concentrations of Cr, Ni, Co, Sr, Nb and La in the first group of samples are the highest. The second group contains very high values of Sb, As, Hg and Y but only moderate concentrations of Cr, Ni, Co and La. The concentrations of the two sets of elements mentioned above are the lowest

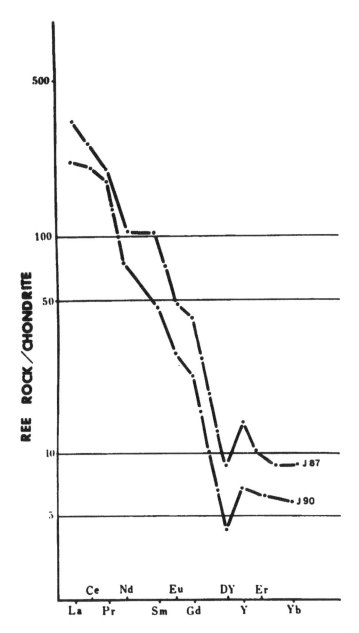

Figure 2.10. The distribution curves of the rare earth elements of two samples in No. 51 kimberlite pipe in Liaoning Province [103].

in the third group. This shows that there exists a unique association of anoma- lously high concentrations of Hg, Sb, As and Y, with moderate concentrations of Cr, Ni, and Nb in productive breccia associated with kimberlite. Figure (2.11)

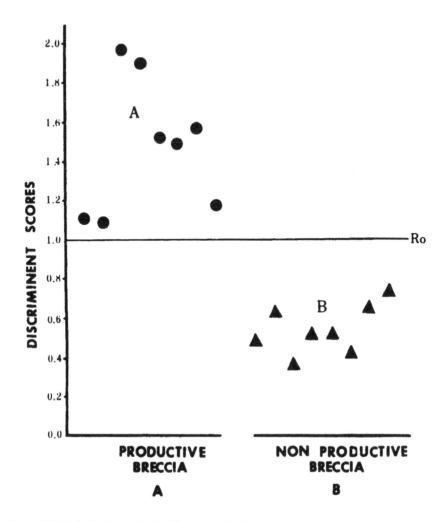

Figure 2.11. Discriminant analysis of known productive breccia and non-productive breccia [103].

shows the results obtained from the geochemical survey of (Cr × Nb), Nb and (As × Sb × Hg) in the productive and non-productive breccia. By using the critical value Ro = 1.08, the two groups, A and B, can apparently be distinguished.

Micaceous kimberlites are characterized by higher concentrations of Nb, Ta, Zr, Hf, Sc, P, Sr, REE, Th and U in comparison with non-micaceous kimberlites. The formation of stable complexes in the presence of ions such as OH^-, CO_3^{2-} and F^- is considered to be a possible explanation for these high concentrations.

REE in South African kimberlites show a different degree of enrichment and LRE/HRE (heavy rare earth elements) fractionation previously only reported in carbonatites. REE in whole rock patterns in kimberlites are influenced by mineralogical effects.

Indications of diamondiferous kimberlites

Primary outcrops of kimberlite bodies or related rocks may be detected as a result of prospecting work, including heavy concentrate sampling, and lithochemical survey by secondary haloes and dispersion trains. Related rocks include picrite-porphyrites and meimechites without the inclusions of deep-seated pyrope-bearing rocks, their minerals and diamonds. It is known that kimberlites are the basic sources of diamonds. Until recently, all porphyritic ultrabasic rocks filling diatremes were considered to be kimberlites, and were therefore subjected to expensive sampling with the aim of detecting diamonds.

The differences in the given types of kimberlites are demonstrated in Table (2.7). Kimberlites differ from allied rocks by a lower content of silica, alumina, ferric and ferrous iron, and sodium oxide.

Table (2.7) illustrates that picrite-porphyrites go beyond the borders of the kimberlite family, with the exception of rocks of the Ingiliysk complex.

Comprehensive differences exist for meimechites, which are poorer than kimberlites in alkalis, calcium, and carbon dioxide, but are richer in iron, aluminum, titanium and chromium, i.e. elements that are found in their basic mass.

Diatremes of trappean composition, confined to the system of deep faults, differ from kimberlites by petrochemical parameters. Traps are characterized by a reduced content of SiO_2, and increased content of the sum of alkalis, as well as magnesium, titanium, and phosphorus. Kimberlites differ from allied rocks by the composition of admixture elements (Fig. (2.12)).

When detecting kimberlites, their potential productivity is determined. The main criterion of diamond content in kimberlites is the presence of associations of mantle (extra-kimberlite) minerals of the diamond–pyrope facies. The mineralogical criteria of diamond content in kimberlites have been developed by N. V. Sobolev and A. D. Kharkiv. The types of paragenesis of minerals associated with diamonds are given in [108]. Analysis of our own material and available data on the mineral composition of kimberlites indicates that garnets, chrome-diopsides, chrome-spinellids, and ilmenites with an increased content of chromium form under conditions of stable diamond crystallization. The stated minerals are characteristic accessories of diamonds. Garnets are divided into [58]:

(1) high-chromic and low-calcic garnets with a knorringite component;

(2) high-chromic and high-calcic garnets of uvarovite-pyrope composition, forming at lower thermodynamic parameters than garnets of the first type; and

(3) garnets with a relatively low chromium content, moderate calcium content, and increased iron content.

Garnets relating to the first and second groups occur in highly diamondiferous pipes and form under conditions that are favourable for the formation of diamonds. A low calcium content in garnets is typical of parageneses, formed at early stages of recrystallization, i.e. of an olivine-garnet composition without clinopyroxene [58]. A direct correlation is observed between the content of chromium, magnesium, and calcium. Magnesian varieties contain up to 19 per cent Cr_2O_3,

Table 2.7.

Chemical composition of kimberlites and picrite-porphyrites, per cent [100]

Oxides	Borders of kimberlite family	Complexes of picrite-porphyrites		
		Ziminsk	Ingiliysk	Chapinsk
SiO_2	19.76–37.44	37.5 ± 0.00	29.74 ± 1.44	31.23 ± 0.87
TiO_2	0.24– 5.37	1.06 ± 0.13	5.33 ± 0.35	2.80 ± 0.26
Al_2O_3	1.33– 7.77	8.02 ± 0.07	5.53 ± 0.29	9.00 ± 0.74
Fe_2O_3 + FeO	3.88–14.08	11.37 ± 0.19	14.74 ± 0.52	15.74 ± 3.09
MgO	14.00–33.06	18.25 ± 0.28	10.06 ± 1.01	10.32 ± 1.54
Na_2O	0.00– 0.81	0.25 ± 0.03	0.45 ± 0.04	0.20 ± 0.00
K_2O	0.00– 3.59	1.82 ± 0.30	1.04 ± 0.10	0.71 ± 0.17

and from 0.6 to 7 per cent CaO. Garnets found in the concentrates from diamondiferous kimberlite pipes are characterized by an increased content of chromium, while the knorringite component in them reaches 40 per cent. An admixture of sodium is typical of garnets (from 0.1 to 0.25 per cent by weight). Alongside garnets of the pyrope-almandine series, with a low calcium content, chromium pyropes have been detected as inclusions in diamonds. The inclusions of garnets from a diamond found in one of the pipes are characterized by a high content (up to 36 per cent) of the calcium component. They contain, (percentages): pyrope 57.1, almandine 5.8, spessartine 0.6, grossular 7.0, andradite 5.9, uvarovite 23.6. The $Cr/(Cr + Al)$ ratio in these garnets reaches 25 per cent [109].

The typomorphic indication of garnets is their colour, whose intensity is conditioned by various concentrations of the knorringite component. In garnets with a Cr_2O_3 content of between 0.2 and 13.0 per cent, the colour varies from light-orange to dark-violet, while Cr_2O_7 content is less than 2.5 per cent in light pink and red varieties, which contain more than 0.5 per cent TiO_2. The presence of green garnets (uvarovites enriched in calcium) is not a mandatory indication of diamondiferous kimberlites.

Chrome-spinellids occur often in kimberlite bodies. Their content is low (up to 0.5 per cent of the kimberlite mass). They occur in the form of solid inclusions in diamonds and growths within diamond aggregates. The former are distinguished from the growths by a lower ferruginosity. Their common specific feature is an increased content of chromium (up to 62 per cent Cr_2O_3) [109]. Chrome-spinellids are characterized by a high $Cr/(Cr + Al)$ ratio, low TiO_2 content, and low Al_2O_3 content. They are clearly differentiated even from those varieties of ultrabasic rocks that are most rich in chromium. Chrome-spinellids occur in the latter rocks with an admixture of ulvöspinels (up to 28 mol per cent) and magnetitic component (up to 30 mol per cent).

X-ray microanalysis and infra-red spectroscopy are used to determine the composition of single grains of chrome-spinellids. Traditional chemical analysis is time-consuming and demands a comparatively large amount of simple material, which is not always available. The application of infra-red spectroscopy is sufficiently simple and gives high performance; the method is based on the linear

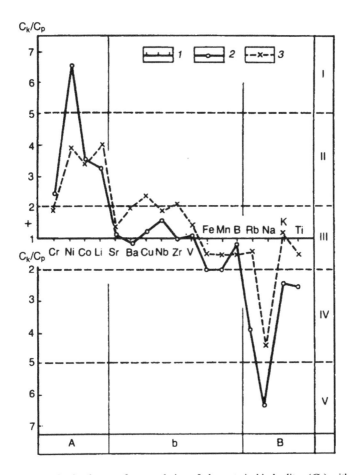

Figure 2.12. Change in the degree of accumulation of elements in kimberlites (C_k) with respect to their content in picrites (C_p) [45]. Concentrations of elements: 1 — in picrite (adopted as unity); 2 — in diamondiferous kimberlites; 3 — in barren kimberlites. A — elements of kimberlite magma; B — elements of both magmas; C — elements of picrite magmas. Changes in the concentrations: I — appreciable accumulation in kimberlites; II — fractionation in kimberlites; III — no changes; IV — fractionation in picrites; V — appreciable accumulation in picrites.

relationship between the parameters of the elemental cell of chrome-spinellids and the value of the absorption band maximum in the range 685–617 cm^{-1}. According to infra-red spectroscopy, a clearly expressed bias of the stated diagnostic absorption band is observed toward the long-wave region of the spectrum with an increase of Cr_2O_3 concentration as found in chrome-spinellids.

The specific features of clinopyroxene composition in kimberlites are the substantial structural admixture of potassium (up to 0.4–0.8 per cent of K_2O by weight), the enhanced role of Al^{VI} in diamondiferous parageneses, and the association with high pressures [109]. Chromium-rich clinopyroxenes with an $Na_2Cr_2O_6$ content of up to 44 per cent have been detected in some pipes, which is possible at

Table 2.8.

Characteristics of kimberlitic zircons [64]

	Kimberlites	
Index	Diamondiferous	Barren
Shape	Rounded grains	Euhedral grains
Cleavage	Common	Absent
Zr/Hf ratio	38.2	53.3
U content, %	Average 16.5×10^{-4}	Average 213×10^{-4}
Coating	Very common	Absent
Fluid inclusions	Common	Absent
Solid inclusions	Kimberlite minerals	Tubular inclusions

high pressures. Orthopyroxenes are represented mainly by enstatites ($f = 5-15$ per cent), containing admixtures of aluminum and chromium. The enstatite inclusions in diamonds contain more chromium, and less aluminum and calcium than peridotite inclusions in kimberlites; an admixture of Na_2O (up to 0.3 per cent) has been identified in enstatites of deep parageneses.

Ilmenites relate to intermediate formations in the series ilmenite–geikielite, with $f = 50-75$ per cent; the minerals contain up to 40 per cent Fe_2O_3, 10 per cent Cr_2O_3, and 1.5 per cent Al_2O_3 [109]. Picroilmenites are distinguished by a wide range of compositional variation. Ilmenite, encountered in diamonds in the form of inclusions, is characterized by an extremely low Fe^{3+} and MgO content (0.11–0.14 per cent). The increased content of manganese in picroilmenites is evidence for the absence of diamonds in kimberlites [74].

The high content of chromium in garnets, clinopyroxenites, chrome-spinellids and ilmenites is associated with the fact that the low oxidation states of chromium are stable at high pressures and temperatures in the upper mantle. Chromium enrichment of the minerals of the early stages of formation is associated also with the effect of the crystal field. Experimental data confirm the theory that the role of chromium increases as pressure rises, and that of high-chromic minerals occur in the deepest parts of the upper mantle.

Zircon is a characteristic accessory mineral of kimberlites, the average zircon content being 1 g tonne. A comparison of the characteristics of zircons from Tanzania and the Republic of South Africa is given in Table (2.8).

Estimation of the level of erosion of kimberlite outcrops

An experiment into the effectiveness of the geochemical method was carried out in the known kimberlite region of Liaoning and Guizhou Province, China [103]. The application of the geochemical method to prospecting for kimberlite, including the use of rare earth element zoning to estimate the level of erosion of kimberlite outcrops was shown in the deposits of China. Samples were collected from different levels. Kimberlite pipe (No. 51) is comparatively rich in diamond. The ICP spectrographic method was used to study the distribution of the rare earth elements — light rare earth elements (LRE) and heavy rare earth elements (HRE) (Fig. (2.13)).

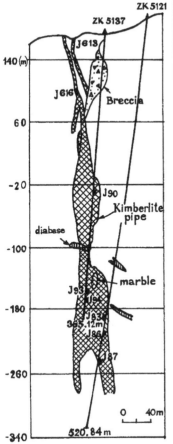

NO.	LRE product	HRE product	TRE product
J613	2.4×10^5	2.1×10^2	5.1×10^7
J616	6.1×10^7	2.8×10^2	1.7×10^{10}
J90	6.8×10^8	3.6×10^2	2.4×10^{11}
J93	7.5×10^8	4.7×10^3	3.6×10^{12}
J94	1.5×10^{10}	8.1×10^3	1.2×10^{14}
J83	6.3×10^8	2.5×10^3	1.6×10^{13}
J86	8.1×10^8	8.4×10^2	6.8×10^{11}
J87	7.2×10^9	6.0×10^3	4.3×10^{13}

Figure 2.13. Zoning of the multiplicative element values of No. 51 kimberlite pipe in Liaoning Province, norhteastern China [103].

From the Fig. (2.13) we could see that the product in the uppermost part of the pipe (J613) is of the order of 10^5, and the total rare earth (TRE) product 10^7. The LRE in the upper part of the pipe (J90–J93), which is rich in diamonds, is of the order of 10^8, and the TRE, of the order of 10^{11}–10^{12}. The values reach a maximum at J94, decrease through J83 and J86 and increase again at J87. This could be explained by [103]: 1) there is symmetrical zoning — the values of the LRE and TRE products, after reaching a maximum in the middle part of the pipe, decrease toward its root; or 2) the values will be highest in the root of the pipe — the fluctuation of the LRE and TRE products at J83–J87 shows that these samples are not taken from the root — and the pipe will extend downwards.

Methodology of evaluating diamond content in kimberlites

Kimberlite bodies, found in the process of prospecting, are graded on the basis of general geological and petrogeochemical materials, and then those with the best prospects are explored by mineralogical methods. The location and facial specific features of the kimberlites are studied. Pipes of primary deposits are most favourable for detection when they are situated in the central parts of the provinces and are characterized by low akalinity and ferruginosity, but an increased content of nickel, chromium and magnesium. The presence of mantle material, evacuated by the kimberlites in the pipe, as well as minerals and xenoliths of the faces of diamondiferous pyrope peridotites are the major factors conditioning the potential presence of diamonds in the kimberlites. Mineralogical analysis of pounded samples makes it possible to grade kimberlites and determine diamond-containing pipes in the total mass. Commonly 300–400 kg samples are taken from workings on a grid of 20×20 m. The samples contain material from all the varieties and zones of the kimberlites. Lump samples are taken simultaneously with the aim of detecting xenolites for correct calculation of the number of mineral phases. More trustworthy data are obtained in this case on the mineral composition of the kimberlite rocks. Crushing of the samples and the extraction of diamonds is performed in two stages:

(1) crushing by stages, with sieving after each stage and isolation of the following classes: $-20 + 8$, $-8 + 4$, $-4 + 2$, and $-2 + 0$ mm;

(2) crushing to 1 mm and concentration in a concentrator.

Representative samples for mineralogical analysis are taken from the enriched concentrates. It is necessary to sample an appreciable amount of rock by this method of preliminary evaluation of the diamond content of the pipes. However, the volume of the samples should ensure the extraction of at least 100 grains of accessory minerals. To this end, it is sufficient to crush samples with a mass of 5–10 kg and to isolate the monofractions of garnets, chrome-spinellids, and zircons. The differences in the species and chemical composition of the accessory minerals serve as a criterion to establish the productivity of the kimberlites. The typomorphic features of garnets serve as the mineralogical criterion of the diamond content. About 100 garnet grains are selected from the concentrates of kimberlite pipes [58]. Their index of refraction and the parameters of the unit cell are determined. An advantageous indication of diamond content of kimberlites is the presence of garnets with indices of refraction ranging from 1.7720 to 1.8002, the parameters of the unit cell being a size of $11\,606 \times 10^{-4}$ to $11\,678 \times 10^{-4}$ cm, the presence of varieties with a violet colour and without indications of dichroism, a $Cr/(Cr + Al)$ ratio of 25, and a Cr_2O_3 content of between 6 and 16 per cent by weight. A high content of Cr_2O_3 is characteristic of chrome-spinellids in diamondiferous kimberlites. Chromites of diamond association are isolated using data from infra-red spectroscopy on chrome-spinellids of other parageneses. An advantageous indication of diamondiferous kimberlites is the development of garnets therein, enriched in chromium; and an unfavourable indication is the presence of minerals of the titanium group. The prospectiveness of kimberlite bodies can be evaluated by calculating the indices of garnetization of the chromium and titanium groups of accessory minerals. To this end, according to the recommendations of V. S. Rovsha, a correlation is determined of the weight amount (percentage) of

Table 2.9.

Coefficient of garnetization

Index	Degree of diamond content in various regions				
	1	2	3	4	5
gr(pyrope)xr	4	4	9	190	55
TiO_2 gr/TiO_2 il (10^{-4})	—	45	770	31	33

1, 2 = low diamond content.

3, 4, 5 = high diamond content.

garnets (gr), chrome-spinellids (xr), and the ratio of titanium levels in garnets to Ti content in ilmenite (TiO_2 gr)/(TiO_2 il) (Table (2.9)).

The kimberlite bodies most likely to have a high diamond contents are those wherein increased concentrations of accessory minerals of the chromium group are observed, with a predominance of associations arising at high pressures (pyrope), and a relatively low amount of chrome-spinellids.

It should be noted that determination of the degree of diamond content in kimberlite pipes by means of mineralogical and geochemical criteria presupposes the isolation of pyropes and chrome-spinellids from pounded samples enriched in chromium. The element is determined in a microanalyser, or by infra-red spectroscopy; trustworthy data on the degree of diamond content in kimberlite pipes are obtained in this case. The typomorphic specific features of zircons can also be used to evaluate kimberlite bodies and establish the origin of these minerals. Highly diamondiferous kimberlites contain zircons, characterized by perfect cleavage, resulting from tectonic shoves, as well as by a reduced content of uranium $(17 \times 10^{-4}$ per cent), and the value of Zr/Hf, which is equal to 38 [64]. The most significant indication of highly diamondiferous kimberlites is the presence of zircons, containing liquid inclusions and whitish zones, confined to healed fissures, as well as characteristic coatings that are easily determined in the field.

Thus the criteria for diamond content in kimberlites are the presence of: (a) garnets, enriched in chromium, poor in iron and titanium; (b) clinopyroxenes with a high content of chromium, potassium, and sodium; (c) chrome-diopsides with a high chromium content; and (d) zircons with well-manifested cleavage and containing liquid inclusions. The mineralogical indication of the Precambrian primary sources is the presence of boron minerals, phosphorus, fluorflorencite, gorceixite, tourmaline, etc.

To enhance the reliability of the conclusions, the mineralogical indications of the depth of kimberlite bodies should be considered simultaneously with the character of element distribution. To this end, diagrams and graphs are used (Figs (2.14)–(2.17)) [45]. The kimberlites being evaluated are compared in these graphs with the standard ones (diamondiferous, poor, and barren). Minimal contents of key elements, namely, Fe, Ti and Al, are typical of kimberlites in diamondiferous fields. Dots in the central part of the diagrams correspond to barren regions (except the correlation Al–Ti), and the highest concentrations of elements are characteristic of picrite-porphyrites. The most advantageous content ranges of

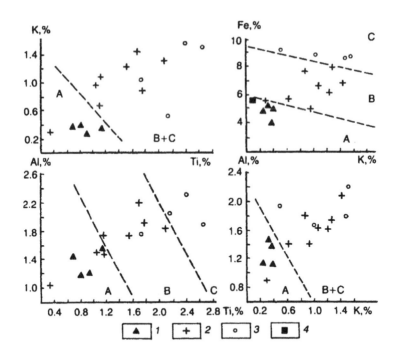

Figure 2.14. Correlation of key elements in kimberlites of various fields in Siberia [45]. 1 — average data for diamondiferous kimberlites; 2 — data for barren kimberlites and kimberlites poor in diamonds; 3 — data for picrite–porphyrites; 4 — data for pyrope lherzolite. A — field of diamondiferous kimberlites; B — field of barren kimberlites and kimberlites poor in diamonds; C — field of picrite–porphyrites.

Fe−K−Al−Ti, within which diamondiferous kimberlite bodies occur, are (percentages): 6 Fe — 0.5 K; 1.8 Al — 0.5 K; 1.2 Ti — 0.5 K.

Estimation of the quantitative content and determination of the diamond quality is very important. Diamonds that do not contain nitrogen (type II) are very rare. They crystallize at a great depth, in the zones of the mantle wherein the magma is not enriched in volatile components. Some type II diamonds have served as centres for the crystallization of type I diamonds. The ratio of the number of paramagnetic centers in colourless diamonds (adopted as a standard) to the total content is a constant value; changes in the conditions of their formation are judged by deviation of this value.

'Nitrogen-free' diamonds are determined by electronic paramagnetic resonance or by the following simple method: the diamonds are placed on the emulsion side of a photoplate sensitive to ultraviolet radiation, and light from a quartz-mercury source or a spectrograph is directed perpendicular to the plate. The blackening of the photoplate indicates diamonds that contain nitrogen.

Stages and sequence of prospecting work

Airborne magnetic survey and interpretation of aerial photographs are performed at the first stage of geological and prospecting work at a scale of 1 : 200 000 to

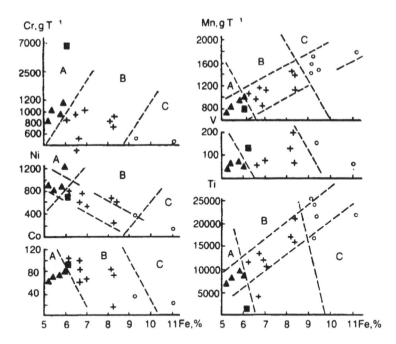

Figure 2.15. Correlation of the ferric group elements in kimberlites of various fields in Siberia [45]. See Fig. (2.14) for symbols [45].

detect faults and kimberlite bodies. Sampling and prospecting are carried out along river dispersion trains. Rapid prospecting of a territory is ensured at a comparatively low expenditure. The presence of diamonds, pyropes and chrome-spinellids in the heavy concentrates, and an increase in the size of their grains, suggest the presence of diamondiferous bodies in the territory being explored. Heavy concentrate sampling of small brooks is performed around anomalies that are remote from traps. Electrical and gravity prospecting are applied to determine the nature of magnetic anomalies.

At the second stage, kimberlite prospecting is performed on the basis of geological surveys carried out at a scale of 1 : 50 000 in conjunction with heavy concentrate sampling, detection of secondary dispersion haloes of indicator elements, and geophysical work. The main purpose is to detect specific zones and areas that are promising for prospecting and exploration for diamonds.

Heavy concentrate sampling of alluvial and deluvial deposits (transition from trains to dispersion haloes), and highly accurate airborne magnetic survey at a scale of 1 : 25 000 are performed in the course of prospecting-surveys during the first field season.

Areas of developed kimberlite bodies are contoured during the second and third field seasons from the data of lithochemical surveys and heavy concentrate sampling of deposits on the slopes, which is followed by ground magnetic survey. In regions where kimberlites outcrop to the surface, or are overlain with a shallow layer of loose formations, magnetic prospecting is applied in conjunction with

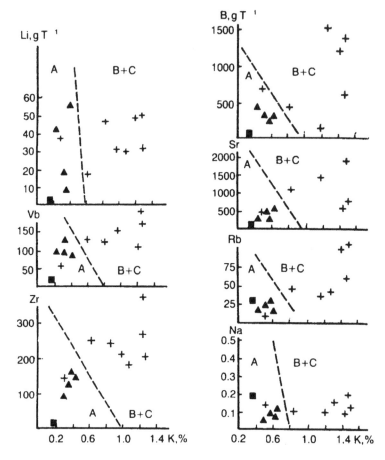

Figure 2.16. Correlation of trace elements in kimberlites of various fields in Yakutia [45]. See Fig. (2.14) for symbols.

mining and drilling operations. Exploration is completed with evaluation of the diamond content of the kimberlites.

Traditional and new types of diamond-bearing rocks and methods for their estimation

Diamond-bearing rocks comprise the following types: kimberlites, lamproites, impactites, eclogite-amphibolites and eclogite-gneiss metamorphic complexes, ultrabasic (ultrabasites) and basic (basaltoids) rocks.

Diamond in kimberlites is a polygenetic mineral and is genetically related to ultrabasic and basic mantle rocks. Kimberlitic magma transported diamond and other minerals of the deep-seated rocks (products of the disintegration of the mantle substratum). These conclusions are substantiated by radiological and mineralogical data, by crystal inclusions with an octahedral cut (negative crystals) present

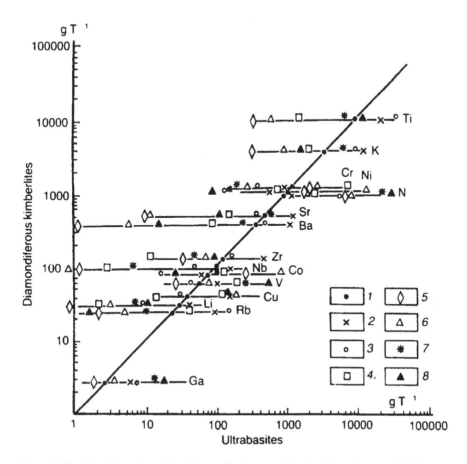

Figure 2.17. Correlation of geochemical specific features of rocks of mantle genesis [45]. 1 — diamondiferous kimberlites; 2 — barren kimberlites; 3 — picrites; 4 — pyrope peridotites; 5 — geosynclinal harzburgites; 6 — chondrites; 7 — continental tholeiites; 8 — oceanic tholeiites.

in diamonds, whose deformation occurred over a long period of geological time (a billion years and even more), and by the presence of diamond in deep-rock xenoliths of various compositions. The characteristics of the composition of kimberlite are governed, as a rule, by the structure of the crystalline basement.

Natural geothermometers, the nature of the dislocation structure of crystals, and the forms of nitrogen show that the crystallization of diamonds occurred at a temperature of 1200–1500 °C, the depth of formation being as much as 200 km. These data necessitate important conclusion:

– the main techniques applied in the search for diamond-bearing kimberlites are mineralogical and geophysical methods based on revealing the structural, compositional and physical inhomogeneities of the kimberlite fields; the division of kimberlites into productive and unpromising cannot be implemented by pertochemical and geochemical methods.

An urgent problem is that of the complex estimation of kimberlites and the subsequent extraction of gold, platinum and precious coloured stones — pyrope, chrysolite, chrome-diopside. Diamonds can be preserved during rock processing on the basis of blending.

Large, rich diamond deposits abroad are associated with lamproites. The characteristic features of lamproites are their Mg, K, Ti, Zr and Ba enrichment by light rare earth elements, the availability of olivine, phlogopite and richterite in the rocks and the presence of such potassium minerals as leucite, wadeite, jappeite and shcherbakovite as well as zircon, barite, garnet (of the pyrope series with low titanium contents), chrome-diopside and chromite. The geological mapping of lamproites and the identification of proper olivine varieties among tuffs are necessary for predictive work. The latter may be carried out in the following sequence: geophysical research (magnetic prospecting, ondometric methods), geological mapping, geochemical work (detection of potassium and barium aureoles), drilling, mine-sinking, large-scale bulk sampling with a view to producing the required quantity of diamonds, and the evaluation of their content and qualities. Large diameter wells are drilled with bulk sampling of the material (sample mass — 40 kg).

The methods currently in use, based on the presence of sub-calcic peridotitic garnet, high Na_2O in eclogitic garnet, and high Cr–Mg chromites and ilmenites, markedly underestimate the grade of the Australian lamproites. This underestimation is partly a result of poor representation in the concentrates of the eclogitic paragenesis which dominates the lamproite diamonds. It also results partly from the peridotitic diamond component present being dominated by the lhersolitic rather than the harzburgitic paragenesis. A further problem for lamproite is the absence or scarcity of megacrystal ilmenite. The common groundmass ilmenite present is at least partly of secondary origin and, therefore, not suitable as an indicator of diamond preservation.

Most lamproites from the West Kimberley and Argyle contain little mantle-derived garnet (Fig. (2.18)).

The lherzolitic mineral suite is prominent within the peridotite diamond paragenesis in Australian lamproites, unlike the African and Siberian kimberlites in which the harzburgitic suite is pre-eminent. This is demonstrated for the lamproites by frequent chrome diopside inclusions within diamond [144] and indirectly confirmed by concentrate mineralogy which, while containing lherzolitic garnets, commonly lacks the sub-calcic varieties although G10 garnets do occur as rare inclusions in the diamonds. As yet, no sub-calcic garnets have been found in Argyle concentrate the world's richest diamond-bearing lamproite (Fig. (2.18)) [145].

The poor representation of eclogitic minerals in concentrates restricts their use in predicting grade and belies the importance of the eclogitic diamond paragenesis at Ellendale and especially at Argyle. Carbon isotope and diamond inclusion work by [144] show that loughly 96% of the Argyle stones and nearly half of those from Ellendale are of eclogitic origin. Analysed Na_2O levels on Ellendale and Argyle eclogitic garnets from heavy mineral concentrates are < 0.05% and therefore not encouraging for the presence of diamonds.

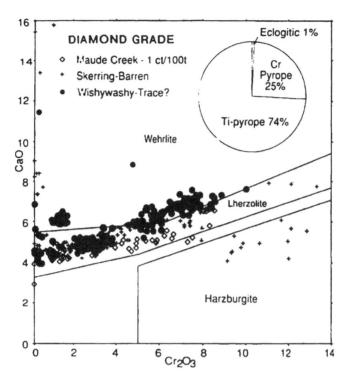

Figure 2.18. Garnets form the Argyle lamproite pipes: CaO versus Cr_2O_3 plot, and pie diagram showing relative abundance of garnet types in concentrate [157].

Most ilmenites recovered from lamproite concentrates (Figs (2.19) and (2.20)) are Mg-poor Mn-bearing groundmass types which cannot be used in diamond grade predictions; a few have moderate to high MgO and moderate Cr_2O_3 but do not have megacryst morphology and therefore have doubtful validity as a predictor of moderate diamond preservation [140, 145, 155].

One problem with lamproites is the paucity of eclogitic minerals in the concentrates which prevents proper assessment of the contribution from the dominant eclogitic paragenesis of the Australian lamproite diamonds [145, 157].

In assessing the diamond content in metamorphic rocks the following characteristics are taken into account as criteria of their productivity:

- wide variations in the mineral and chemical composition of rocks, the frequent alternation of very contrasting Precambrian geological formations metamorphosed under the conditions of the amphibolitic faces of metamorphism;

- the development of a complex and variegated paragenesis of minerals in the zones of potassium metasomatism: graphite, quartz, feldspar, forsterite, biotite, phlogopite, muscovite, spinel, clinohumite, tourmaline, pyroxene (diopside, omphacite), garnet (pyrope-almandine, pyrope-grossular), dolomite, calcite.

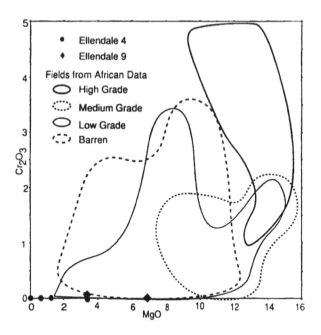

Figure 2.19. Ilmenites from Ellendale lamproites; Cr_2O_3 versus MgO plot [157]. Data from [140].

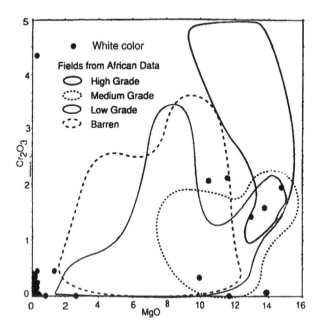

Figure 2.20. Ilmenites from the White color lamproite pipe: Cr_2O_3 versus MgO plot [157]. Data from [155].

Diamonds contained in metamorphic formations are usually fine and green-yellow. Mineralogical (typomorphic signs of garnet, pyroxene) and geochemical features (by potassium aureoles) are used in the search for diamonds.

2.2. BERYL, TOPAZ AND TOURMALINE

Beryl and topaz are gemstones whose transparent, colourless or beautifully coloured varieties are cut and used widely in jewellery. The following types of deposits are known: (a) pegmatitic; (b) greisen; and (c) hydrothermal. Five types of pegmatitic deposits are of major commercial interest, including:

- miarolitic chambered microclinal pegmatites;
- miarolitic cavity microclinal pegmatites (Brazil, Madagascar);
- miarolitic microclinal and microcline–albite pegmatites substituted by cleavelandite and lepidolite (Brazil);
- muscovite–microclinal (beryl–muscovite) pegmatites; and
- beryl-bearing muscovite–topaz–quartz greisens.

Placers, containing the stated minerals, are formed as a result of destruction of the primary deposits.

Shows of beryl and topaz are associated with holodifferentiated albite microclinal chambered pegmatites. Beryl forms in products of the final stage of granite crystallization, especially in pegmatitic melts. Beryl and topaz occur in all the varieties of granite pegmatites, crystallizing in geological phases D–E (yellow and green beryl, aquamarine), F (white beryl, pink vorobyevite with an admixture of alkalis), black schorl, blue indigolite, green wehrdelite, and light blue or pink rubellite. Topaz originates in geological phases E and F (in miarolitic microclinal pegmatites). Migration, redistribution, and concentration of beryllium and boron occur as a result of the activity of fluorine-bearing post-magmatic gaseous–liquid and liquid solutions. Prospecting indications of beryl and topaz deposits are as follows:

(1) Development in the region of pegmatites, occurring in complex multiphase shallow granite intrusives, having a hypsabyssal habit.

(2) Presence of leucocratic, alaskite, or biotite granites with a coarse- or medium-grained structure.

(3) Development of holodifferentiated pegmatites with a clearly visible quartz core, feldspathic, pegmatoid, and aplitic zones. The pegmatitic bodies have a stock-like, tubular, or irregular shape.

(4) Finds of beryl, topaz and their fragments in alluvial and eluvial–deluvial deposits. Beryl and topaz, resistant to chemical weathering, are accumulated in deluvial and alluvial formations in the course of destruction of the indigenous sources of gemstones. Their presence in heavy concentrates presupposes the occurrence of a nearby indigenous source, and the parageneous associations make it possible to determine the type of primary deposit.

The characteristic series of beryl and topaz accessories in pegmatites of the first genetic type are: morion, smoky quartz, albite (basic minerals), ilmenite, zircon, fluorite, magnetite, pyrite, tourmaline, garnet, molybdenite, cassiterite, epidote, apatite, phenacite, arsenopyrite and columbite. The indicator elements of productive mineralization are: F, B, Cl, K, Na, Li, Rb and Cs.

Morion, smoky quartz, microcline, albite (basic minerals), muscovite, veined berthite, spodumene, tourmaline, kaolinite, apatite, columbite, cordierite, desmine, rutile, zircon, amazonite, garnet and monazite are associated minerals of pegmatites of the second type. Microcline predominates over albite in pegmatites of this type. The indicator elements of mineralization are: K, Na, Li, Rb, Cs, F, Be, B, P, Nb, Ta, Ti, Zr, Bi, Zn, Pb and Cu.

Smoky quartz, microcline, aquamarine, topaz, cleavelandite, lepidolite, elvanite, tourmalines, muscovite, cassiterite, columbite and garnet are the most typical paragenous mineral associations of pegmatites of the third type. The indicator elements are: F, Cl, K, Na, Li, Rb, Cs, B, Nb, Ta and Sn.

Quartz, muscovite, citrine, amethyst, columbite, monozite, almandine-spessartine, ilmenorutile, fluorite, apatite and tourmaline are associated minerals of voidless pegmatites of the fourth type. The indicator elements are: F, B, K, Na, Li, Rb, Cs, P, Ti, Nb and Ta.

Quartz, ferberite, cassiterite, hematite, molybdenite, arsenopyrite, bismuthine, fluorite and siderophyllite occur in greisens in association with topaz and beryl. The indicator elements are: Be, B, F, K, Na, Li, Rb, Cs, Cl, P, Sn, W, Mo, As, Bi, Cu, Pb and Zn. Calcite, fluorite, garnet and hematite occur in hydrothermal (postvolcanic) deposits in association with beryl and topaz. The indicator elements are: F, B, Be, Cl and Hg.

Let us consider the geochemical criteria of prospecting for gemstones conforming to each type of deposit.

Deposits of the first commercial-genetic type (holodifferentiated
albite−microclinal miarolitic chambered pegmatites)

Rich concentrations of beryllium, fluorine, boron, lithium and rubidium in pegmatites are a favourable precondition for applying geochemical methods of prospecting for zones containing gemstones. Experimental work by different methods has demonstrated the possibility of applying hydrogeochemical, lithogeochemical and decreptophonic methods of prospecting for pegmatites. Hydrochemical haloes of fluorine and boron have been detected in water, which propagate in the form of a train up to 200 m from the pegmatitic bodies in the direction of the water flow.

Pegmatites occur also with a fluorine halo, propagating over a distance of 5 m and more in different directions from the pegmatite. The fluorine content in the halo is 0.1−0.3 per cent, and the contrast = 0.19. The configuration of the halo in the general form repeats the shape of the pegmatite. The dimensions of steaming haloes reach 10−40 m.

Areas prospective for topaz and beryl are defined by analysing the results of prospecting and geological surveys. Particular attention is given to testing the petrochemical characteristic properties of granites. The following petrochemical properties of the mother granite constitute the prospecting indications of gemstones: a high content of alkaline elements (8.4–9.2 per cent) with the predominance of potassium over sodium; and an increased value of the agpaite ratio (0.8–0.9) with a silica content of 70 per cent and more. These criteria assist in identifying the most prospective magmatic complexes in the total mass of granitoid rock.

Lithogeochemical surveys on a grid of 500×50 m (scale 1 : 50 000) and 100×20 m (scale 1 : 10 000), oriented transverse to the prevailing strike of the orebearing structures, are advisable under conditions of poor exposure. It is advisable also to orient the surveys so that they cross the most likely migration paths of the elements in zones with displacement of the dispersion haloes, or if the latter are characterized by an isometric outline. The lines of the profiles should be drawn parallel to the contour lines of the topography. When prospecting for deep-seated pegmatitic bodies, it is advisable to sample water courses with a maximum discharge of 0.2 $m^3 s^{-1}$ (streams with a maximum length of 4 km) by two to three samples in the mouth, middle and downstream parts. The volume of the water sample should be 0.5 l. The samples are used for detecting fluorine, boron and alkaline elements directly in the field.

Soil sampling is performed in a representative horizon (usually A or A_0) using a furrow cutting the entire thickness of the eluvial–deluvial formations. Fractions of 1 mm are used to detect F, B, K, Na, Li, Rb, Cs and Be.

Reliable grading of the haloes is accomplished by heavy concentrate–geochemical methods. to this end, 2 kg samples of broken rock and fine-grained material are taken from holes 0.5 m in depth. Batches of 50 g each are sieved out of the sample for spectral analysis, and a 10 g sample is washed on the tray for mineralogical analysis. Before grinding and burning, the heavy concentrate fraction is subjected to mineralogical analysis.

Areas with spatial superposition of the mineralogical and geochemical haloes are the most prospective ones and ought to be subjected to detailed survey. The grid of reconnaissance surveys is made denser to define the zones of mineralization (considering that the dimensions of the pegmatitic bodies are usually 100×10 and 80×5 m), and shallow workings are made. The gemstones are fully extracted from the detected cavities, and the samples are documented and sent to the laboratory for a conclusion on the quality of the raw material. The primary haloes of the elements are studied during the prospecting survey. At the stage of detailed prospecting–exploratory work particular attention is given to geochemical investigation of the lower sampling intervals in the holes in order to detect the anomalies of the deep-seated vein bodies. The enclosing rocks should be examined in a similar way when making workings. In homogeneous enclosing rocks, wherein only the pegmatites proper can be the marker horizons, endogenic haloes are the most reliable prospecting indications of veins, especially under conditions of tectonic splitting of the rock mass. Geochemical samples are taken along the profile (bedrocks at the surface, walls of workings, core samples from holes) at 3–5 m intervals. Five on six chip samples of rocks with a size of 3–4 cm^2 across are combined into one sample from an interval of sampling. The samples are disintegrated in a jaw crusher to 1 mm. This fraction is used for decrepitation analysis. After crushing and quartering, a part of the material is subjected to grinding and analysis. The samples are analysed for a wide range of elements (F, B, K, Na, Li, Rb, Cs, Y and Yb). Multiplicative haloes (Na × K × Li × Rb) are constructed to increase the contrast of endogenic anomalies and to obtain more objective information, and decrepitation analysis of bedrock samples is performed for detecting steaming haloes.

The mineralogical and geochemical zonality is studied. The following zonality (from the periphery to the centre) is a favourable prospecting indication: (a) zone of graphic pegmatite; (b) zone of pegmatoid and block pegmatite; (c) quartz core; and (d) area of free growth. The following are the mineralogical and geochemical indications of prospecting for zones with gemstones:

- presence of coarse flake muscovite in macrocrystalline pegmatite of the axial parts, with biotite in the peripheral parts of the body;

- development of albitization processes, availability of albite pseudomorphs replacing the microcline; and

- increased concentrations of fluorite, garnet, muscovite and hydromica, as well as berylium, fluorine, boron, lithium and rubidium in the pegmatites.

It should be considered that negative and positive anomalies of fluorine and sodium are observed directly near the nests, and muscovite is enriched in lithium in the direction from the periphery toward the pegmatite centre.

Combined application of mineralogical, geochemical, and geophysical methods (radiowave sounding) ensures reliable detection of pegmatitic bodies and makes it possible to correct the direction of prospecting.

Similar techniques of geochemical survey may be used on pegmatitic and greisen-vein fields of other genetic types. Some specific methodological features of this work are discussed below.

Deposits of the second genetic type (zonal albite–microcline and albite–amazonite–microcline miarolitic pegmatites)

Petrochemical specific features of granites associated with pegmatites are as follows: increased silica content (about 70 per cent), moderate alumina content (about 14 per cent), relatively high alkalinity (the sum of alkali content is about 8 per cent) with an insignificant predominance of potassium over sodium, and an average value of the agpaite ratio (about 0.75).

Zonal pegmatitic veins, containing gemstones, form under calm tectonic conditions in supra-intrusive zones and occur in mica-enriched rocks. The occurrence of amazonite is an indirect indicator of jewellery topaz, while rose quartz in the core indicates a possible occurrence of beryl. Ore-free microclinal and plagioclase–microclinal pegmatites, lacking miarolitic cavities, differ from pegmatites with gemstones by their considerably lower content of rare alkaline elements and high K/Rb values, drawing these pegmatites closer to mother granites.

The following fluctuations in the element concentrations (g tonne^{-1}) [115] are found in potash feldspar of magmatic pegmatites, crystallyzing at an insignificant temperature and pressure drop in the absence of screens for the accumulation of volatiles. Lithium, 14–22 (average 18±3); rubidium, 260–620 (average 526±55); cesium, 11–30 (average 26 ± 4). The values of K/Rb and Rb/Cs change very little. The variance is characterized by low values. Miarolitic cavities with gemstones are not found in these veins. The concentration of rubidium and cesium increases to 950 g tonne^{-1} and 60 g tonne^{-1}, respectively, and the values of K/Rb and Rb/Cs decrease during autometasomatic reworking of potash feldspars from pegmatites. Miarolitic cavities with rubellite, lepidolite, and fluorite form in the hanging walls of veins.

The geochemical features of pegmatites with gemstones are conditioned by postmagmatic processes. Miarolitic cavities form under the effect of solutions enriched more in Rb, Li and Cs than in potassium.

The variance of the Li, Rb, and Cs contents increases greatly in veins with superimposed metasomatic processes, while the K/Rb and Rb/Cs values experience considerable fluctuations as compared to the primary magmatic bodies. Superimposition of several successive stages of metasomatic mineralization over the primary mineral parageneses intensifies the contrast of mineral distribution. The average content of rare elements in potash feldspars also increases.

Rubidium and cesium, as related to potassium, are also accumulated in biotites and muscovites of pegmatites in the later stage of the postmagmatic process. The value of K/Rb decreases in the following succession: lepidolite–rubellite, muscovite–beryl–topaz, and muscovite–wehrdelite pegmatites. The parameters of distribution of rare alkaline elements in potash feldspars, muscovites and biotites are the criteria of the presence of miarolitic cavities to be considered during prospecting and evaluating work. The major index of pegmatite productivity is the variance of element concentration, which is specified by the character and intensity of post-magmatic mineral formation in the pegmatites (geochemical factor). The values of K/Rb, K/Cs and Rb/Cs are the most effective indicators when evaluating pegmatites for gemstones.

Deposits of the third genetic type (zonal microcline–albite with spodumene miarolitic pegmatites with intensive development of the lepidolite–cleavelandite complex)

Many petrochemical and mineralogical–geochemical prospecting criteria of such deposits are the same as for the second type deposits. Miarolitic pegmatites with comprehensive lithium mineralization are found in rocks enriched in magnesium and iron, i.e. in ultrabasites, basic rocks and amphibolites. The evolution of lepidolite–cleavelandite substitutions with the development of coloured tourmalines is characteristic of miarolitic zonal rare-metal pegmatites.

Deposits of the fourth genetic type (beryl–bearing muscovite-microclinal pegmatites without miarolitic cavities)

These deposits resemble micaceous pegmatites by the geochemical conditions of formation and are therefore not discussed in detail here. Pegmatites of muscovite composition (large plates of green muscovite) with tabular and lath-like albite are prospective for jewellery beryl.

Deposits of the fifth genetic type (beryl-bearing muscovite–topaz–quartz greisen)

The formation of thick haloes of hornfels with an increased concentration of Sn, W, Be and F are characteristic of exocontact zones of greisen rocks. Granites, associated with greisens, are notable for a high content of silica and alkalis. The specific feature of productive zones is the presence of an argillomicaceous mass, wherein topaz and beryl crystals are concentrated.

The types of miarolitic pegmatites differ in the content of Li, Rb and Cs in K-feldspars. In crystal-bearing (commercial for quartz crystals) pegmatites, there are relatively low levels of Rb and Cs, but, in topaz–beryl types the levels are substantially higher. The maximum content of Li and Cs is in K-feldspars from the tourmaline type of miarolitic pegmatites.

Miarolitic pegmatites from the south–western Pamirs are a source of coloured tourmaline. In pegmatite fluids, potassium predominates over sodium and the content of rare elements is close to the clarke with the pronounced enrichment of the volatile materials with boron. Therefore, the pegmatites are essentially orthoclase in composition. Within fields of coloured tourmaline-bearing miarolitic pegmatites, the composition of K-feldspar and plagioclase can be used to assess pegmatite productivity (Table (2.10)). Commercially productive veins contain K-feldspar and plagioclase with Li and Cs contents 6–8 times higher than in non-productive veins; Rb is about 2–2.5 times higher. Even weakly productive veins can be distinguished from either non-productive or commercially productive ones.

A high content of rare alkalis in K-feldspars from spodumene — pegmatites has been found not only in Siberia, but also in many regions of the world [104]. The petalite type of Siberian pegmatites displays similar concentrations of Rb and Cs in the K-feldspars as those from the Bernic Lake (Tanco) deposits in Canada [119]. The Ba/Rb ratio in K-feldspars from complex pegmatites is as low as 0.002–0.004, about five orders of magnitude lower than it is in muscovite pegmatites. As was shown for one example of the spodumene-type deposits in Siberia, it is possible to use the correlations of rare alkalis in K-feldspar to carry out mineralogical mapping in the pegmatite field and to forecast the commercial importance of individual pegmatite bodies among a cluster. The pegmatitic bodies

Table 2.10.

Average content of Li, Rb and Cs (ppm) in feldspars in miarolitic pegmatites with different tourmaline production capacity [104]

Pegmatites	K-feldspars			Plagioclases			Region
	Li	Rb	Cs	Li	Rb	Cs	
Non-productive	27	662	36	21	5	2.3	Malkhan field
Weakly productive	70	880	103	64	11	7.0	Transbaykal
Commercially productive	210	1180	211	179	13	10	

of deposits are accompanied by primary haloes of a wide range of elements. The negative haloes are formed by Na, Ti, Cu, Ba and Sr, and the positive ones are formed by F, K, Li, Rb, Cs, As, W and Hg.

The size and nature of the primary haloes around pegmatites are determined by the type of pegmatite, its degree of replacement, the type of the host rocks and the size of the pegmatitic body.

It has been ascertained that multiplicative haloes have hardly any gaps and are more unbroken than monoelement haloes. Geochemical investigations of wall-rock aureoles in proximity to pegmatite bodies have been extensively described by [11, 104]. From these works, it is evident that certain types of pegmatites cast aureoles enriched in elements diagnostic of the type (B, Li, Rb, Cs, Be, F) and that such elements may migrate several hundreds of feet from their source, guided largely by non-diastrophic and diastrophic fabrics present in their host rocks.

A zonation aureole can be used to estimate the depth of burial of a pegmatite body, and accordingly, if the body has a dip or plunge component, in which direction is occurs (Fig. (2.21)). Accordingly, it can be seen that the lithium data decrease uniformly as the depth of burial of the pegmatite increases from the south-central area of the pegmatite body. The rubidium data behave similarly to those of lithium, but decrease much more rapidly and hence uotline a smaller portion of the pegmatite. Cesium data are non-existent over pegmatite and hence suggest that the body is buried too deeply for detection.

The zonal constitution of the elements is expressed most obviously by the $Li^2/(Rb \times Cs)$ ratio, which decreases gradually when the depth increases (e.g. in the deep direction of a pegmatitic vein). At the same time, the significant ratio $(Ba \times Sr \times Pb)/(Zn \times Cr \times Cu)$ decreases in the negative haloes. The zonal constitution of the positive haloes formed by F, Cl and Hg has a uniform character not only in a qualitative but also in a quantitative respect. Thus, in any haloes the Cl/F ratios are similar at the same level.

According to our study the following methods and geochemical operation can be recommended for prospecting purposes:

(1) Express spectral analysis of the whole volume of the geochemical samples for a wide range of elements, including Ti, Cu, Ni, V, Mo and Na. Its aim is to reveal negative anomalies, some of which may be connected with ore bodies;

(2) Quantitative analysis of a number of samples taken from the location of a negative anomaly in Li, Rb, Cs, F and Cl. Its aim is to select anomalies caused by commercially productive pegmatites.

(3) Semi-quantitative analysis of some selected samples for Hg and forms of mercury. The number of forms of mercury is greater and its thermal spectrum wider in the supra-ore zone than in the zone beneath the ore.

(4) An effective method of obtaining wider and more obvious geochemical haloes consists in analysing the distribution of different heavy mineral fractions of geochemical samples instead of the same samples as a whole.

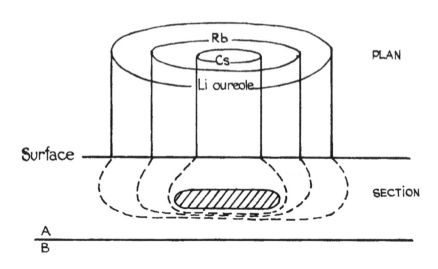

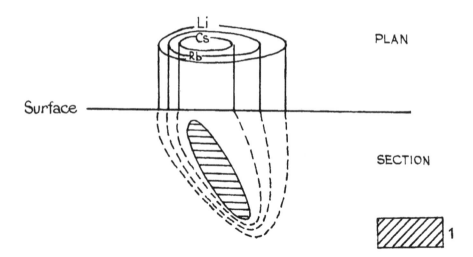

Figure 2.21. Schematic representation of geochemical wallrock aureoles of Li, Rb and Cs developed about horizontal cylindroidal pegmatites (A) and plunging cylindroidal pegmatites (B) [119]. 1 — Pegmatite.

(5) The geochemical estimation of the mineral potential of the deep-seated levels and the nearest flanks by prospecting with a different degree of detail on the basis of the zonality of negative haloes and Cl and F haloes. The use of specific features of Li, Rb and Cs haloes to clarify the nature of anomalies.

(6) The geochemical determination of the present level of erosion of the anomaly, based on the zonal constitution of Cl, F and Hg.

2.3. EMERALD

Emerald, a dark-green chromium-bearing variety of beryl, relates to the group of gemstones of the first class. It occurs in three genetic groups:

(1) pegmatite (miarolitic microclinal pegmatites);

(2) pneumatolytic–hydrothermal (glimmeritic greisens); and

(3) hydrothermal (telethermal calcite and pyrite–albite veins).

Deposits, associated with pegmatites and occurring in basic and ultrabasic rocks, are small in size. High-quality jewellery emerald is encountered very seldom (USA, Norway). Deposits of the second type are the major source of jewellery varieties of emeralds. They are localized in ultrabasites, diorites, and phlogopite glimmerites that are associated genetically with acid and ultra-acid granites. The formation of these deposits is conditioned by pneumatolytic–hydrothermal processes in ultrabasites as a result of which thick phlogopite and biotitic margins have formed along the selvage of quartz–plagioclase veins.

The most favourable conditions for the formation of emeralds have appeared as a result of the effect of aggressive beryllium-bearing components of granite melts, enriched in fluorine, chlorine, and boron, on chromium-bearing ultrabasites (up to 15 per cent Cr_2O_3). Glimmerites (concentrating chromium-bearing varieties of beryl) are associated with pegmatites, albites, quartz–feldspar and quartz veins. Emerald mineralization, localizing only in glimmerites, were formed at the final stage, following the formation of phlogopite. Multistage formation of mineral phases is clearly manifested at emerald deposits of the glimmerite type. The injection of pegmatitic veins occurred at the first stage, accompanied by weak manifestation of hydrothermal changes. The second, productive stage resulted, through repeated tectonic movements, in the origination of emerald-bearing formations due to the effect of mineral-forming solutions on the substratum rocks and pegmatitic veins, accompanied by the formation of phlogopite–biotitic glimmerites. Chlorine, F, CO_2, H_2O and Hg were introduced from the depths, whereas the basic mass of glimmerite formations was derived from the enclosing rocks. Muscovite–fluorite and muscovite–albite veins formed at the beginning of the third hydrothermal stage, and quartz and quartz-sulphide veins with calcite and zeolites formed at the last stage.

Telethermal deposits are associated with quartz, calcite and albite veins, occurring in shales, aleurolites and limestones. The authors believe that the basic mass of alkali and volatile components, as well as beryllium, have been introduced from the depths. The effect of aggressive solutions on the enclosing rocks caused mobilization of such elements as Fe, Mg, Ca, Si, Cr, V and Sc. Mineral parageneses of emerald-bearing zones depend on the composition of the enclosing rocks.

The following indications are used in prospecting for deposits of emerald-bearing pegmatites:

• presence of microclinal pegmatites and granites, occurring in ultrabasites and mica schists;

• exposures of pegmatites, containing miarolitic cavities.

The following indications are used in the search for deposits of beryl (emerald-bearing) greisens:

(1) Pegmatites and veins of granite among serpentinous ultrabasites, as well as at the contact of the latter with diorites and amphibolites.

(2) Presence of glimmerites, accompanied by pegmatites, albitites, quartz–feldspar and quartz veins.

(3) Direct finds of glimmerites.

For telethermal deposits, the presence of shales and arkoses at the contact with quartz–calcite and albite veins in carbonate rocks; and direct finds of calcite, dolomite–calcite and pyrite–albite veins forming stockworks, are indications used in prospecting.

The mineral composition of emerald-bearing zones is quite diverse. Microcline, quartz, tourmaline, biotite, albite, beryl and kaolin are most widespread in endogenic deposits of the pegmatitic type. Deposits of the pneumatolytic hydrothermal type are characterized by diverse lithological–petrographic composition of the enclosing rocks (shales, amphibolites, diorites, ultrabasites) and, accordingly, of the mineral composition of emerald-bearing zones: phlogopite (biotite), apatite, alexandrite, tourmaline, fluorite, chrysoberyl, topaz, garnet, muscovite, clinohumite, molybdenite, chalcopyrite, pyrite, zinwaldite, margarite, hornblende, tremolite, talc, chlorite, calcite and zeolites. The basic mass of emerald gravitates to the phlogopite zone. Phlogopite of glimmerites contains inclusions of sphene, apatite, zircon, ilmenite, rutile and pyrrhotite. Green and grey glimmerites occur in emerald bodies [129]. Grey glimmerites appeared later and replace the green ones. Emeralds occur more frequently in grey glimmerites. Zones of altered rocks (actinolite, actinolite–chlorite, chlotite and talc) develop around glimmerite veins, occurring in serpentinites. When glimmerites form at the contact of serpentinous ultrabasites with diorites or amphibolytic gneisses, the latter are affected only by biotitization; if tremolitic rocks are the enclosing ones, then phlogopite–tremolite, tremolite–chlorite–talc, and chlorite–tremolite zones develop around the glimmerites.

Aqueous extracts of emerald green and beryl take the following composition (in equivalent per cent): Na (54–64), K (24–31), Li (2–5), Mg (1.5–3), Cl (7–19), F (1–10), bicarbonate (62–92), and sulphate ion (up to 8). Emeralds are associated in some deposits (Zimbabwe) with pegmatites localized in intensively altered ultrabasites. The pegmatites contain lepidolite, spodumene and tantaloniobates. Tremolite, garnet, haematite, magnetite, muscovite, biotite, limonite and tourmaline occur in emeralds as inclusions. The typical accessories of emerald of the telethermal type are pyrite, calcite, dolomite, aragonite, fluorite, barite, albite, goethite, apatite, parisite and graphite.

The data stated above, and the results of studying the mineralogical composition of glimmerite zones, indicate that the most effective indicator elements of emerald mineralization are fluorine, chlorine, boron, beryllium, potassium, sodium, rubidium, lithium, chromium and titanium. The most typical indicator elements of the deposits of the first genetic group are boron, fluorine, beryllium, potassium and sodium. Deposits of the second genetic group contain admixture-elements, characterized by endogenic introduction (fluorine, beryllium, boron, chlorine and potassium were introduced at the pegmatitic and pneumatolytic stages, whereas mercury and carbon dioxide came in at the hydrothermal stage), and by migration in the zones (impoverishment of the zones of productive mineralization in chromium, titanium, sodium, copper at the pneumatolytic stage, and in vanadium at their mobilization from the enclosing rocks). Fluorine, sulphur, carbon dioxide, potassium and beryllium were introduced into telethermal deposits from the depths, whereas calcium, iron, magnesium, chromium, thorium, vanadium, selenium, copper and nickel were derived from the enclosing rocks. The major element responsible for the colour of emeralds is chromium, whose content in glimmerites ranges from 0.01 to 0.47 per cent, whereas maximum concentration of the element is characteristic of magnesian varieties [129]. Titanium is present along with chromium in essentially aluminiferous glimmerites [129]. The following correlations are observed between the elemental contents: negative (chromium–titanium, aluminum–iron) and positive (silicon–magnesium). In the process of potassium–fluoric metasomatosis, chromium goes into solution and migrates with other elements, which results in the formation of emeralds. Thus, migration and redistribution of elements is observed were there are signs of emerald mineralization as well as at other types of hypogenic deposits, that condition the potentiality of using geochemical methods of prospecting. Secondary dispersion haloes of the stated elements surrounding glimmerite zones are the basis for using geochemical prospecting for endogenic emerald-bearing bodies. Reduced concentration of chromium and titanium in glimmerites are related to prospecting prerequisites, associated with the specific features of formation of emerald mineralization. The plotting of graphs of chromium distribution from the data of sampling loose deposits and as a result of profile traverses of bedrocks has demonstrated that glimmerite zones are identified by negative anomalies of chromium. Gradual reduction of the chromium content in the direction from the ultrabasites to the central zones of glimmerite complexes is conditioned by the removal of chromium. An increase in the fluorine and potassium concentration is observed in the same direction. Experimental work has proven the applicability of magnetic and electrical prospecting and neutron logging in the search for emerald-bearing zones.

Currently, geochemical methods are used most extensively at the stage of prospecting for emerald-bearing zones. These methods allow detection of productive blind bodies that do not outcrop to the surface, and they allow evolution of the vertical zonality of the deposits with the aim of identifying the present level of erosion of the ore-bodies and the extension of mineralization. Geochemical methods make it possible to grade glimmerites into productive and unproductive ones, thereby reducing the volume of expensive sampling.

The indicator elements of emerald deposits of different types are ore-forming and associated elements with universal distribution, that is to say Be, F, Na, K, Li, Rb, Cs, Mo, Pb, Zn, Cu, Ag and Hg.

Deposits of different compositions are formed in different geological conditions, and this is reflected in the composition of the associated primary geochemical haloes. F, K and Na are typomorphic elements of high-temperature deposit haloes

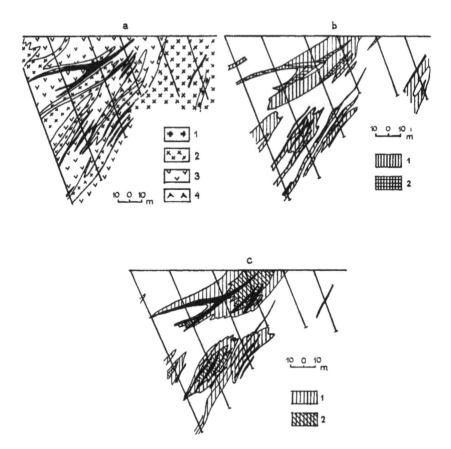

Figure 2.22. Primary haloes around emerald veins. Black bodies from glimmerites. (a): 1 — Plagioclasites, 2 — diorite, 3 — serpentinites, 4 — amphibole-rock. (b): Cesium, 1 — 100–1000 g/t, 2 — 1000 g/t. (c): Rubidium, 1 — 100–1000 g/t, 2 — 1000 g/t [78].

related to acid intrusions; rare elements and sometimes Li and Rb are typomorphic elements of hydrothermal deposits.

Hydrothermal deposits of emerald are accompanied by primary geochemical haloes of a complex composition which are considerably larger than the ore bodies and develop more than 150–200 metres above the level of the ore bodies (Fig. (2.22)). A close spacing and close genetic relationships have been established between primary and secondary haloes of emerald mineralization. As the secondary haloes inherit the quantitative relationships of the elements characteristic of the primary haloes, it is possible to use the above criteria to interpret and assess the secondary haloes.

The characteristic feature of haloes of emerald deposits is vertical zonality, displayed contrastingly in the direction of dip of the ore-bearing zone and expressed by the following series (from top–bottom): Hg–As–Sb–F–Cl–Li–Na–K–Rb–Ag–Pb–Zn–W–Cu–Ni. Any member of the series may be excluded or reduced in relation to the deposit type, but the above vertical variation in element associations is maintained.

These regularities in the distribution of the indicator elements in the primary geochemical haloes have made it possible to develop criteria for the geochemical forecasting of emerald deposits. Special coefficients of formation type (c) have been proposed for the interpretation of the formation type of mineralization [59]:

$$C_1 = (F \times Li)/Pb^2 \quad \text{and} \quad C_2 = K^2/(Li \times Rb).$$

Data on the degree of complexity of halo composition and the value of the zonality coefficient of type $C_3 = (Li \times Cl \times F)/(Cu \times Ni \times Co)$ are utilized to assess the level of erosion of the geochemical anomaly.

The variety of glimmerites is caused by the mineral composition of the host rocks and the temperature and chemical activity of the solutions. Accumulation of emeralds up to industrial concentrations occurs with the repeated redistribution of Cr. The final low-temperature stages of metasomatic rock transformations are the most favourable for the migration and concentration of Cr. To appreciate the potential ore content of glimmerites, we applied a group of geochemical methods of investigation, which makes it possible to recognize the geochemical signs of metasomatite emerald content.

According to the above methods, the geochemical signs of commercially productive zones of emerald-bearing glimmerites are:

1. Commercially productive zones of glimmerites as against ore-free zones are characterized by higher levels of Cr, F, Li, B and Hg, and decreased concentrations of Ti.

2. A relatively increased content of disseminated mercury is peculiar of ore-bearing glimmerites (5×10^{-6} — in ore-bearing, 1×10^{-6} — in ore-free).

3. Increased content of minerals in productive zones of glimmerites.

The presence of emeralds in glimmerites is conditioned by a combination of a whole series of interrelated factors: temperature, acidity and alkalinity of mineral-forming solutions, and the composition of the enclosing rocks, which is manifested in the specific features of the chemical composition of glimmerite complexes and quantitative relationship of such elements, as potassium, sodium, chromium, titanium and fluorine. The following factors are favourable for the formation of glimmerites: (a) presence of enclosing rocks rich in chromium (ultrabasites); (b) increased content of potassium and fluorine in the mineralforming fluids, conditioning the aggressiveness of the solutions and mobilization of Cr from the substratum rocks; and (c) optimal ratio of chromium and titanium, because the latter forms solid mineral phases with chromium Cr_2TiO_5 at high titanium concentrations, thereby removing chromium from the system, and preventing the colouring of beryl. The process of potassium–fluoric metasomatosis of the ultrabasites results in changes is the overall concentration and correlation between the oxide and silicate chromium in favour of the latter, while the entire chromium in the glimmerite zone is silicate [129].

Until recently, assessment of the productivity of glimmerites due to non-uniform distribution of emerald mineralization has been performed by means of bulk sampling. Glimmerites were removed from the total mass of rocks by hand. Considerable volumes of underground workings (up to 70 per cent) were made in glimmerites containing no emeralds.

Evaluating criteria for determining the content of emeralds by geochemical indications have been elaborated by comparing the chemical composition of micas and glimmerites with the data of bulk sampling of the latter [129]. Highly productive glimmerites contain, per cent: $K_2O > 7.2$; $F > 2.8$; $0.2 < Na_2O < 0.05$; $TiO_2 < 0.15$; $0.8 < Cr_2O_3 < 4$. High concentrations of fluorine (> 4 per cent) are necessary for the formation of emeralds at low concentrations of beryllium in the system (< 0.01 per cent) [129].

Methods of work

Loose formations are sampled along the profiles, oriented crosswise the probable strike of the ore-bearing structures, to detect and contour secondary haloes in those regions, where the bedrocks are overlain by eluvial–deluvial deposits. Prospecting by lithogeochemical dispersion haloes is performed usually within prospective areas, detected during preceding geological surveys. Areal (at $1 : 50\,000$ and $1 : 10\,000$ scales) geochemical and mineralogical sampling of loose formations is advisable on promising sites at the prospecting stage in semiclosed regions. During geochemical exploration samples with a mass of $200–300$ g are taken from the horizon of eluvial–deluvial formations, which is representative under given conditions. A horizon is considered to be representative when the specific features of distribution of the indicator elements manifest most close correlation with the anomalies in the bedrocks: the primary source of hypergene anomalies. After grinding a fine fraction (< 0.1 mm) of the samples is subjected to analysis.

The identification of emerald-bearing bodies by geochemical methods consists in detecting anomalies of potassium, fluorine, chlorine, beryllium and boron around glimmerite bodies. Prospecting for the latter is impeded by the fact that similar anomalies may be produced by pegmatitic veins and granite dykes. When grading anomalies, it is necessary to use additional geochemical and mineralogical criteria:

(1) presence of haloes, caused by introduction and removal of chromium, titanium and sodium, whose combination indicates the epicenters of glimmerites;

(2) presence of direct indicators of emeralds, i.e. beryllium, fluorine, boron and chlorine, associated with the pneumatolytic (productive) stage;

(3) composite character and contrast of anomalies, and superposition of haloes of different elements in space.

Combined application of mineralogical and geochemical methods is quite effective. The heavy concentrate method gives good results in prospecting for emerald deposits. When applying this method, it is necessary to consider the character of the emerald accessories, which can indicate the type of primary deposit. The finds of emerald tourmaline, apatite, beryl, topaz, fluorite, phlogopite and lithium micas in the heavy concentrates are favourable prospecting prerequisites. Heavy concentrate and geochemical samples are taken in regions with thick loose formations (tens of metres) from boreholes tapping bedrocks. When selecting sampling points, it is necessary to consider geological factors. Samples from deluvial formations are taken along the prospecting lines, arranged perpendicular to the direction of the assumed dispersion train (by contour lines along the slope). In this case, the distance between the sampling points along the lines should be less than between the lines. Samples are taken on a square grid in the case of a gently sloping topography in the region being explored.

Areal geochemical sampling of bedrocks along profiles, oriented across the strike of assumed mineralized zones, is performed in open regions at the stage of prospecting within promising areas. The sampling grid is 250×25 m (scale $1:25000$), and 100×10 m (scale $1:10000$). In some cases, it is 50×5 m (scale $1:5000$). Bedrocks are sampled by the dotted furrow method. Five or six rock chips (at equal spacings) are taken from the sampling interval ($3-5$ m) and are combined into one sample with a mass of $150-200$ g. The pegmatites, glimmerites and fractured zones are sampled separately. The sampling interval is changed with the alteration of rocks. Each sample should comprise chips of only one variety of rocks. It is advisable to take samples of 5 kg from these rocks, which are crushed, and then are concentrated with the aim of studying the heavy minerals of the bedrocks. After crushing (to a size of 1 mm) the sampled material is quartered: one half is preserved as a duplicate, and the other is subjected to analysis after grinding. Geochemical anomalies are contoured by the value of the minimal anomalous contents calculated with a 5 per cent level of significance (unilateral 2.5 per cent). The method of constructing multiplicative haloes ($Li \times Rb \times Cs$) is advisable to enhance the anomalies. Such haloes are characterized by considerable size, uniform distribution, and absence of background sites within them. The results of areal sampling of bedrocks are plotted on graphs in the form of geochemical maps (arrangement of anomalies). The results of mineralogical explorations are considered when interpreting geochemical anomalies. In particular, indices, recommended by A. I. Sherstyuk, are recorded. Namely, the correlation of ultrabasites and diorites (coefficient $U/D-I$) and the index of heterogeneity, and alternation of rocks ($C_2 = 0.3$), indicating the presence of emerald-bearing zones.

With the aim of quantitative estimation of vertical zonality, it is advisable to calculate the indices of zonality and the change in the multiplicative indices of the haloes for the following elements; Ba, Sr, As, Sb, Hg, Ga, Cu, Co, V, B and F. Increased concentrations of Ba, Sr, As, Sb, Hg and Ga are characteristic of supra-ore zones, and of Co, V, B, Ni and Cu for zones beneath the ore.

The present level of erosion of the anomalies is estimated by comparing the ratio of the average contents and productivities of element pairs with those in haloes of the known ore bodies. Upon detection of glimmerites, they are graded into highly, weakly and unproductive ones. Geochemical studies can be applied to isolate the emerald-bearing zones from the total mass of glimmerites.

Geochemical criteria for detecting most productive glimmerite zones within ultrabasites are associated with the specific features of the composition of these rocks. It is necessary to consider in this case the increased content of characteristic (typomorphic) elements: potassium, rubidium, fluorine, chlorine, beryllium and boron.

A geochemical diagram is used to evaluate the emerald content of glimmerites, as well as a graph of the glimmerite composition (Figs (2.23)–(2.25)). They are used to grade glimmerites into highly, weakly and unproductive ones. It makes it possible to discriminate the most prospective zones, deserving detailed exploration and bulk sampling. However, certain difficulties arise in some cases at interpreting geochemical data, owing to the fact that the chemical composition

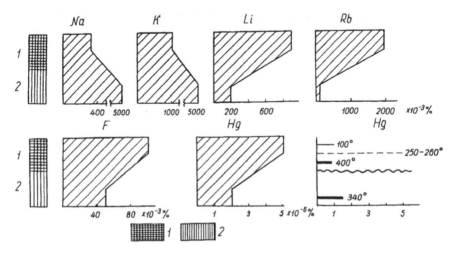

Figure 2.23. Content elements in glimmerites: 1 — highly productive with emerald; 2 — nonproductive glimmerites.

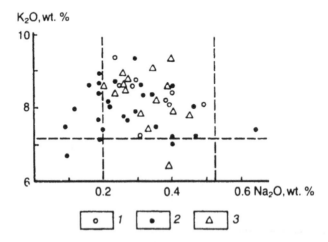

Figure 2.24. Graph of dependence of bulk composition of potassium and sodium in glimmerites [57]. 1 — highly productive glimmerites; 2 — glimmerites that have not been characterized by bulk sampling; 3 — low productive glimmerites.

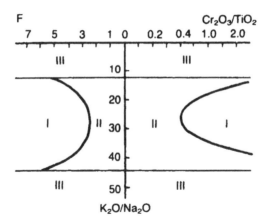

Figure 2.25. Geochemical diagram for evaluating emerald content in glimmerites [57]. Types of glimmerites: I — highly productive; II — low productivity; III — unproductive.

of glimmerites depends on the composition of the substratum. When glimmerites are localized in crystalline schists, the potassium/sodium ratio is greater than in diorites. The following factors should be considered to be sure of reliable grading of the glimmerites:

- the presence of indicators that are characteristic of the metasomatic process in glimmerites, replacing ultrabasites: fluorine, chlorine, beryllium, rubidium, boron and mercury, whose contents are independent of the composition of the enclosing rocks;
- polymorphism of the phase states of mercury in the rocks;
- presence of negative anomalies of chromium, which in the process of potassium–fluoric metasomatosis is removed from the enclosing rocks and settles with other elements, thereby causing the formation of emeralds; and
- high radioactivity of glimmerites (0.07 A kg^{-1} at a background of 0.001 A kg^{-1}).

The typomorphic specific features of phlogopites are taken into consideration. This last one among glimmerites, substituting ultrabasites, is characterized by low indices of refraction and a greyish-brown colour. Emerald-bearing sites are contoured in the dispersion halo of accessory beryl by the green colour of the latter, as well as by the presence of chrome spinellid grains. It should be taken into account that prospective apoultrabasite glimmerites contain such accessory minerals as chromite and pyrrhotite. The chromium–titanium ratio therein is $1-2.7$ (1.0 on the average), and $0.02-0.08$ in apodiorite glimmerites (0.07 on the average).

During prospecting work special attention is given to detecting grey varieties of glimmerites, which are the most productive ones. Intermediate varieties of glimmerites differ very little owing to the gradual alteration of their properties. When grading glimmerites, the latter are subjected to the effect of oxalic acid. Productive grey glimmerites are loosened (in mass) along the intergrowth planes of phlogopite flakes in 5–10 days, whereas the green ones are loosened in 3–4 weeks [129]. When processing emerald-bearing formations, it is advisable to use the chemical method to extract emeralds from the glimmerites.

2.4. CORUNDUM, SAPPHIRE, RUBY AND SPINEL

The basis of this mineral group is corundum, whose chemical composition meets the formula Al_2O_3. Ruby, a transparent variety of corundum, is of light-red or dark-crimson colour. Light-blue, blue, green, yellow, violet and pink varieties are called sapphires. Transparent varieties are used in jewellery. Commercial deposits of the latter relate to the exogenic type. They are formed as a result of destruction of basic sources.

Natural corundum, α-Al_2O_3, often colourless but allochromatically coloured by different elements, is found as a highly prized gemstone in three varieties: ruby, sapphire and corundum. As in violet corundums (Table (2.11)) low contents are also detected in blue, green and yellow corundums by microprobe analysis.

The yellow colour with weak pleochroism caused by lattice imperfections in sapphires from Sri Lanka results with the added presence of Cr^{3+}, in the rare orange corundum which is named padparadscha and — like the irradiated yellow sapphire — is only known in Sri Lanka.

The following primary deposits of corundum are known:

(1) magmatic;

(2) contact–metasomatic;

(3) skarn;

(4) pegmatitic;

(5) greisen; and

(6) metamorphogenetic.

They differ by the petrographic composition of the enclosing rocks, the shape and dimensions of the ore bodies, the relationship with the enclosing rocks, and the alteration of the latter around the ore.

Table 2.11.

Chemical data of natural corundums (electron microprobe) [101].

Location	Colour	SiO_2	TiO_2^*	Cr_2O_3	V_2O_3	$Fe_2O_3^{**}$
Sri Lanka	yellow	0.005	0.013	0.009	0.007	0.03
Sri Lanka	light blue	0.02	0.03	0.003	0.006	0.11
Sri Lanka	dark blue	0.02	0.03	0.007	0.004	0.17
Australia	bluish-green	0.005	0.08	0.09	0.001	1.04
Thailand	bluish-green	0.005	–	0.09	0.006	1.40
Umba, Tanzania	violet	0.01	0.001	0.15	0.03	0.41
Longido, Tanzania	red	n.d.	0.07	1.08	1.22	0.07
Tsavo Park, Kenya	red	n.d.	0.05	1.37	0.34	0.27

n.d. = not determined.
* total titanium as TiO_2
** total iron as Fe_2O_3.

Deposits of the first genetic type are associated with alkali-basalts and vein basaltoids (basic lamprophyres)

Two viewpoints exist on the genesis of gem-quality corundum:

(a) corundum realises excessive alumina at the stage of magma protocrystallization; and

(b) the mineral crystals are xenogenic, entrained by the magma from the wall rocks at its intrusion. The characteristic series of corundum accessories is: spinel, magnetite, zircon, garnet, apatite, zeolite and cancrinite.

Deposits of the second genetic type

The deposits are associated with massifs of secondary quartzites, occurring at the contact of acid-effusive rocks with granites. Effusive formations (lava, pyroclastic rocks) with the ratio of molecular amounts $(Na_2O + K_2O + CaO)/Al_2O_3 \leqslant 1$ are most favourable for the formation of alumoquartzites. Andalusite, haematite and muscovite are associated minerals.

Deposits of the third genetic type

These are associated with calc-magnesian and silicate skarns. In the first case, the enclosing rocks are represented by dolomitic marbles and calciphyres, that occur at the contact with granites. The associated minerals of ruby are spinel, chondrodite, forsterite and diopside. The granites are rich in alkalis and silica, but poor in calcium and magnesium. In the second case, the enclosing marbles and calciphyres occur at the contact with syenites. Nest-like accumulations of sapphire are associated with oligoclase, andesine, ruby, phlogopite, scapolite, apatite, forsterite and pyrite.

Deposits of the fourth genetic type

Deposits of the fourth type are associated with corundum pegmatites, occurring in zones of development of granite pegmatites, accompanying the intrusions of granites or alkali syenites. Occurrences of corundum pegmatites are divided into two subtypes: corundum syenite–pegmatites and corundum plagioclasites rich in manganese. The first are closely associated with alkali and nepheline syenites, localizing within their limits or amongst the syenite enclosing rocks, but near the contact with intrusives. The enclosing rocks near the vein bodies are rich in biotite. Sapphire, occurring in veins, is associated with magnetite, apatite and phlogopite. Corundum plagioclasites (plumasites, kyschtymites, borzovites) occur in the form of dykes among ultrabasites and their serpentinous varieties. Altered rocks, represented by biotitized, phlogopitized, chloritized, vermiculitized and talc varieties, occur near productive bodies. The associated minerals of corundum are green spinel (pleonaste) and tourmaline. Occurrences of gem-quality sapphire at the contact of dolomites with pegmatites are known (India). The associated minerals are actinolite, tremolite, tourmaline and garnet.

Deposits of the fifth genetic type

These are confined to glimmerite greisens in ultrabasic rocks. The enclosing serpentinous ultrabasites are intruded by gangue rocks (granite–aplites and pegmatites). Ruby and (rarely) sapphire, occurring in biotite–phlogopite glimmerites, are associated with plagioclase, chromite, talc, chlorite and serpentine.

The structure of altered rocks enclosing ruby mineralization in one of its occurrences, demonstrates mineralogical geochemical zonality, i.e. alteration of the plagioclase zone to micaceous and amphibolic occurs from the centre to the periphery. An introduction of potassium is observed in the productive micaceous zone; the content of this element decreases in amphibolites, but still remains high as compared to the original rock. Strong variations in the sodium content have been found at the contact with glimmerites. The formation of metasomatities is accompanied by redistribution of chromium in dunites. Low concentrations of chromium have been recorded in the centre of the plagioclase core; its content increases from the micaceous to the amphibolitic zone and at the contact of the latter with dunites. The content of chromium reaches 30.9 per cent at the contact of amphibolites, dunites and talc chlorite chromite formations. The content of Cr_2O_3 diminishes to 0.3 per cent in altered dunites in the contact zone. The zone of contact of dunites and metasomatites is characterized by the introduction of volatile elements (F, Cl). A gradual increase in the Mg, Cr, and Fe content is observed in dunites at the transition from altered to unaltered varieties. Ruby is formed without the introduction of aluminum but during its redistribution in the process of metasomatic alteration of the basic rocks (gabbro).

Deposits of the sixth genetic type

Such deposits are confined to metamorphosed, alumina-rich formations, and are divided into four subtypes:

(1) ruby-bearing marbles;

(2) magnetite–haematite, chloritoid and margarite emeries in marbles;

(3) spinel emeries in metamorphic schists, gabbro-norites and granites; and

(4) corundum rocks in crystalline schists. A series of ruby accessories is quite characteristic of the first subtype: phlogopite, garnet, forsterite, diopside, actinolite, tremolite, chondrodite, scapolite, spinel, graphite, and pyrite.

A geochemical diagram Figs (2.26) and (2.27) to evaluate the ruby-bearing marbles and content of Mg and Sr in calcite of ruby-limestone.

Emery is formed during metamorphic processes in the conditions of amphibolitic facies of metamorphism under the effect of alumina-bearing mineralforming solutions on carbonate formations. Ore bodies are localized among old metamorphic rocks (marbles and gneisses). They are usually stock-shaped. The accessory minerals of productive mineralization are corundum, chloritoid, pyrite, magnetites, rutile, diaspore, apatite, biotite and sericite.

Deposits of the second subtype, occurring among crystalline limestones and represented by sheet lenses, are accompanied by pyritized and graphitized rocks.

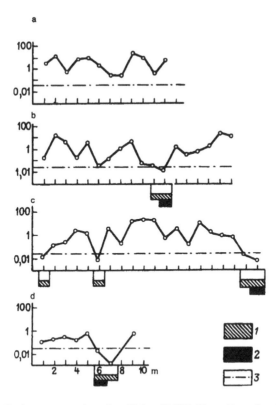

Figure 2.26. Distribution concentration of coefficient K-(100 Mg × Fe × Sr ×H$_2$O) around ruby marbles. a — non-productive rocks; b — productive rocks; 1–2 — marbles: 1 — with ruby; 2 — little ruby types rocks; 3 — anomalous level.

Emeries are known that are enriched in sulphides subjected to limonitization, and easily detectable at the surface. Spinel emeries contain increased concentrations of magnetite and, therefore, can be detected by magnetic prospecting. Corundum is associated with spinel, magnetite, plagioclase and sillimanite.

Deposits of the fourth subtype form lens-shaped bodies among enstatite sillimanite–cordierite, cyanite schists and plagioclase gneisses. Corundum is associated with muscovite, chlorite, cyanite and sillimanite.

The heavy concentrate method is the most effective one in prospecting for corundum and its gem-quality varieties. The shape of the corundum crystals, the degree of their roundness, and the character of paragenous associations (which may indicate the genetic type of the primary sources) should be considered. It is advisable to apply the heavy concentrate-geochemical method:

- in territories devoid of loose cover, or in the case of its poor development, by sampling natural exposures;
- at the edges and lower levels of the deposits, by taking samples from workings and boreholes; and

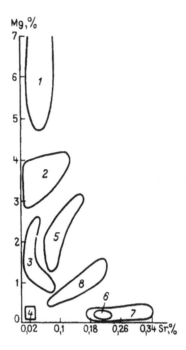

Figure 2.27. Content of Mg and Sr in calcite of ruby-limestone rocks. 1–7 marbles: 1 — magne-sians; 2–3 — calcites; 4 — with ruby; 5 — contact with pegmatites; 6 — marbles and skarn with phlogopite; 7 — metasomatic rocks; 8 — vein of phlogopite–calcite.

- at sites overlain with allochthonous formations, by sampling the core from shallow test wells. The mass of the samples should be 5–10 kg. The samples are first subjected to crushing (to 1.00 mm). The crushed sample is then washed on a pan to the 'grey' heavy concentrate. The varieties of the corundum are determined by mineralogical analysis.

Residual secondary dispersion haloes form over the shows of corundum, brought by tectonic and denudation processes to the present level of erosion. A totality of chemical elements, occurring in the ore bodies in increased or decreased concentrations compared to the enclosing rocks, is characteristic of each type of corundum deposit. The typomorphic indicator elements of the deposits of the first type are Al, K, Na, Zr and F; of the second type, K, Na, Li and Rb; of the third type, K, Na, Cl, B and P; of the fourth type, F, K, Na, P and B; of the fifth type, Cr, Ni, F and Cl; and of the sixth type, Al, P and F.

Corundum minerals in the weathering zone are the most stable ones. Their stability is conditioned by the properties of the component elements, represented mainly by hydrolysers (Al, Ti, Zr, Fe). Corundum migrates into secondary dispersion haloes without essential changes during the weathering process. The entire excessive aluminum in the secondary dispersion haloes (as compared to the background) is in a free state.

Therefore, lithogeochemical survey is performed with the aim of detecting the secondary dispersion haloes of the main corundum-forming elements. Samples should be taken from a depth of 30–40 cm from a horizon transient to the parent rock. The sampling step is 20 m, which should be reduced to 10 m over corundum deposits. Fraction of 1 mm, taken from the BC horizon, is subjected to analysis. The samples are analysed for Al, Fe, Si and Ti by the X-ray spectral fluorescence method. In addition, a wide range of elements are determined by means of the express spectral analysis.

Secondary dispersion haloes of Al (as well as K, Na, P, Cl and F), associated genetically with shows of corundum, contain an excessive amount of these elements as compared to the enclosing rocks. To enhance the trustworthiness of the geochemical data, the results of lithogeochemical surveys should be presented in the form of maps of isolines of complex indices:

(1) $n = (\text{Al} \times \text{Fe})/\text{Si}$;

(2) $n_1 = (C_i \times \text{Al}_2\text{O}_3)/(C_i \times \text{SiO}_2)$;

(3) $n_2 = \dfrac{C_i}{C_b}\text{Al}_2\text{O}_3 + \dfrac{C_i}{C_b}\text{Na}_2\text{O} + \dfrac{C_i}{C_b}\text{K}_2\text{O}$;

(4) $n_3 = (\text{F} \times \text{Cl})$,

where C_i is the concentration of an element in an individual sample; C_b is the background concentration of an element; n is an index indicating the rock enrichment in aluminum and iron; n_1 is an index of rock enrichment in aluminum; n_2 is the summary coefficient, considering the content of aluminum and alkali elements; and n_3 is the multiplicative index of the halogen concentration.

Other elements, accompanying corundum, can also be used in lithogeochemical prospecting, especially in the case of their appreciable concentration in mineralized zones. However, it is very difficult to detect secondary dispersion haloes in deposits of this type of raw material owing to their weak contrast.

The geochemical distribution of rubidium and strontium in the central granulite belt of Sri Lanka, where many of the gem deposits (corundum, spinel, tourmaline, zircon) are found, has been described in the paper by Gamage (1992) [31]. This paper discusses the use of Rb–Sr ratios in stream sediments and rocks to delineate areas. The gem minerals found in stream sediments are derived from the adjacent bedrock that forms the hillsides of the alluvium-filled valleys. The seven areas, shown in Fig. (2.28) represent different levels of known gem potential, as follows [31].

(1) Areas of known high potential (Opanayake and Hattota Amuna); extensive gem production in these areas is derived from stream sediments.

(2) Areas of moderate potential from which some gem occurrences have been reported (Bogawantalawa and Pubbiliya).

(3) Areas of low, very low or no potential (Nilambe and Ridigama).

The analytical results showed that the concentrations of Rb and Sr in the sediments varied significantly from area to area. In the first area high Rb–Sr ratios were observed for the gem-bearing zone west and southwest of Opanayake. Most of the prominent gem-bearing locations had an Rb–Sr ratio of around 1.0 or more,

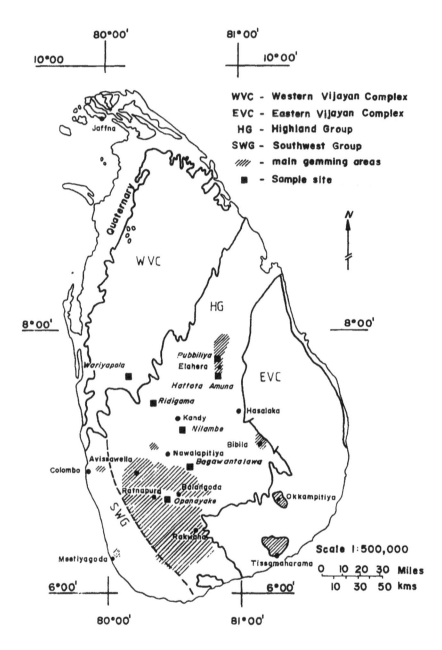

Figure 2.28. Map of Sri Lanka showing main geological groups, gemming areas and sample sites [31].

particularly in the gem pits. In the stream sediments, a value of approximately 0.4 could be used as a rough guide in the delineation of target areas for detailed gem exploration. Figure (2.29) illustrates the Rb–Sr plots for areas with different gem potential. It is observed that even though there is a certain degree of

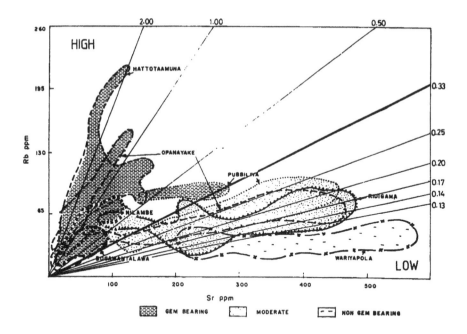

Figure 2.29. Rb–Sr plots for the stream sediments [31].

overlap between the fields, there is a clear variation in the Rb–Sr ratios from high gem-bearing to low gem-bearing regions. The plots of Rb–Sr (Fig. (2.30)) show that there is a wide scatter for the metamorphic rocks (gneisses, charnokites, marbles, quartzites, granites). Even though the potential for using the Rb–Sr ratios of the different rocks in the main gem-bearing and non gem-bearing rocks may appear to be limited, an understanding of the role of magmatic rocks in the genesis of gem minerals may enhance their use. Figure (2.30F) illustrates this point clearly. The Rb–Sr ratios obtained for the granitoid intrusives in the Highland and South-western Group (known to be gem-bearing) and the Western Vijayan Complex (known to be non gem-bearing) are markedly different. We consider the concentration levels of Sr and Rb in the rocks to be the key geochemical tracer of the degree of development and the extent of mobilization of ore-formation.

The method of the Sr–Rb classification of geological formations was developed on the basis of prospecting studies. The method makes possible the reliable and express estimation of the potential ore content of complexes at the stage of prospecting and surveying and the delimitation of the ore productive areas at the stage of prediction and evaluation.

The rubidium content of the rocks is the most effective indicator in the lithium-fluoric type of rocks. Sr and Rb can be considered a universal indicator of vein productivity for the whole series of granulitic types.

The origin of precious spinel deposits is different. The geological reasons for prospecting for and predicting deposits are based on their connection with defi-

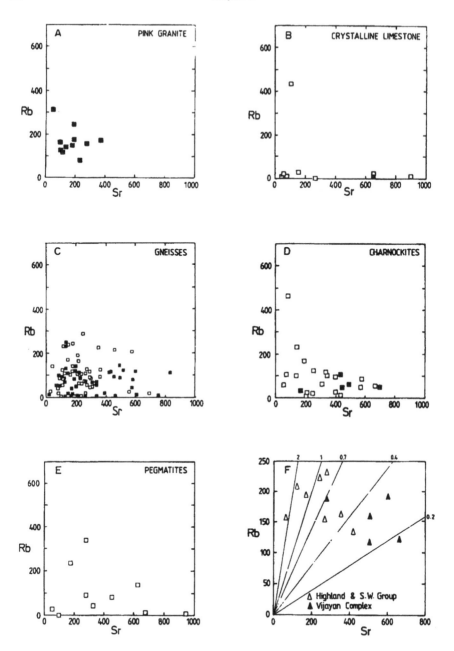

Figure 2.30. Rb–Sr plots for rocks [31].

nite geological formations, taking into account the lithological, structural tectonic and mineral–geochemical peculiarities of these formations. Of main interest are magnesian skarns, glimmerites and plagioclastites in crystalline shields and central massifs.

Table 2.12.
Average abundance of the elements in spinel (%).

Number	Elements									
	SiO_2	TiO_2	Al_2O_3	FeO	MnO	MgO	Cr_2O_3	V_2O_5	H_2O^+	Σ
1	1.35	0.01	70.00	1.59	0.03	27.15	0.02	0.02	0.23	100.38
2	0.70	0.01	70.25	1.80	0.05	27.20	0.01	0.01	0.30	100.33

The evaluation of deposits is based, first and foremost, on the quality of the mineral products and the scale of mineralization. The prospecting methods depend on the nature of the distribution of the useful component: 1) sharp irregular pockets; 2) irregular-impregnated; 3) relatively uniform and compact, continuous or interrupted. The necessary volume of bulk sampling decreases from the first group to the third and the role of drilling increases. The composition for some typical spinels is shown in Table (2.12).

In prospecting for concentrates, samples enriched with heavy minerals (concentrates) are produced by sieving and classifying 10 litres of book sediment. The production and processing of the samples are very expensive, but in the case of some elements it is only possible to obtain enrichment up to the range of detectability by classification. The grain-size fraction greater than 0.15 mm is investigated mineralogically, the fraction smaller than 0.15 mm, by spectral analysis. As a

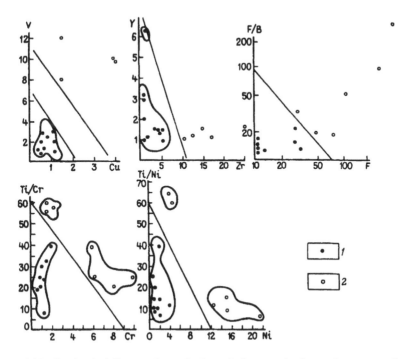

Figure 2.31. Geochemical diagrams for evaluating spinel content in skarns: 1 — non-productive skarns; 2 — productive skarns.

result, we can mark off the catchment areas of anomalous samples. On the basis of the stated association of minerals, the method makes it possible to reach conclusions about the corresponding assemblages.

In spinel deposits of economic importance there are distinct, contrasting and large primary haloes of direct (K, Rb, Sr) and indirect indicator elements. In prospective areas of less than economic value, the haloes of indicator elements are small in size and loose. Deposits are characterized by a more distinct zonality of the haloes, the redistribution of elements and the presence of leaching haloes; as a rule, there is a positive correlation between K and Rb, Li and Hg.

Hydrothermal spinel deposits are usually associated with the passage of large volumes of fluid through the rocks, and so it might reasonably be expected that in zones of recrystallization, near to the focus of mineralization, this action would lead to an increase in the abundance of fluid inclusions in the associated rocks. Studies of spinel deposits have revealed that the frequent presence of halite-bearing fluid inclusions can be correlated with the areas of mineralization.

A geochemical diagram used to evaluate the spinel content of rocks is shown in Fig. (2.31).

2.5. RHODONITE

Rhodonite is a gemstone whose basic component is a mineral of the same name, belonging to the wollastonite group [Mn, Ca] Si. Commercial deposits of rhodonite occur rarely, mainly in association with manganese mineralization. The following types of rhodonite deposits are distinguished: (1) rhodonite skarns on polymetallic ore fields; and (2) regionally metamorphosed manganese-bearing igneous sedimentary strata.

Deposits of the first type are localized in carbonate formations, wherein skarn bodies with rhodonite form deposits of irregular shape. Rhodonite rocks, containing various solid mineral admixtures, are usually characterized by poor ornamental properties and low value. The associated minerals are calcite, rhodochrosite, Mn-garnet, quartz, bustamite, manganosalites, spharelite, galenite, wollastonite, hedenbergite, ilvaite and epidote. The indicator elements of rhodonite skarns are manganese chlorine, fluorine, lead, zinc, copper, molybdenum, strontium, barium and mercury. The main structures controlling the arrangement of rhodonite skarns are fan-shaped dislocations with a break in continuity, developing along the line of contact of limestones with small intrusions of granodiorite-porphyries and quartz monzonite-porphyries. Rhodonite makes up low-temperature skarn deposits formed during bimetasomatic and infiltrating processes in the fractures of limestones, or at the contact of the latter with granitoids. This pattern may be a prospecting indication of rhodonite. It should be taken into account that carbonate rocks are favourable substratum for the formation of skarns, and the major factor for the formation of rhodonite is the presence of manganese-rich rocks. The manifestation of rhodonite in silicic-slate formations and deposits of rhodonite in jasper formations are distinguished among deposits of the second type. Rhodonite deposits in silicic-slate formations, relating to the largest ones, are localized in quartzites, quartz–coaly–sericite and quartz–sericite slates. The productive zones of rhodonite are small (thickness 3–5 m, length 10–15 m), lens-shaped concordant bodies. The indicator minerals are manganese garnet, bustamite, tephroite, rhodochrosite, piedmontite, manganese and iron oxides, hausmannite and spessartite. The indicator elements are Mn, Pb, Cu, Zn, Ba, Cl and Hg. The ornamental properties of the raw material are very good. Rhodonite, quartz–rhodonite, and rhodochrosite–rhodonite form aggregates of bright pink, red, crimson, and spotted colours.

Rhodonite deposits associated with jasper formations are localized in jasper, tuffite, tufogenic argillite, and flinty and mica slates. Productive bodies have a lens-shaped, stratified form, and extend over several hundred metres along the strike, reaching a thickness of up to 10 m. The accompanying minerals are garnet, manganocalcite, tephroite, rhodochrosite, piedmonite, braunite, quartz, calcite, barite, hematite, pyrite, chalcopyrite and chalcedony. The indicator elements are Mn, Cu, Pb, Zn, Ba and Sr.

The contents of elements (Fig. (2.32)) show that there is a wide scatter for rhodonite in various types of rocks.

Rhodonite deposits of these groups relate to metamorphogenetic formations that have developed in the process of transformation of primary sedimentary or igneous (exhalation)-sedimentary and carbonate manganese ores under the effect of regional metamorphism.

Magmatic, lithological formation, and metamorphic factors are considered when prospecting for rhodonite deposits. Rhodonite deposits of the metamorphogenetic

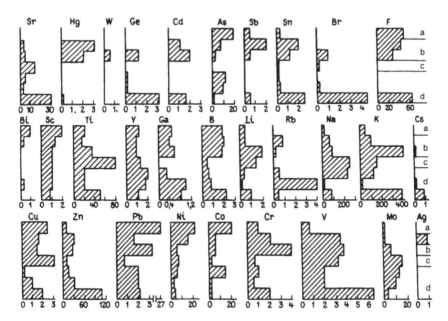

Figure 2.32. Content of elements in rhodonite from various formations. Formations: (a) — rhodonite on polymetallic ore fields, (b) — skarns, (c) — jaspers, (d) — silicic-slates. Content of the elements $n \times 10^{-3}\%$ and (Hg, Mo, Br) $n \times 10^{-6}\%$.

type are associated with igneous–siliceous formations and the spilitic–keratophyric type of volcanism in geosynclinal regions. The concentration of manganese in igneous–siliceous rocks (jasper, quartzitic-like rocks), formed as a result of underwater effusive activity of the spilitic type, is the geochemical criterion of prospecting for rhodonite. Rhodonite occurrences in calcic skarns lack any considerable practical significance and are characterized by small concentrations of brightly coloured mineral (in pegmatites and hydrothermal veins, in manganese-bearing formations). They are usually explored alongside with the ore content.

Rhodonite showings are localized in rocks that were metamorphosed in greenstone slate, or epidote–amphibolite facies. Silicic formations, that are spatially separated from carbonate (calcic) rocks, are most favourable for detecting deposits of highly ornamental varieties of semi-precious rhodonite.

Figure (2.33) illustrates the compositions of some typical rhodonite from various formations.

Mineralogical and geochemical criteria ('manganese' caps of considerable size and characteristic black colour, as well as the presence of skarns enriched in lead and zinc) play a certain role in prospecting alongside with direct most reliable prospecting indications (finds of rhodonite in exposures and in disintegrated blocks piled up on mountain slopes). High manganese concentrations in pyroxenes are favourable indications for finding rhodonite. Associations of accessory minerals and, first of all, of silicates are indirect prospecting indications for finding rhodonite mineralization. Thus, the presence of high concentrations of braunite or rhodochrosite are favourable factors for the development of rhodonite mineralization and a wide thermal spectrum of mercury sublimation in altered rocks in the presence of low-medium- and high-temperature forms (Fig. (2.34)).

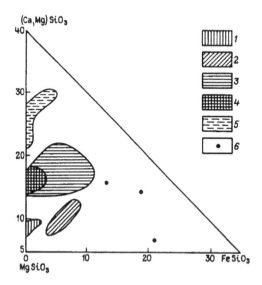

Figure 2.33. Graph of variation in composition of rhodonite from various formations. Formations: 1 — pegmatites; 2 — silicic-slate; 3 — skarn deposits; 4 — jaspers; 5 — gondite series rocks; 6 — rhodonite on polymetallic ore fields.

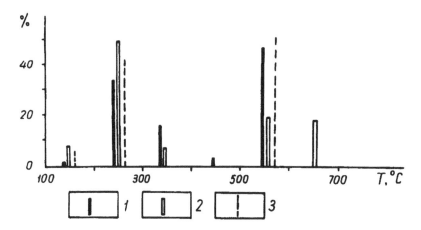

Figure 2.34. Regularities governing the changes in the forms of mercury in rhodonite from deposits: 1 — Ural; 2 — Uzbekistan; 3 — Siberia.

Prospecting for rhodonite at a scale of 1 : 200 000 is carried out in regions with igneous–sedimentary strata, as well as in areas with silicic slate or jasper formations. Primary attention is given to exploring potentially productive formations (with geochemical sampling) along regional dislocations with a break in continuity by rare crossings transverse their strike.

When prospecting at a scale of 1 : 50 000, the work is concentrated in areas with silicic slate or silicic formations. Manganese bearing horizons of rocks are traced and explored. Geochemical samples are taken to detect manganese, as well as barium, strontium, elements of the ferric group, and mercury. Geochemical surveys can be carried out in closed regions to detect dispersion haloes of manganese in eluvial–deluvial formations. Most prospective anomalies in siliceous rocks are stripped by shallow workings (trenches, strippings). Lumps of rhodonite varieties are sampled for quality evaluation of the raw material. The ornamental and artistic qualities of the raw material are determined on the basis of examining the lump samples (the mineral composition and colour of the rocks are determined). Shows of rhodonite of bright pink colour, relating to the type of regionally metamorphosed manganese deposits, are the most prospective ones.

2.6. GRAPHITE

Graphite is one of the mineral varieties of carbon. Rock is also called graphite if its main components are graphite and other mineral admixtures. Magmatic, contact-metasomatic, pegmatitic, high-temperature hydrothermal, and metamorphogenetic types of deposit are known. The first types, that are usually insignificant in scale, are associated with intrusive, dyke, and effusive rocks, where flakes and, infrequently, dense accumulations of tight crystalline graphite occur. The second types form at the contact of carbonate and igneous rocks. Graphite with large flakes is dispersed in skarns; it concentrates sometimes in the form of deposits of irregular shape. The latter are characterized by small sizes and low quality of the raw material.

Rare and small deposits are confined to pegmatites. Deposits of the fourth type, characterized by high-quality raw material, occur in intensively metamorphosed carbon-bearing sedimentary formations. Vein deposits of graphite are localized in gneisses and crystalline schists. Deposits of the fifth type form in the process of metamorphism of coal or sedimentary rocks that are rich in carbon. Microcrystalline (amorphous) varieties of graphite (up to 95 per cent) are found in the form of veinlets. Large deposits, that are shaped like irregular strata and lenses, form in ancient metamorphic rocks.

Metamorphic and high-temperature hydrothermal deposits are of greatest economic interest. The basic mass of carbon was derived from the enclosing sedimentary rocks (organic remains, carbonates). Magma (CO and CO_2 gases contained in it) could be a partial source of carbon. Transformation, mobilization and redistribution of carbon in sedimentary rocks occurred as a result of regional and contact metamorphism. Thus, metamorphism of graphite gneisses in the Ukrainian shield occurred at 720–750 °C and a pressure of 750 MPa, i.e. under conditions that were transient from the amphibolitic to the granulitic facies of metamorphism.

Favourable indications used in prospecting for graphite are:

(1) spread of carbon-bearing sedimentary rocks (limestone, coaly bituminous shale, coal) in folded areas with active magmatic activity;

(2) presence of intrusive rocks of various composition that are in contact with carbon-bearing sedimentary formations;

(3) distribution of gneisses and crystalline schists with intercalations of graphite-bearing limestone and dolomite in ancient complexes.

Accessory minerals of graphite of magmatic genesis are nepheline, microcline, aegirite–augite, albite, cancrinite, calcite and ilmenite at the occurrence of alkaline and nepheline–syenite in limestones. Parageneses that are typical of skarns (diopside, wollastonite, amphiboles, garnet, scapolite, forsterite, phlogopite, apatite, sphene, serpentine, tremolite and vesuvianite) develop in contact–metasomatic deposits with limestones. Epidote, zoisite, diopside, amphibole, antigorite, sericite, calcite, wollastonite and fluorite are abundant in the graphitic zone at the contact of limestones with gabbroids. Pegmatites, occurring in limestones, contain impregnated flakes of graphite with pyroxene, amphibole, garnet and quartz. The characteristic series of graphite accessories in deposits of the fourth genetic group includes quartz, biotite, orthoclase, calcite, pyrite, pyroxene, magnesite, rutile, orthite, zeolites, forsterite, tremolite, phlogopite, wollastonite and scapolite. In Mexican shows, graphite is associated with tourmaline, hematite, pyrite, biotite and gypsum. In metamorphogenetic deposits, localizing in gneisses and schists, graphite is associated with feldspar, quartz, calcite, pyrrhotite, pyrite, sphalerite, galenite and biotite. The presence of limestone or dolomite in the enclosing rock is characteristic of these types of deposits. The accessories of graphite in this case are minerals that are characteristic of skarns. Owing to its high rigidity, graphite preserves well in eluvial–deluvial formations in weathering zones. The presence of graphite in loose formations in the weathering zone of the bedrock is a major prospecting indication of deposits. Data on the mineral and chemical composition of the enclosing rocks indicate that carbon, vanadium, copper, germanium, phosphorus, potassium, manganese and fluorine are typomorphic indicator elements of graphite deposits.

Gently sloping contacts of contrasting rocks, differing sharply in petrographic composition (gabbroids and limestones, gneisses and dolomites) that are clearly different by the content of admixture elements, are favourable pathways for the penetration of hydrothermal solutions. Tectonic shoves, indicated by mercury anomalies, occurred along the contact of these rocks. We have determined the following admixtures from the data of spectral analyses of graphitic zones at one of the deposits ($n \times 10^{-3}$ per cent): Cu-200; Zn-50; Pb-3; Ni-10; Co-3; Cr-10; V-15; Mo-40; Ag-4; Sn-2; Ba-40; Ti-500; Sc-1; Li-2; Be-0.04; Y-3; Zr-15; P-50; Ge-2; Ga-3; and B-5. An increased content of vanadium (up to 0.1 per cent) is observed in hydrothermally modified rocks near graphitic veins. This evidence all points to redistribution of the elements at active participation of the organic substances contained in graphite, creating a reducing situation favourable for the deposition of vanadium from the mobilizing solutions. This element is associated mainly with graphite and sericite. It substitutes isomorphically for aluminum in the latter, and in graphite it is in the form of sulphides, oxides and organometallic compounds. Graphite rock is enriched in molybdenum in some regions. The isotopic composition of carbon, determined by mass spectrometric methods, is one of the criteria used in prospecting for graphite deposits. Graphite in commercial deposits is enriched by 0.12–1.36 per cent in the light isotope of ^{12}C (in values $\delta^{13}C$) as compared to light graphite that is dispersed in the rocks [73]. Thus $\delta^{12}C$ for graphite from the Ukraine changes from -0.30 to -3.76 per cent (from carbonate formations -0.3 to 1.6 per cent and gneisses 1.8 to 3.7 per cent). The general trend, consisting in a gradual increase of the light isotope ^{12}C content in the graphite in the direction from carbonate rock to gneisses, is preserved permanently [73]. As a result of experimental work, the principal potentiality has been proved of discovering graphite deposits by geochemical and geophysical methods. The specific features of the composition of the primary rocks and the subsequent processes of their metamorphism are the prerequisites for applying geochemical methods for discovering graphite deposits. In this case, increased concentrations of phosphorus, fluorine, copper, and vanadium are determined in graphite-bearing bodies. The maximum content of CO_2 above the graphite deposits is 5.6 per cent (at a 0.2 per cent background). Negative anomalies of the natural electric field potentials are found around graphitized rocks and bodies of monomineral graphite.

Methods of work

It is advisable to carry out geochemical exploration in *a fortiori* graphite-bearing regions. Under conditions of poor exposure, it is necessary to explore secondary haloes, carry out atmochemical surveys, apply electric prospecting, and the method of the natural electric field. Samples are taken from deluvial formations from a depth of 0.3–0.4 m on a grid of 250×25 m (scale $1:50\,000 - 1:25\,000$), and

80×5 m (scale $1:10\,000$) at a thickness of the overburden up to 5 m. The following indicator elements are determined: fluorine, mercury, antimony and arsenic (by special methods), and copper, vanadium, strontium, barium (by spectral analysis). Weak anomalies of elements, masked by noises, are enhanced by multiplication. When the overburden is more than 5 m thick, the subsurface atmosphere is explored in the process of atmochemical survey in specially drilled holes 5 m in depth. In this case, gas samples with a volume of 0.5 l are taken and the complex of components (CO_2, O_2, CH_4, H_2) therein is determined by the chromatographic method. Air samples are taken near the bottom of the hole by means of a special sampler, a packer-sonde, which is a hollow perforated pipe 0.4 m long, whose upper end is inserted into a rubber chamber (packer). Simultaneously with sampling, it is possible to determine the CO_2 content directly in the field by a mine interferometer, ShI-10. The content of mercury vapour is determined in the subsurface atmosphere by means of an atomic absorption photometer. The volume of control exploration during gas survey should be 10 per cent as compared to common survey. Gaseous haloes are divided by composition into the following anomalies: methane and hydrogen; methane, hydrogen and carbon dioxide; methane, hydrogen, carbon dioxide, and oxygen; and hydrogen and carbon dioxide. The detected coinciding gaseous and lithochemical anomalies are spot-sorted after verification by control-detail sampling, and prospecting-estimation holes are drilled at the most promising ones.

The variety of the graphite-like substance is determined by its chemical properties (yield of volatile components, solubility in organic solvents, electric resistance). The genetic nature of the graphite is determined by examining the stable isotopes of the carbon. If the graphite-bearing zones are located among metamorphic rock, determination of productive mineralization is possible by geophysical methods (radio instrumentation, electromagnetic sounding).

It is advisable to explore the primary haloes at the stage of detailed prospecting and exploration when estimating the size of the graphite deposits, the depth of erosional truncation and identification of blind bodies. To this end, sampling of the bedrock along the profiles is performed within the prospective areas. The sampling grid is 250×25 m (scale $1:25\,000$), 100×10 m (scale $1:10\,000$). Samples are analysed to determine P, V, Pb, Zn, Cu, Mo, Ag and Ba, whose haloes are contoured by the results of the analyses with an accuracy that is necessary for prediction.

It is advisable also to apply electrical exploration at the reconnaissance stage. The position of individual bodies in plan, as well as their depth, angle of dip, and thickness are determined according to amplitude–phase anomalies.

The following main problems can be solved through interpreting geochemical and geophysical anomalies:

(a) determination of the formational nature of the anomalies (presence of mineral and geochemical associations that are characteristic of individual types of deposits); and

(b) estimation of the present level of erosion of the anomalies (by vertical zonality of the haloes).

The latter problem is the most important one. A universal series of indicator elements of vertical zonality is considered in this case (supra-ore indicators: Ag, Ba, Sr, Hg, As, Sb; and sub-ore indicators: Mo, V, Ni, Co, Cr). The level of erosional truncation of the anomalies is estimated by comparing the values of the parameters with those of haloes of known graphite bodies. It is most effective to use in this case the parameters of multiplicative haloes, which are the most contrasting ones.

Graphite is a widely occurring mineral in the metamorphic terrains of Precambrian shields, central massifs and folded zones. It is found in various gneisses, crystalline schists, quartzites and marbles, confined to definite stratigraphic levels.

The scale of graphite mineralization increases substantially in the parts of initial graphite-bearing rocks with imposed hydrothermal alterations, especially in zones of tectonic disturbances. Here, graphite content reaches 40–45 per cent locally. The process is frequently followed by the intensive sulphide mineralization of graphite ores, forming bed-like, lens-shaped, vein and stockwork bodies. In the zones of hydrothermal alteration there are various graphite morphologies: flaky, grained and fibrous. The high graphite content in the zones of initial graphite-bearing rocks with imposed hydrothermal alteration is probably due to the reduced conditions which promote graphite mineralization. The carbon of these graphites is commonly enriched by the light isotope ($\delta\,^{13}C = -33.00 + 49.90\%o$) [71].

If during the early stages of exploration the anticipated types of deposits show certain characteristic geophysical properties, the relevant exploration method (magnetic, electromagnetic, gravity, resistivity, etc.) will be applied to the area to try to identify these properties.

Graphitic rocks with a certain electromagnetic character are easily observed. The target is a graphitic conductor which can be successfully located using TDEM (Time Domain Electromagnetic) techniques with a computer programme.

2.7. ASBESTOS

Asbestos is the genetic name given to a group of fibrous mineral silicates found in nature. They are all incombustible and can be separated by mechanical means into fibre of various lengths and cross sections. However, each differs in chemical composition from the others.

It is generally recognized that there are two main groups of asbestos. The first group contains the fibrous serpentine called chrysotile and comprises about 94 per cent of the world production of asbestos. The second group contains five minerals in the amphibole series — crocidolite, amosite, anthophyllite, tremolite, and actinolite.

Table (2.13) shows the range of compositions for the most important components in commercial asbestos. In addition, there will be many trace elements present but these are rarely present in amounts greater than 0.20 per cent.

Chrysotile constitutes about 94 per cent of the current world production of asbestos and, of this amount, all but a fraction of a per cent is derived from

Table 2.13.

Compositions and cell-contents for some typical chrysotiles. (Analyst: D. G. Hiscock, Cape Asbestos Fibres, Ltd.) [43]

	Thetford, King Beaver Mine	British Kolumbia, Cassiar	Russia, Asbest	Rhodesia, Shabani	Swaziland, Havelock Mine
SiO_2	38.75	40.75	39.00	39.70	39.93
FeO	2.03	0.28	1.53	0.70	0.45
Fe_2O_3	1.59	0.44	0.54	0.27	0.10
Al_2O_3	3.09	3.37	4.66	3.17	3.92
CaO	0.89	0.35	2.03	1.08	1.02
MgO	39.78	41.28	38.22	40.30	40.25
MnO	0.08	0.03	0.11	0.26	0.05
Na_2O	0.10	0.07	0.07	0.04	0.09
K_2O	0.18	0.04	0.07	0.05	0.09
H_2O^+	12.22	12.86	11.37	12.17	12.36
H_2O^-	0.60	0.78	0.77	0.64	0.92
CO_2	0.48	0.44	1.83	2.13	1.04
Total	99.63	100.37	100.20	100.51	100.24
Mg	5.79	5.70	5.60	5.68	5.70
Fe^{2+}, Mn	0.12	0.01	0.12	0.07	0.04
Na, K	0.04	0.02	0.02	0.02	0.02
Al	0.05	0.23	0.26	0.23	0.28
Total	6.00	5.96	6.00	6.00	6.00
Si	3.80	3.85	3.86	3.88	3.84
Al	0.30	0.15	0.28	0.13	0.31
Total	4.1	4.00	4.14	4.01	4.15
O	10.07	9.91	10.49	8.65	0.53
$(OH)_6$	5.92	7.07	5.93	6.49	6.66
$(OH)_2$	2.01	1.02	1.58	1.43	0.80

deposits whose host rocks are ultrabasic in composition. Chrysotile is the result of hydrothermal reactions under certain conditions within serpentine rock, and it is only natural that there will be imperfections within the crystal structure and the inclusion of impurities between fibre and at the ends of the fibre near the wall rock. The other fraction of chrysotile production is derived from serpentinized dolomitic limestone. Among the other varieties of asbestos, amosite and crocidolite are found in certain metamorphosed ferruginous sedimentary formations and, together, account for some 5.3 per cent of world production. Tremolite and anthophyllite make up the balance of production and are generally found in association with highly metamorphosed ultrabasic rocks.

Crocidolite and amosite occur in metamorphosed siliceous–ferruginous sediments. The fibre are often comparatively long, ranging up to several inches. These long lengths have no advantage for most uses and may detract from their value, since the greatest use of asbestos is in asbestos cement products where fibre lengths of 1/8 to 3/8 inches are more suitable.

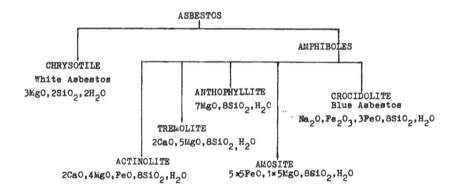

Figure 2.35. Varieties of asbestos [43].

Anthophyllite has a low ignition loss comparable to that of tremolite. It is never known to occur with fibre sufficiently flexible to be spun nor with the tensile strength required to make it of value to the asbestos cement industry.

Richterite asbestos is a rare type of a fibrous potassium (potassium-bearing) richterite, which forms is metasomatically altered sedimentary dolomites on the contacts of alkaline rocks and intensively in the zone of super-carbonate fenites. These deposits are characterized by a great amount of asbestos and geological reserves.

Figure (2.35) shows a classification of asbestos, together with its approximate chemical composition.

Table (2.13) contains some analyses of chrysotiles from various sources together with their cell contents, and Table (2.14) contains similar data from more recent analyses of amphibole asbestos. Regardless of source there is a marked uniformly among the compositions of the chrysotiles, and one has to look hard for differences which might explain variations in other chemical and physical properties. The analyses disclose the presence of two impurities, CO_2 being linked with calcium–magnesium carbonates, while FeO and Fe_2O_3 arise from an impurity of magnetite. Given a sufficiently wide range of analyses it is not difficult to distinguish between the various kinds of amphibole asbestos. For example, the most useful marker elements among the crocidolites are Na and Mg. Amosites have a high ferrous-iron content, approaching 40 per cent FeO.

Chemical analysis will give some indication of the degree of natural weathering to which an amphibole asbestos has been exposed, a question which particularly applies to amosite and crocidolite. Amosite should contain ferrous iron only, and ground surface weathering or subsurface leaching shows up in increasing amounts of ferric iron on analysis. Fresh crocidolite has a distinct excess of FeO over Fe_2O_3 in analysis, and when FeO is equal to or less than the Fe_2O_3 content, the crocidolite cannot be considered to be fresh.

A comparison of the general properties of the six varieties of asbestos is given in Table (2.15).

Table 2.14.

Compositions and cell-contents of some typical amphibole asbestos varieties. (Analyst: D. G. Hiscock and staff of Cape Asbestos Fibres, Ltd.) [43]

	Koegas, Cape Prov.	Kuruman, Cape Prov.	Pomfret, Cape Prov.	Malipsdrift, Transvaal	Cochabamba, Bolivia	Hammersley Range, W. Australia
SiO_2	50.90	50.70	52.00	59.41	55.65	52.85
FeO	20.50	17.50	17.65	15.11	3.84	14.94
Fe_2O_3	16.85	18.30	16.05	14.03	13.01	18.55
Al_2O_3	nil	0.70	nil	nil	4.00	0.18
CaO	1.45	1.30	1.20	0.49	1.45	1.07
MgO	1.06	3.05	4.28	3.53	13.09	4.64
MnO	0.05	0.06	tr.	tr.	tr.	tr.
Na_2O	6.20	5.30	6.21	4.63	6.91	5.97
K_2O	0.20	tr.	0.06	0.28	0.39	0.05
H_2O^+	2.37	2.53	2.43	2.07	1.78	2.77
H_2O^-	0.22	0.29	0.26	0.14	tr.	0.22
CO_2	0.20	0.45	0.09	0.09	tr.	0.23
Total	100.1	100.18	100.23	99.78	100.35	101.47
Na, K	1.92	1.59	1.86	No data		1.91
Ca, Mn	0.21	0.13	0.12			0.13
Total $2M_4$	2.13	1.72	1.98	No data		2.04
Fe^{3+}	1.98	2.13	1.85	No data		2.00
Total $2M_2$	1.98	2.13	1.85	No data		2.00
Fe^{2+}	2.68	2.27	2.26	No data		1.90
Mg	0.25	0.69	0.93			1.06
Total $2M_1 + M_3$	2.95	2.96	3.19	No data		2.96
Si, Al	7.97	8.00	7.96	No data		7.98
22(O)	22.05	22.03	22.06			21.82
2(OH)	1.95	1.97	1.94			2.18

Minerals of the serpentine group (chrysotile, crocidolite, amosite) and amphiboles (anthophyllite, rezhikite, rhodusite, tremolite, actinolite) that separate into thin fibres under mechanical effect, relate to asbestos. Commercial deposits of chrysotile asbestos are associated with serpentines that have formed as a result of metamorphism of ultrabasites or dolomitized limestones under the impact of hydrothermal processes. Two types of deposit are distinguished, apoultrabasite and apocarbonate. Deposits of the first type include those of chrysotile asbestos (associated with serpentinization of dunites) pyroxenites, and peridotites, which are divided into three sub-types: Bazhenovsky, Labinsky and Karachayevsky. These sub-types are distinguished by the morphology of the asbestos deposits and the kinds of ores. Asbestos mineralization of the Bazhenovsky subtype is associated with partially or fully serpentinized peridotites, while the Labinsky and Karachayevsky subtypes are associated with fully serpentinized ultrabasites. The

Table 2.14.

(Continued)

	Amosite Penge, Transvaal	Amosite Weltevreden, Transvaal	Anthophyllite Paakkila, Finland	Tremolite, Pakistan	Prieskaite Koegas, Cape Prov.
SiO_2	49.70	51.30	57.20	55.10	53.80
FeO	39.70	35.50	10.12	2.00	25.30
Fe_2O_3	0.03	0.90	0.13	0.32	1.90
Al_2O_3	0.40	nil	–	1.14	1.20
CaO	1.04	0.95	1.02	11.45	10.20
MgO	6.44	6.90	29.21	25.65	4.30
MnO	0.22	1.76	–	0.10	0.40
Na_2O	0.09	0.05	–	0.14	0.10
K_2O	0.63	0.51	–	0.29	0.40
H_2O^+	1.83	2.31	2.18	3.52	2.60
H_2O^-	0.09	0.05	0.28	0.16	nil
CO_2	0.09	0.25	–	0.06	0.20
Total	100.53	100.54	100.48	99.93	100.40
Fe^{2+}	1.78	1.28	–	0.23	–
Ca, Mn	0.23	0.33	0.16	1.71	1.71
Mg	–	–	2.08	0.31	–
Na, K	0.07	0.12	–	0.11	0.13
Total $2M_4$	2.08	1.73	2.24	2.36	1.84
$Fe^{2+}(Fe^{3+}, Al)$	1.39	1.35	–	–	1.72
Mg	0.61	0.65	2.00	2.00	–
Total $2M_2$	2.00	2.00	2.00	2.00	1.72
Fe^{2+}	2.09	2.03	1.15	–	2.00
Mg	0.91	0.97	1.85	3.00	1.00
Total $2M_1 + M_2$	3.00	3.00	3.00	3.00	3.00
Si, Al, Fe^{3+}	7.99	8.17	7.79	8.00	8.26
22(O)	21.98	21.96	22.00	22.05	22.46
2(OH)	2.02	2.04	2.00	1.95	1.54
H	–	–	0.68	–	0.45

second type of deposit, resulting from serpentinization of dolomitic limestones, is of minor commercial significance.

Amphibole asbestos mineralization is localized in ultrabasites, occurring among gneiss–migmatite formations. Metamorphic (regional metasomatic) and contact-reaction (bimetasomatic) deposits are distinguished. In the first type, productive mineralization is associated with dunite–harzburgite or gabbro–pyroxenite–dunite formations. A high degree of metamorphism in rocks enclosing ultrabasites that meet the conditions of the amphibolite facies, is characteristic of anthophyllite asbestos deposits. High pressure (700–1000 MPa) at 630–660 °C contributed to the formation of anthophyllite, which later turned into asbestos. Anthophyllite asbestos formed at the final stages of regional metamorphism when there was a high CO_2 content in the fluid [74].

Table 2.15.
Properties of asbestos fibers [43]

Property	Actinolite	Amosite	Anthophyllite	Chrysotile	Crocidolite	Tremolite
Structure	Reticulated long prismatic crystals and fibre	Lamellar. Coarse to fine fibrous and asbestiform	Lamellar. Fibrous asbestiform	Usually highly fibrous fibres fine and easily separable	Fibrous in ironstones	Long. Prismatic and fibrous aggregates
Mineral association	In limestone and in crystalline schists	In crystalline schists, etc.	In crystalline schists and gneisses	In altered peridotite adjacent to serpentine and limestone near contact with basic igneous rocks	Iron rich silicious agrillite in quartzose schists	In Mg limestones as alteration products of highly magnesian rocks. Metamorphic and igneous rocks
Origin	Results of metamorphism	Metamorphic	Metamorphic. Usually from olivine	Alteration and metamorphism of basic igneous rocks rich in magnesian silicates	Regional metamorphism	Metamorphic
Veining	Slip or mass fibre	Cross fibre	Slip. Mass fibre unoriented and interlacing	Cross and slip fibre	Cross fibre	Slip or mass fibre
Essential composition	$Ca \times Mg \times Fe \times$ Silicate water up to 5%	Silicate of Fe and Mg higher iron than anthophyllite	Mg silicate with iron	Hydrosilicate of magnesium	Silicate of Na and Fe with some water	Ca and Mg silicate with some water
Crystal structure	Long and thin columnar to fibrous	Prismatic. Lamellar to fibrous	Prismatic. Lamellar to fibrous	Fibrous and asbestiform	Fibrous	Long and thin columnar to fibrous
Crystal system	Monoclinic	Monoclinic	Orthorhombic	Monoclinic (pseudo-orthorhombic?)	Monoclinic	Monoclinic

Table 2.15.
(Continued)

Property	Actinolite	Amosite	Anthophyllite	Chrysotile	Crocidolite	Tremolite
Color	Greenish	Ash gray or brown	Grayish white. Brown-gray or green	White. Gray. Green. Yellowish	Lavender-blue. Metallic-blue	Gray-white. Greenish. Yellowish. Bluish
Luster	Silky	Vitreous. Somewhat pearly	Vitreous to pearly	Silky	Silky to dull	Silky
Hardness	6±	5.5–6.0	5.5–6.0	2.5–4.0	4	5.5
Specific gravity	3.0–3.2	3.1–3.25	2.85–3.1	2.4–2.6	3.2–3.3	2.9–3.2
Cleavage Optical properties	110 perfect Biaxial negative extinction inclined	110 perfect Biaxial and positive extinction parallel	110 perfect Biaxial positive extinction parallel	010 perfect Biaxial positive extinction parallel	110 perfect Biaxial ± extinction inclined	110 perfect Biaxial negative extinction inclined
Refractive index	1.63± weakly pleochroic	16.4±	1.61±	1.51–1.55	1.7 pleochroic	1.61±
Length	Short to long	2 to 11 in., varies	Short	Short to long	Short to long	Short to long
Texture	Harsh	Coarse but somewhat pliable	Harsh	Soft to harsh. Also silky	Soft to harsh	Generally harsh. Sometimes soft
Specific heat, Btu per lb per °F	0.217	0.193	0.210	0.266	0.201	0.212
Tensile strength, psi	1000 and less	16 000 90 000	4000 and less	80 000 100 000	100 000 300 000	1000 8000

Table 2.15.
(Continued)

Property	Actinolite	Amosite	Anthophyllite	Chrysotile	Crocidolite	Tremolite
Temperature at maximum ignition loss, °F	—	1600 to 1800	1800	1800	1200	1800
Filtration properties	Medium	Fast	Medium	Slow	Fast	Medium
Electric charge	Negative	Negative	Negative	Positive	Negative	Negative
Fusion point, °F	2540	2550	2675	2770	2180	2400
Spinnability	Poor	Fair	Poor	Very good	Fair	Poor
Resistance to acids and alkalies	Fair	Good	Very good	Poor	Good	Good
Magnetite content	—	0	0	0–5.2	3.0–5.9	0
Mineral impurities present	Lime, iron	Iron	Iron	Iron, chrome, nickel, lime	Iron	Lime
Flexibility	Poor	Good	Poor	High	Good	Poor
Resistance to heat	—	Good. Brittle at high	Very good	Good. Brittle at high	Poor, fuses	Fair to good

Rhoducite asbestos deposits are confined to sandy–clayey rocks. Rezhikite asbestos, forms in zones of talcatization and carbonatization of serpentinites as well as in the strata of calcareous–dolomitic marls, alternating with sandstones. These two deposits both derived the main components for their formation from the enclosing rocks. This is confirmed by the fact that mineralization does not extend beyond these surrounding rocks.

The following prospecting indications of asbestos mineralization are common.

(a) For chrysotile asbestos deposits:

- Presence of ultrabasites of the dunite–harzburgite formation.
- Tectonic dislocations in ultrabasites that are controlled by veined granites, talc–carbonate rocks, and schistic serpentines.
- Presence of fan-shaped faults and series of jointed dislocations.

(b) For anthophyllite asbestos deposits:

- Development in the axial zones of ultrabasites of the dunite–peridotite and gabbro-pyroxenite formations, occurring in rocks with manifestations of granitization, boudinage, and dome-shaped structures; abundance of granite dykes that cleave the serpentinites.
- Shows of coarse-grained anthophyllite–talc ultrabasites, confined to zones of gneiss contact with amphibolites, as well as migmatitic formations.
- Metamorphism of the ultrabasite enclosing rocks in the amphibolitic and granulitic facies with shows of several stages.

The mineral composition of chrysotile asbestos deposits is quite diverse. Three groups of minerals are distinguished:

(a) primary: serpentines (lizardite, chrysotile, antigorite), talc, carbonates (ankerite, magnesite, calcite, breunnerite), brucite and magnetite;

(b) secondary: bastite, nemalite, garnet, actinolite, tremolite, chlorite, pyrite, chalcopyrite, pyrrhotite, hematite, pyroaurite, ophite and chromite; and

(c) accessory and rare: hydromagnesite, hydrotalcite, kerolith, brugnatellite, josephinite awarcite and graphite.

Diopside, forsterite, garnet, tremolite, vesuvianite, scapolite, ophicalcite and chlorite have been found in apocarbonate deposits of chrysotile-asbestos.

The mineral composition of anthophyllite asbestos is also quite diverse. The most widespread minerals therein are:

(1) primary: anthophyllite, tremolite, actinolite, phlogopite, chlorite, talc, bastite, nontronite, vermiculite, calcite, magnesite and dolomite;

(2) secondary: olivine, enstatite, chrysotile, lizardite, magnetite and limonite; and

(3) accessory: zircon, sphene, garnet, cyanite, epidote, apatite, barite, spinel, chromite, ilmenite, rutile, corundum, pyrite, pyrrhotite, galenite, hematite quartz, chalcedony, psilomelane, graphite and malachite.

Prospecting indications of chrysotile asbestos and anthophyllite asbestos deposits also include the presence in the river alluvium of asbestos-bearing rocks, and accessory minerals: magnetite, lizardite, antigorite, talc, brucite and carbonates. Shows of multistage transformation of serpentinites are prospecting indications of chrysotile asbestos. The products of serpentinization of primary dunites, harzburgites, and pyroxenites, distinguished by a generally simple composition (Σ MgO + SiO_2 + Fe_2O_3 + FeO $\approx$ 99 per cent), are quite diverse:

(1) loop-shaped, transverse-fibrous serpentine (α-chrysotile), replacing olivine at the initial spatial serpentinization;

(2) thin-foliated lizardite, forming under the effect of superposed processes at an early stage of serpentinization under conditions of insignificant inflow of H_2O and removal of MgO and FeO from the ultrabasites;

(3) chrysotile asbestos associated with magnetite, antigorite, talc and carbonates, forming at the final stage.

Determination of the stated paragenous associations makes it possible to specify the prospecting indications. The prospects of detecting chrysotile asbestos diminish appreciably in the case of intensive manifestation of anti-goritization and listvenitization, and, vice versa, the presence of the chrysotilization stage indicates a potential asbestos occurrence. The dependence is established of the refraction indices on the chemical composition of anthophyllite and anthophyllite asbestos from ultrabasites of the dunite–harzburgite and gabbro-pyroxenite–peridotite formations. The stated ratio is expressed by the regression equation:

(1) $Ng = 1.6073 + 0.01727$ F, $Np = 1.585 + 0.00166$ F;

(2) $Ng = 1.613 + 0.00136$ F, $Np = 1.592 + 0.001213$ F [96]

Anthophyllite asbestos from ultrabasites of the dunite–harzburgite formations is characterized by a high content (in per cent) of SiO_2 (2.1), and MgO (3.04); and a lower content of FeO (2.43), Fe_2O_3 (0.99), Al_2O_3 (50), and CaO (60); and a lower value (5.6) of the total iron content as compared to anthophyllite from ultrabasites of the gabbro-pyroxenite–peridotite formation. Table (2.16) presents data on the changes in the content of the basic oxides in anthophyllite and anthophyllite asbestos. The specific and quantitative composition of the minerals in the rocks of the substratum change in the process of development of anthophyllitic asbestos mineralization. The amount of enstatite and olivine decreases at transition from serpentinites to anthophyllite zones, whereas the content of chrysotile, actinolite, talc, chlorite, calcite, and dolomite increases in this direction.

Amphibole asbestos mineralization is confined to certain facial rock types which differ noticeably from non-asbestos-bearing types in their petrographic composition and petrochemical characteristics.

The temperature conditions of the formation of deposits correspond to certain stages in the alteration of the enclosing rock: 40–120°C for rhodusite–crocidolite–asbestos, 60–230°C for actinolite asbestos, 540–590°C for anthophyllite–asbestos. Asbestos proper is only formed under low temperature conditions, either at the beginning of the process, when the initial gel is in its most fluid state and there is a considerable predominance of the liquid phase in solutions, or else, towards the end of the process, when the processes of hydration and decomposition take

Table 2.16.

Fluctuation limits of content of basic oxides and cations in anthophyllite (numerator) and anthophyllite asbestos (denominator) from ultrabasites weight per cent [96]

Components	SiO_2	Al_2O_3	Fe_2O_3	FeO	MgO	CaO
Dunite–harzburgite formation						
Basic oxides	56.17–59.14	1–0.94	0–2.16	5.13–12.35	26.77–31.73	0–0.96
	56.70–60.13	0.03–1.55	0–2.83	5.05–11.15	26.56–30.51	0–0.45
Cations	7.76–8.00	0–0.32	0–0.28	0.58–1.44	5.45–6.25	0–0.14
	7.79–8.00	0–0.26	0–0.28	0.66–1.28	5.21–6.15	0–0.06
Gabbro-pyroxenite–peridotite formation						
Basic oxides	54.21–56.75	0.45–2.78	0.65–3.32	10.08–13.51	24.04–26.37	traces–1.32
	56.26–60.29	0–0.84	0–0.22	7.05–11.76	24.92–28.56	0–0.39
Cations	7.51–7.97	0.15–0.46	0.07–0.34	1.17–1.58	5.01–5.50	0–0.27
	7.71–7.98	0–0.14	0–0.14	0.78–1.32	5.04–5.94	0–0.05

place in the crystalline amphibole rock and ore [10, 43]. The production of amphibole asbestos other than crocidolite and amosite includes significant amounts of anthophyllite, a small quantity of tremolite, and token amounts of fibrous actinolite.

The anthophyllite occurs as a series of lenses of amphibolitized and serpentinized ultrabasic material originally thought to be dunitic is composition. These lenses appear to have become detached from larger bodies during an early period of severe deformation. The serpentinized amphibolite lenses occur in a biotite gneiss which is intruded by granites and pegmatite. Exploration for these isolated bodies is difficult and often costly as the area is covered by drift and the ore is processed at a centrally located mill.

Although some occurrences of fibrous actinolite are reported, production is extremely limited and is of negligible value.

Both crocidolite and amosite occur in bedded metamorphosed ironstones (ferruginous quartzite, iron-rich silicified argillite) (South Africa). Cross-fibre veins are found as closely spaced ribbons roughly conformable with the bedding, which in some localities is distorted and steeply dipping.

Associations of the following minerals are characteristic of the manifestations of rezhikite asbestos: talc, carbonate, actinolite, alkali amphibole, and quartz; and of rhoducite ore bodies: sericite, dolomite, anatase, brookite, hematite, calcite and opal.

Conditions for the application of geochemical methods in prospecting for asbestos

A favourable condition for applying geochemical methods in the search for asbestos is the fact that asbestos-bearing serpentinites are characterized by a more diverse, specific, and higher quantitative content of minerals and elements than barren ones. This principle is based on data relating to the conditions of asbestos formation. It is generally recognized that the formation of chrysotile asbestos occurred in two stages. At the first stage, uniform serpentinization of ultrabasites resulted from autometamorphism under the effect of solutions that were originally contained in the ultrabasitic magma, or were assimilated by the latter from the enclosing rocks. At the second, productive stage, asbestos-bearing bodies formed as a result of serpentinization of ultrabasites by hydrothermal solutions. The processes of serpentinization proceeded during the period of intensive asbestos formation at relatively long-term and various thermodynamic levels in preliminarily altered ultrabasites. As a result of incomplete serpentinization, the composition of the latter was favourable for the formation of chrysotile asbestos. According to the results of analyses and calculations of the substance balance, low-mineralized solutions containing H_2O, Cl, F, CO_2 and Hg (deep spring) circulated during asbestos formation; the basic mass of components was derived from the substratum rocks (Si, Mg, Fe, Na, Ti, etc.). Chlorine-bearing solutions, contributing to the dissolution of mineral phases, catalysed the reaction of asbestos formation.

Chrysotile asbestos consists mainly of Mg, Si, H_2O (hydrous magnesium silicate), and anthophyllite asbestos is made up of Mg, Si and Fe. The wide occurrence of these elements in rocks containing asbestos-bearing bodies, eliminates their application as geochemical indicators. It is advisable in this connection to use the following admixture of elements in geochemical prospecting for asbestos: Ni, Co, V, Cr, Cu and Ti (magnetite, pyrite, sulphides of Ni, Co, Cu), Sr and Ba (barite, carbonates), Cl and F (serpentines, brucite, apatite), and Hg (sulphides, carbonates).

Migration and redistribution of the elements of the ferric group (Ni, Co, V, Cr, Cu and Ti) in the processes of asbestos formation results in the formation of positive and negative anomalies around productive bodies. Thus, nickel is redistributed in the course of allometamorphic serpentinization and asbestos formation, which is accompanied by the formation of independent mineral phases (native $FeNi_3$).

Chlorine and fluorine are the major indicators of asbestos mineralization. A direct correlation is observed between the content of chlorine in the rocks and the degree of their serpentinization. The content of Cl in serpentinized dunites and fresh peridotites is 0.3 and 0.04 per cent, respectively. The content of Cl in unaltered ultrabasites (~ 5 per cent serpentine) is 0.007 per cent, in medium-serpentinized (10–75 per cent) rocks it is 0.030 per cent, and in intensively altered ultrabasites it is 0.080 per cent [111].

Comparative exploration of chrysotile asbestos and barren rocks indicates the enrichment of the former in chlorine. A high concentration of this element in asbestos-bearing zones is conditioned by the effect of the mineral-forming solutions on the substratum rocks, causing an increase in the amount of chlorine-containing minerals and liquid inclusions therein. The content of Cl in serpentinized veinlets (Quebec, Canada) is 0.5 per cent, while near the contacts of chrysotile asbestos veins it is 0.8 per cent, and in the olivine its not more than 0.001 per cent [111]. The presence of chlorine in asbestos-forming solutions had an appreciable effect on the distribution of iron between brucite and silicates, causing the enrichment of brucite in this element [111]. The chloride composition of the hydrothermal solutions explains an increased concentration of mercury (from 18×10^{-6} to 20×10^{-6} per cent) in chrysotile asbestos. It has been proved experimentally that mercury can be transported in the form of chlorine complexes.

The presence of fluorine in serpentinizing solutions is proved by the results of chemical analyses: 0.001 per cent in unaltered ultrabasites, and 0.15 per cent in the serpentinized ones. Considerably higher concentrations of F are observed in anthophyllitized ultrabasites, wherein fluorine-bearing amphibole asbestos is found.

Carbon dioxide is an important component of asbestos-forming solutions, which is proved by the results of chemical analyses of ultrabasites and by the composition of the liquid inclusions. The content of CO_2 in serpentinized ultrabasites is 0.2–0.4 per cent, which is 100–200 times more than in unaltered, or in slightly altered rocks. The content of CO_2 ranges from 0.4 to 7.9 per cent in anthophyllitized ultrabasites.

The difference in the distribution of the elements in unaltered ultrabsites depends on the quantitative combinations of the minerals and concentrators therein. The redistribution of elements during autometamorphic is considerably greater than during allometamorphic serpentinization. An introduction of K, Na and Si into ultrabasites, and the removal of Fe, Mg and CaO (anthophyllite asbestos deposits) occurs at the first stages of metamorphism, and a considerable introduction of CO_2 takes place at the stage of anthophyllitization (and asbestos formation). Migration and redistribution of the elements condition the potentiality of halo formation around asbestos-bearing zones. Cl, F, SO_2, Hg, Sr and B, elements of the ferric group, can be typomorphic indicators of chrysotile asbestos deposits, while F, K, Na, P and Cl can indicate anthophyllite asbestos.

Positive anomalies of Cl, F, SO_2, Hg, Sr, Ba and B, and negative anomalies of Ni, Co, Cu, Ti and V, are indicators of chrysotile asbestos mineralization. A combination of positive and negative anomalies and a sharp variance in the content of such elements as Ti, Cr, V and Ni are observed near asbestos bodies because the necessary components were derived from the enclosing rocks in the process of asbestos formation. Haloes of the chloride (220°C) and sulphide forms of mercury develop around chrysotile asbestos zones.

Fluorine anomalies are observed at rhoducite asbestos deposits that lack an established association with granite (Fig. (2.36)).

Geochemical criteria of prospecting for anthophyllite asbestos are associated with the specific features of the composition of the primary ultrabasites and the

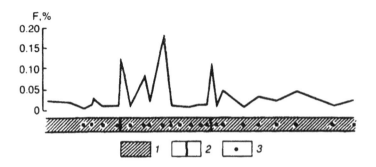

Figure 2.36. Fluorometric profile across the strike of a rhoducite asbestos zone [60]. 1 — marls, sandstones, and aleurolites; 2 — rhoducite asbestos ore bodies; 3 — sampling points.

subsequent process of their metamorphism. The increased content of Cr, Ni, Co and Cu therein is taken into consideration. Endogenic haloes of these elements develop around ultrabasites, containing anthophyllite asbestos. Endogenic haloes of increased concentrations of Cr, Ni, Co and V, and negative anomalies of Ti have been detected at the Sysertsky anthophyllite field. A combination of positive and negative anomalies of Ti, confidently indicating the epicentre of productive mineralization, is an important indication for discovering anthophyllite asbestos zones.

The experimental application of geochemical methods of prospecting for anthophyllite asbestos at several ore fields has demonstrated their effectiveness and advisability. Productive asbestos bodies, encountered in ultrabasites, were accompanied, according to A. P. Bachin, by Ni, Co and Cr haloes with a content of 0.01–0.05 per cent (up to 0.2 per cent) compared with background values of 0.005 per cent. Positive anomalies of Ni, Cr, Mo and Co, and negative anomalies of Ti have been detected above anthophyllite asbestos deposits in loose formations.

Geochemical methods can be applied on a wide scale and effectively in prospecting for asbestos deposits in open regions. It is possible to perform in this case: (a) determination of the parent rocks that asbestos mineralization is associated with; (b) detection of blind and deep-seated asbestos bodies; (c) determination of the volume of the bodies and quality of the raw material; and (d) determination of concrete areas for prospecting by establishing the formation affiliation of ultrabasites.

Massifs of the dunite–harzburgite formation, consisting of peridotite and harzburgite, wherein the weight ratios of $Mg/Si \approx 1.25–1.29$, are close to those in chrysotile asbestos, and the content of pyroxene varies within the range of 10–20 per cent, are favourable for the formation of asbestos-bearing zones. The presence of asbestos is conditioned also by the character of the subsequent alteration of these rocks. Comparatively small-scale deposits are associated with rocks of the pyroxenite–peridotite formation (Karachayevsky type). Intrusions of the dunite–clinopyroxene and dunite–pyroxene–gabbro formations are unpromising for asbestos formation by their chemical composition and physical properties. The degree of serpentinization of ultrabasites, as a process of hydration of primary anhydrous silicates of magnesium and iron, has a considerable effect on the formation of asbestos and is an important evaluating criterion. Large commercial deposits of chrysotile asbestos are usually localized in partially serpentinized ultrabasites. Irregular and incomplete serpentinization (from 40 to 60 per cent), as well as manifestation of the chrysotilization stage, are characteristic of the latter. Intensively manifested processes of pre-ore serpentinization (autometamorphism) and intensive manifestation of antigoritization (final stages) are unfavourable for the formation of chrysotile asbestos.

The character and degree of serpentinization are manifested, fist of all, in the alteration of the paragenous associations of minerals that are characteristic of various stages of the process:

(1) primary association of ultrabasites (olivine, enstatite, chrome-spinellid);

(2) association of early autometamorphic serpentinization (chrysotile, lizardite, brucite, cohenite), magnetite is unstable; and

(3) products of superposed allometamorphic serpentinization (chrysotile asbestos, maghemo-magnetite, brucite), developing under the effect of hydrothermal solutions at the site of early associations [46]. Formations that are originally unfavourable for the formation of asbestos become favourable during serpentinization of the substratum rocks (reduction of the Mg/Si value with introduction of silica into the dunites).

Zonality is a key evaluating indication of chrysotile asbestos and anthophyllite asbestos deposits. A certain regularity is observed in the change of the degree of ultrabasite serpentinization in the vertical section (a decrease with depth). Local changes in the degree of its intensity, associated with tectonic shoves, are observed as the background of a common pattern. On the whole the degree of serpentinization is of a continuous–discontinuous character. Predominant serpentinization of ultrabasites near the surface and its decrease with depth are associated with the fact that the intrusions in the near-surface zones are subjected to most intensive influence of volatile and acidic components. It is significant that the intensity of serpentinization correlates directly with the degree of iron oxidation. The maximum values of the degree of oxidation are typical of chrysolite serpentinites whose formation is characterized by a decrease in ferruginosity caused by the removal of iron (F 0.04–0.05) [46, 111].

The central parts of anthophyllite asbestos deposits are composed of talc–anthophyllite–carbonate rocks with serpentine relics that are replaced at the periphery by talc–carbonate–anthophyllite and anthophyllite–talc zones [96]. Commercial concentration of asbestos is confined mainly to the anthophyllite–talc zone. Anomalies of U, Mo and Zr are observed in enclosing rocks containing asbestos; Cr, Ni, Co and Ti haloes are found in tremolite and anthophyllite rocks. Distinct differences in the physical properties of asbestos-bearing formations and their enclosing rocks is a favourable indication for applying geophysical methods.

Geochemical methods in prospecting for asbestos

Lithogeochemical survey and examination of the petrochemical features of ultrabasites are most effective in prospecting for asbestos deposits. Since the dispersion trains are formed by the material transported to the bed of the watercourse from its vast catchment area, an analysis of samples of bottom sediments gives an impression the geochemical characteristics of a sizeable territory and can point to the presence of ultrabasites. An increased content of magnesium in the water and the presence of chlorine therein can serve as criteria of asbestos bodies occurring at depth. Alongside sampling bottom sediments to detect dispersion trains, it is advisable to apply also the method of analysing the composition of alluvial formations, allowing accurate recording and determination of the character of finds of serpentinites with asbestos veinlets. Asbestos-bearing formations differ in appearance from rocks of other composition and are easily discovered in the field. The distance of transportation from the original source is determined by the roundness of the fragments. The affiliation of the serpentinite pebbles and boulders to one of the varieties of asbestos (chrysotile or anthophyllite) is determined and the direction of prospecting is then corrected.

Prospecting by secondary haloes is performed within prospective areas, which have been detected on the basis of geological and geophysical indications, at a scale of $1:25\,000-1:10\,000$. Exposed secondary haloes, reflecting hypergene destruction of serpentinites, contact zones of various rocks, as well as asbestos-bearing bodies, form under conditions of semi-closed regions. Survey of secondary haloes of Cl, F, Hg, Ni, Co, Ti, Cu, Sr, Ba and other elements makes it possible to contour ultrabasite massifs, tectonic dislocations, zones of development of talc serpentinites, as well as asbestos mineralization. The contacts of ultrabasite intrusions are detected by anomalies of Ni, Co, Cr and Cu elements as compared to the enclosing rocks. Geochemical and geophysical data are compared when the position of the ultrabasite massifs is insufficiently determined by the secondary dispersion haloes (thick overburden, gradual transition to enclosing rocks). Talc–carbonate and anthophyllite formations near asbestos-bearing zones are characterized by low concentrations of elements of the ferric group as compared to ultrabasites, and, conversely, by high concentrations of Cl, F, Sr, B, SO_2, Pb, Zn, etc. Endocontacts of serpentinites with ultrabasites dissected by granite dykes and indicated by anomalies of Hg, Cl, Ba, Sr, Zn, As and Sb, are favourable for localization of asbestos. Chrysotile asbestos bodies are indicated by haloes of Cl, F and SO_2, anthophyllite asbestos by haloes of F, Na, K, and rhoducite asbestos ones by haloes of F. Eluvial–deluvial formations are sampled along profiles that are oriented across the strike of the assumed ultrabasite massifs to detect and contour secondary dispersion haloes. The ranging poles and grid stations that are used in geophysical work, can be used in geochemical surveys. Samples of 200–250 g are taken from a representative horizon of loose formations.

The prospecting technique varies at chrysotile asbestos deposits, depending on their genetic type.

The following observations concerning exploration and evaluation apply mainly to chrysotile deposits and, in particular, to those which occur in ultrabasic rocks. These observations are essentially a review of current methods employed in Canada.

In Africa a large part of all chrysotile is mined by underground methods. The ore bodies are generally tabular in shape with a pronounced dip. Several underground methods have been used, including cut and fill.

In the case of asbestos-bearing zones occurring in partially serpentinized peridotites, a sampling grid of 250×25 m (scale $1:25\,000$) is sufficient to contour and detect perspective zones. In the case of shows in fully serpentinized ultrabasites, distinguished from the previous ones by smaller sizes, the sampling grid should be 100×10 m (scale $1:10\,000$), or 150×20 m. The sampling grid should be 50×5 m (scale $1:5000$) when detecting single deposits that are rarely combined into zones of bodies that are associated with small massifs of serpentinites, or horizons of serpentinized carbonate formations. When prospecting for deposits of chrysotile asbestos formed during serpentinization of dolomitic limestone, special attention should be given to the presence of forsterite, diopside, tremolite, garnet, scapolite and ophicalcite in carbonate rocks. Attention should also be paid to the rocks in the vein system (granite, diabase dikes) that transform under the influence of asbestos-forming solutions into characteristically coloured (yellow, pink, green-white) chlorite, carbonaceous and pyroxene garnet formations, occurring at the contact of asbestos and enclosing rocks. The prospecting indication for asbestos is the presence of garnet, chlorite and vermiculite around the veins.

A mineralogical indication of the potential occurrence of anthophyllite as-bestos, is the presence of secondary coarse-grained orthorhombic pyroxenes (en-statite, hypersthene), stellar aggregates of anthophyllite, and chlorite (pennine, clinochlore, prochlorite, chrome-bearing clinochlore). When prospecting, special attention should be given to the occurrence of talc and talc–chlorite formations. Sometimes outcrops of bedrock are found, and also areas that are overlain with eluvial–deluvial formations. In these cases, samples are taken from the bedrock and loose formations with consideration of the ratio of exposed and overlain areas. When the thickness of the loose formations is up to 10 m, it is possible to apply geobotanical methods and carry out deep lithogeochemical survey with sampling of mud from auger holes.

The use of petrochemical criteria is most advisable in prospecting aimed mainly at detecting asbestos-bearing massifs. Determination of the variety of ultrabasites, prospective for asbestos, is carried out on the basis of formation analysis and petrochemical indications. It is advisable to single out three formation groups on the maps: (1) ultrabasic proper (dunite harzburgite), (2) basic, and (3) ultrabasic alkali. The following petrochemical indices can be used to determine the primary composition of serpentinized rocks.

1. Determination of the M^1/S ratio [56]. The sequence of calculating opera-tions is as follows: (a) the content of chemical components in the rocks is brought to 100 per cent; (b) the molecular amounts of oxides are calculated; (c) all the iron is reduced to its monoxide form (FeO); (d) the parameter M^1/S is calculated by the formula: $M^1/S = [MgO + (2Fe_2O_3 + FeO) + MnO + NiO]/SiO_2$; and (e) the calculated parameters are plotted on a graph and the varieties of ultrabasites are determined.

2. Recalculation of chemical analyses of rocks for their mineral composition (olivine = ol,) orthorhombic pyroxene = en, clinopyroxene = di) by the method of N. D. Sobolev [108]. The molecular amounts of the components are used in this case that have been calculated by their weight per cent. Three basic equations are solved:

(1) $di + en + ol = MgO + 2Fe_2O_3 + FeO + CaO + MnO + NiO + CoO$,

(2) $di + en + 0.5 ol = SiO_2$,

(3) $di = 2CaO$.

When adding a respective quantity of bases to silicic acid for the formation of olivine and monoclinic and orthorhombic pyroxene, the sum of the obtained molecular amounts is re-calculated to 100. A correction for the content of chrome-spinellids is introduced for dunites and peridotites. Unfortunately, the method does not determine accurately the composition of the parent rocks and the degree of their serpentinization, but it can serve for approximate calculations. Additional information on the specific geochemical features of the rocks is used to receive more objective data: the values of the atomic quantities of Cr/Al, Cr/Ti, Si/Mg and Si/Mn are determined. It is advisable to calculate the value of the generalized distance D^2 between the compositions of rocks of various intrusions with consideration of the average content of olivine, enstatite, and diopside [6]. The use of this technique makes it possible to determine the degree of proximity of the rock compositions of the surveyed and standard intrusions. The method of plotting graphs of branching bonds is applied (on the principle that the stronger the bond between the elements, the closer the latter are arranged on the plane). Serpentinized peridotites of asbestos-bearing intrusions form the most compact group, which statistically is not different in the content of mineral phases. It is possible to use graphs of the virtual composition for convenient comparison of the intrusions (Fig. (2.37)).

The content of olivine, enstatite, and diopside is plotted on the above graph. The location of the dot on the horizontal axis is determined by the relationship between the first minerals, and the height of the perpendicular indicates the diopside content. When plotting the results of analyses on the graph, it is possible to determine the degree of proximity of the surveyed ultrabasites to the massifs wherein chrysotile asbestos deposits are observed.

3. Determination of the ratio of magnesium to iron, which varies from 9.2 to 15.5 in serpentinized harzburgite, and from 7.6 to 8.9 in fresh dunites. It is advisable to use a special sign to mark on geological maps the serpentinized peridotite intrusions with $M/F > 10$, which are of major interest. Non-commercial

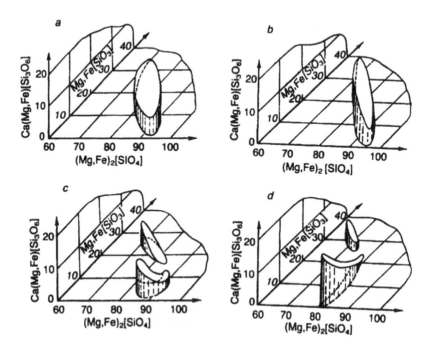

Figure 2.37. Graph of variation in virtual composition of serpentinous ultrabasite massifs in the Urals [6]. Massifs: a — Kimpersaisky; b — Bazhenovsky; c — Buruktalsky; d — Khabarlinsky.

shows of chrysotile asbestos are observed in rocks wherein $M/F > 7$. When interpreting the results of chemical analyses, it is possible to use graphs showing average values of M/F for ultrabasites of the dunite–harzburgite and gabbro-pyroxene–dunite associations on the basis of comprehensive factual material [111]. To this end, it is possible to use data on the absolute ferruginosity of these rocks.

4. The total molecular ferruginosity (F, per cent), and the titanium–iron ($100 \times$ Ti/Fe, per cent) and titanium–magnesium ($100 \times$ Ti/Mg, per cent) ratios are determined to estimate the potential anthophyllite asbestos content of the ultrabasites, and their formational affiliation [108].

Anthophyllite-asbestos-bearing ultrabasites of the dunite–harzburgite formation (most prospective one) are characterized as follows: MgO 35–48 per cent, FeO 0.20–5.41 per cent, Fe_2O_3 0.20–7.60 per cent, TiO_2 0.0–0.1 per cent, CaO 0.1–2 per cent, $M/F > 7$ (the value of $F = 5.1–10$, rarely 11 and 13), $100 \times$ Ti/Fe < 0.5, $100 \times$ Ti/Mg < 25; the chemical composition of rocks of the gabbro-pyroxene–dunite formation is characterized as follows: content of MgO $= 28.72–37.05$ per cent, FeO 2.49–8.47 per cent, Fe_2O_3 3.31–8.16 per cent, TiO_2 0.12–0.61 per cent, $M/F > 7$ ($F = 14–22$, rarely 11), $100 \times$ Ti/Fe $= 1–1.5$ (always more than 1), $100 \times$ Ti/Mg $= 0.3–2.3$ (always more than 0.25) [96].

Examination of the chemical composition and refraction indices of anthophyllite also makes it possible to determine the formational affiliation and prospectiveness of asbestos mineralization. The relative mass of the serpentinized part of the rock, as re-calculated to the anhydrous composition in relation to the mass of the entire rock with subtraction of water, is determined to establish the quantitative indices of the degree of serpentinization [111]. The weight per cent of the products of serpentinization in relation to the entire rock is determined by the amount of water, porosity, and density of these variables. The density is determined by the hydrostatic method, the value of losses on ignition is used instead of H_2O, making the assumption that no other volatile components are present in the sample besides water. It is convenient to determine the degree of serpentinization (volume or mass of the altered part of rock in relation to initial value) by means of a graph, considering the interrelationship of such parameters as density, H_2O^+, and the degree of serpentinization (Fig. (2.38)).

The ratio $(Fe_2O_3 \times 100)/(Fe_2O_3 + FeO)$ is used additionally to determine the degree of serpentinization [108]. Certainly this method is approximate to some extent because it is necessary to compare the chemical analyses at serpentinization with the results of quantitative mineral composition. In addition, the ferruginosity f and the degree of iron oxidation ψ can be determined by the following formulae [111]:

$$f = \frac{(2Fe_2O_3 + FeO + MnO) \times 100}{2Fe_2O_3 + FeO + MnO + MgO} \text{ mol. per cent;} \quad f = \frac{Fe \times 100}{Fe + Mg};$$

$$\psi = \frac{0.9 \times Fe_2O_3}{0.9 \times Fe_2O_3 + FeO} \text{ weight per cent;} \quad \psi = \frac{Fe_2O_3}{FeO + Fe_2O_3} \text{ mol. per cent.}$$

When substituting chrysotile serpentinites for lizardite ones, the degree of iron oxidation decrease in the rocks (4.92, 3.87 and 3.28 correspond to slightly, medium

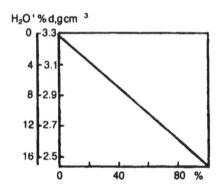

Figure 2.38. Nomograph for determining the degree of early serpentinization of dunites (by density and water content) and harzburgites (by density) [111].

and intensively serpentinized peridotites). This also is true of minerals. The following sequence is observed in the change of the degree of iron oxidation (per cent): lizardite 76–94, chrysotile 68–84; chrysotile asbestos 66–74; and magnetite 69–74 [111]. The thermographic method of surveying ultrabasites and their serpentinized varieties is effective. It is possible to determine by standard curves such minerals as clinochrysotile, lizardite, and antigorite, and to distinguish elastic (strong) chrysotile asbestos from the fragile type.

The method of route traverses of ultrabasites with a sampling interval of 5–10 m is used to determine the alteration of early serpentinization at the present level of erosion. A map of serpentinization is drawn after generalization of the field data and results of analyses. The map shows intervals in the degree of serpentinization as contour lines. A special sign is used to mark areas where serpentinization of the ultrabasites reaches 40–60 per cent, these areas being the most likely for chrysotile asbestos. The frequency of occurrence of serpentinization zones is determined, which, on the whole, can vary from 3 to 100 per cent across the entire massif. Contact zones of fully and partially serpentinized ultrabasites are specially marked.

Shallow workings are dug and holes up to 60 m in depth are drilled for prospecting. Sampling therein is carried out by a continuous point-to-point method: five or six chip samples are taken from each interval at equal distances (3–5 cm), and are combined into one sample with a mass of 200–250 g every 3–5 m. The faults along the section lines and the rock varieties are sampled separately. The sampling interval is different in this case. Anomalies are contoured by the values of minimal anomalous content of indicator elements, which are determined by the basic parameters of distribution of the indicators: mean and standard distribution of the logarithms of element concentrations. Considering that low-contrasting anomalies are located at objects of chrysotile and anthophyllite asbestos, it is advisable to contour them by the values of minimal anomalous concentrations, which are calculated at a 5 per cent level of significance. It is necessary to construct multiplicative haloes of such groups of elements as: $Ba \times Sr \times Hg$; $Cl \times F$; and $As \times Sb \times Ag$.

The data of geochemical surveys are presented in the form of plans and maps with endogenic anomalies, that are divided into groups by the degree of likelihood. It is advisable to discriminate ultrabasites (especially peridotites) by geochemical and petrographic data. In this case, it is necessary to consider the chemical composition of the rocks and the content of pyroxene. Three groups of peridotites are singled out by the quantitative composition of the latter: (a) 5–10 per cent; (b) 10–20 per cent; and (c) 20 per cent. Peridotites containing from 10 to 20 per cent pyroxene, which are most likely for asbestos, are marked off by a special sign. The presence of lizardite, chrysotile, antigorite, and magnetite in the serpentinites is also marked off.

When generalizing the results of mineralogical surveys on maps, it is necessary to mark off by special signs partially serpentinized ultrabasites and essentially lizarditic, chrysotilic, and antigoritic peridotites and dunites. The concentrations of cobalt- and nickel-bearing minerals are marked, especially when crusts of weathering develop on the ultrabasites. Geochemical surveys at the stage of exploration work render comprehensive assistance to prospecting for blind deposits and estimating the quality of the raw material. When carrying out drilling operations, selective abrasion and outwash of the asbestos libre from the core is observed. Therefore, it is advisable to carry out obligatory geochemical sampling of the core to detect occurrences of asbestos bodies at a depth that is realistic for their mining. It is necessary to determine the change in the ratios of the products of linear products of $(Ba \times Sr \times Hg)/(V \times Ni \times Co)$ and $(Ba \times Sr \times As \times Sb)/(Cu \times Ni \times Co \times B)$ to estimate the degree of the erosional truncation, to predict blind asbestos-bearing bodies, and to quantitatively characterize the vertical zonality.

The epicenters of asbestos deposits are detected clearly by the presence of sharp fluctuations in the degree of basicity, oxidation, and coefficient of iron content. To detect asbestos deposits at depth, it is possible to consider the types of the asbestos bodies. Serpentinites with cuts (hair-like veins with a thickness up to 0.5 mm) and rare single asbestos veins usually make up the external zone and, therefore, are of interest as indicators of potential commercial mineralization at depth. Anomalies of such elements as Ba, Sr, As, Hg and Sb have been established for serpentinized peridotites containing veins of chrysotile asbestos. The presence of such veins and endogenic haloes makes it possible to consider the theory that erosion has opened only the upper horizons, and that bodies of asbestos with a large quadrillage should occur beneath them. Maps of deep horizons and graphs of elemental distribution are plotted as a result of these surveys.

The quality of the ore and the asbestos fibre are estimated when carrying out prospecting–exploration surveys. For this purpose, a 12–15 cm core is extracted, and the total width of the cross-sections of the asbestos veins is calculated. The asbestos content of the ore is determined by the relation of the summed width of the asbestos veins to the total length of the core. The grade of the asbestos is determined by calculating the width of the cross-sections of the asbestos veins by classes: −4 mesh (+4.7 mm); −4 + 1 mesh (−4.7 + 1.35 mm); and −1 mesh (−1.35 mm). The content of raw material in each of these classes is determined from the ratio of the total width of the asbestos veins of a given class to the summed width of all the veins. The quality of the fibre, and its strength and flexibility are determined at each new deposit when carrying out geological survey. It is possible to extract individual transverse fibres from the asbestos veins and measure them by means of a scale rule under field conditions. Samples are taken every 3 m, and the results are processed by statistical methods.

The main prerequisite for any fibre development is to have the necessary chemical components for a specific variety of fibre. Without the required chemical constituents, no amount of folding, faulting, or alteration can produce any fibre.

It is interesting to consider this structural origin of chrysotile fibre as well, because the effects of faulting, folding and dilation by shearing or intrusion almost invariably occur in close proximity to occurrences of fibre.

Many of the amosite and crocidolite deposits have been mined from small, narrow open cuts or adits following the fibre-bearing band.

The evaluation of any asbestos deposit entails the determination of its size as well as the grade and quality of its fibre. The dimensions of a mineral zone are established by conventional methods, such as mapping, magnetic surveys, trenching and diamond drilling. The value of the asbestos fibre in the deposit is dependent on numerous physical properties such as fibre length, strength, flexibility, harshness and colour, besides the actual amount of fibre present.

The determination of grade cannot be based on a simple chemical analysis as both the fibre and the wall rock have essentially the same chemical composition. To avoid the complete crushing of the rock and the physical separation of the fibre into different lengths, a method of visual evaluation has been developed which requires a careful enumeration of the total number of fibre veins, together with the average length of fibre in each vein.

Mineralogical–geochemical indications can be used to evaluate the quality of asbestos deposits. The content of magnetite in serpentinites usually increases with the decrease in the asbestos fibre length. Therefore, it is possible to evaluate asbestos quality by the concentration of iron-bearing compounds. To this end, it is necessary to analyse samples of great initial mass, because the distribution of magnetite in rocks is extremely irregular.

The effectiveness of geochemical methods of prospecting is increased by combination with geophysical data.

Stages and sequence of work

Regional geochemical and geophysical surveys are carried out at the first stage at a scale of 1 : 200 000 to determine the geological regularities of location of asbestos mineralization within the territory being explored. It is permissible to carry out geological surveys at a scale of 1 : 100 000 in regions with a very complex geological structure. Prospecting for serpentinite massifs with deposits of chrysotile asbestos can be performed by magnetometric surveys even when they are at a great depth.

In the case of an airborne survey, flight lines are normally spaced at 450 m intervals and flown as close to a 175 m elevation as possible. Where the terrain is rugged, helicopters are used which are capable of maintaining a constant altitude of 100 m above ground level. The spacing between profiles on a ground survey is usually 70–100 m and readings are taken at intervals of 18 or 35 m.

The detected serpentinite massifs are evaluated. The occurrence of ultrabasites of the dunite–harzburgite formation (with a ratio of $M/F > 7$) and ophicalcites is a reliable indication for detecting promising asbestos-bearing regions. Massifs of serpentinized ultrabasites, as well as carbonate formations with the above-stated favourable indications are contoured in the general form. Massifs of ultrabasic rocks are determined from the results of petrochemical, geochemical and geophysical surveys. Space and airborne photogeological surveys are carried out when exploring poorly surveyed areas. Target areas for detailed prospecting surveys are concretized at this stage of work.

Owing to the lack of outcrops within areas of ultrabasic rocks, or within belts in which these rocks are expected to occur, both aeromagnetic and ground magnetic surveys are often employed at the early stage of exploration for asbestos [40]. Ground magnetic surveys may be used to check and clarify anomalies obtained by an airborne survey of a large area, or a ground survey alone may be used for the purpose of exploring a small area.

Magnetic surveys are used to locate and define areas of ultrabasic rocks and, within these, areas which have been subjected to extensive serpentinization. This is possible because this type of alteration produces a higher content of secondary magnetite. Asbestos deposits in ultrabasic rocks are a result of intensive serpentinization, and for this reason asbestos veining is usually accompanied by a higher concentration of magnetite than is normally found in barren serpentine. It follows, therefore, that magnetic anomalies obtained over an area of ultrabasic rocks are favourable places to explore for asbestos. Figure (2.39) shows the results of ground magnetometer surveys carried out over two known ore bodies.

Geological survey is carried out at the second stage at a scale of $1:50\,000$ in the areas that have been chosen according to the data obtained at the first stage. Route and territorial examination of the regions to be evaluated for the abundance of asbestos is the basic technique of geological survey. Geochemical survey is carried out alongside visual prospecting (specification of geological structure and detection of sites of asbestos formation). Lithochemical sampling of loose formations is performed on a grid of 50×50 m in combination with magnetic prospecting. These methods make it possible to contour massifs of serpentinites and to determine therein zones of crushing and cores of slightly altered ultrabasites.

Modern instrumentation has made great strides in recent years, and it is now possible to conduct precise surveys with small lightweight magnetometers, in contrast to the equipment used in the past. Diamond drilling is normally employed to probe beneath the overburden in order to assess and define the limits of an asbestos deposit. Wire-line drilling equipment and the use of non-rotating core barrels are also recommended to minimize fibre loss by grinding of the core during drilling.

Steps are taken to recover the sludge only where core recovery is poor, which is generally the case with an occurrence of slip fibre. Because the fibre tends to fluff

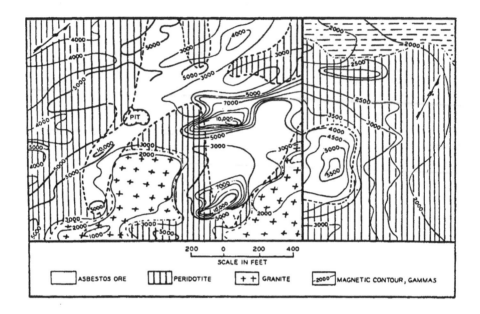

Figure 2.39. Results of ground magnetometer surveys of two known ore bodies [43].

up and remain in suspension, a much greater settling tank capacity is required than is the case when recovering sludge from other minerals. In areas where drilling is impractical, exposure by trenching or exploration beneath the surface by adit or shaft and lateral workings may offer the only means of assessing a deposit.

Special attention should be given to exploring and estimating the occurrences of asbestos (types, size of bodies, fibre quality). A geological–prognostic map of asbestos-bearing areas is the basic result of the second stage.

Geological observations and sampling of ultrabasite intrusions across their strike at 100 m intervals are carried out at the third stage. The route surveys are accompanied by workings, sink to check geochemical and geophysical anomalies. Preliminary prospecting is carried out at shows of asbestos that have been detected during the previous stages and are characterized by favourable indications (extent, quality). Geochemical surveys at this stage make it possible to rarefy the network of holes and workings without worsening determination of the parameters of productive bodies.

Modern power shovels loading into heavy-duty trucks have now supplemented all other loading and transporting equipment where opencast methods are employed. Underground methods which have been used in the past include glory holes, shrinkage and sublevel stopping, and block caving.

Large primary crushers may be located underground. The trend that previously prevailed in Canada, where most of the asbestos was mined by underground methods, has been reversed and now open pit mining prevails.

2.8. MICA—MUSCOVITE

The term 'mica' does not relate to a particular mineral, but to a group or family of minerals of similar chemical composition and to some extent similar physical properties. These minerals are predominantly potassium aluminium silicates with varying amounts of magnesium, iron, and lithium. They have an internal structure of the layered lattice type in which the silicon atoms are in the centre of a tetrahedral grouping of oxygen atoms.

A few of the better known members of the mica group are muscovite, the potash or white mica; phlogopite, the magnesium or amber mica; biotite, the magnesium—iron mica; and lepidolite, the lithium mica. The mineral vermiculite is also a member of the mica group and is treated elsewhere in this volume. Of the known micas, only muscovite, phlogopite and vermiculite find applications in industry.

The mineral mica, which has been known to man since ancient times, has played an important role in the development of our modern electrical industry. Sheet mica can be broadly classified into manufactured and unmanufactured mica [90]. Manufactured mica consists of mica that has been shaped, punched, or otherwise processed into some form suitable for a particular end use. Unmanufactured mica consists of partially hand-trimmed or processed material that has not been prepared for any particular end use. This mica is divided into two commercial classes consisting of sheet mica and scrap and flake mica. Unmanufactured classes of mica differ greatly in their ultimate end use and marketed forms. Sheet muscovite mica can be classified by colour, degree of preparation of the raw material, thickness, size, visual appearance, and electrical quality. In addition, phlogopite mica is also classified according to its thermal stability. Scrap and flake mica consist of mine, trimming shop, and factory scrap that occurs as remnants from mining, processing, and manufacturing operations, and small-particle-size mica which is available from the beneficiation of pegmatites, clays, schists, or other mica-rich host rocks.

Originally scrap mica was derived from the mining and processing of sheet mica and included poor quality sheet mica that did not meet the specifications for size, colour, and quality. The smaller-size mica crystals or flake mica found in coarse-grained, weathered, granitic rocks are known as alaskite.

All commercially valuable muscovites belong to polytype modification 2M, with uncompensated isomorphic substitutions. The majority of muscovite provinces are confined to polymetamorphic complexes, wherein metamorphic cycles occurred before the pegmatites, simultaneously, and after their formation.

Postpegmatitic progressive metamorphism causes dehydration of muscovite and its substitution by parageneses with the participation of disthene, sillimanite and andalusite, and low-temperature postpegmatitic diaphthoresis is accompanied by albitization, neogenesis of spessartine garnet, and sometimes by accessory rare-metal mineralization, that render negative effect on the muscovite quality.

Muscovite-bearing pegmatite bodies concentrate within mica-bearing provinces as regional and local strips of higher mica-bearing zones. Such strips, as a rule, occupy a secant position relative to the rock levels of the metamorphic rock mass. Muscovitization and other metamorphic rock changes take place within them.

The processes of ultrametamorphism and anatexis play an important role in the occurrence of muscovite-bearing pegmatites which were formed at high and extremely high pressures. Mica-pegmatites crystallize at a lower temperature than the other formations of pegmatites.

The differences in mica-pegmatites are caused by the considerable depth and very specific temperature and pressure (PT) conditions of their formation, which determine the constant association of these pegmatites with specific magmatic and metamorphic complexes of wall rocks such as granites, anatectites and disthene gneisses.

Muscovite deposits group into linearly extended belts, having a length of several hundred and a width of several dozen kilometres. They are confined to large fold structures in rocks of the amphibolite facies of the miogeosynclinic group of flysch and flysch-like formations of early and especially median stages of tectonic cycles [74]. The problem of magmatic control has not yet been solved reliably. Many pegmatitic fields are outside the limits of occurrence of large granite massifs and are associated with the latter only by common processes of ultrametamorphism and granitization of rocks.

Plagioclase and microclinic plagioclase (up to purely microclinic) muscovite pegmatites, that give maximum amount of commercial mica, are distinguished. The formation of pegmatites occurred at the background of high-temperature metasomatosis and recrystallization of rocks. The appearance of plagioclase pegmatites is a particular case. The latter are characterized by resemblance of composition and gradual transition to the enclosing rocks, and by presence of skialites and shadow structures. Microclinic and microcline-plagioclase pegmatites crystallized from melt, which is indicated by their independent position in relation to enclosing rocks of any composition, the presence of sharp contacts, graphic granite in the external margins of the bodies, and frequently well-expressed zonal structure of the bodies. Irrespective of their origin, pegmatites were subjected to active reworking by hydrothermal solutions according to the following sequence: early high-temperature alkali stage (absent in plagioclase pegmatites), acid stage, and late medium-temperature alkali stage. Similar transformations are observed in the enclosing rocks.

In Bihar state (India) detailed geological mapping was carried out which covered structural, textural and petrographic analysis and geochemical studies, and included major and trace element analysis of the pegmatites and associated rocks.

The study reveals that [79, 105]:

(1) The pegmatites were formed due to an influx of various chemical components through zones of weakness at the late synkinematic and post-kinematic phases.

(2) The effects of heat and pressure were the main causes of the mobilization of the fluids and volatiles, partly from the anatectic zone and partly from the vapour pressure phase, which activated the process.

(3) The temperature range for the metamorphism of the associated pelitic rocks has been calculated as 522–680°C at 4–6 Kbar pressure.

(4) The pegmatites underwent paracrystalline deformation, as evidenced by the presence of lineations and numerous 'healed fractures' in the constituent minerals of the pegmatites.

Two points of view exist on the genesis of muscovite: crystallization from pegmatitic melt, and metasomatic transformation of feldspars under the effect of hydrothermal solutions. This is evidenced by the dependence between the content of muscovite in pegmatitic bodies, and the composition and character of alteration of the enclosing rocks. The system at that stage was open, which is confirmed by the close contents of the volatiles in the pegmatites and enclosing rocks, alkalinity coefficients of aluminum in hexadic coordination, and lower oxidation of the elements (Fe, Mn) in plagioclase, microcline, biotite, garnet and orthite.

A new type of industrial mica-picrophengite has been found which occurs in both schist and gneiss. In China not only does picrophengite occur in pegmatite, but its crystal size and electrical property also meet industrial requirements.

In the metasomatic column, the composition of (K, Al, Si) in a fluid of pegmatite and the composition of (Mg, Fe) from hornblende (or eclogite) can form picrophengite.

Conditions for the application of geochemical methods in prospecting for and evaluation of deposits

(1) Muscovite pegmatites are localized in the strata, wherein rocks (> 70 per cent) that are favourable for the formation of muscovite are predominant (biotite, garnet– biotite, cyanite–garnet–biotite, bimica plagiogneiss and slate). Pegmatites are confined to that part of the cross-section wherein apoargillaceous rocks are predominant, with an increased content of alumina and constitutional water in the minerals (from 1 to 2 per cent) at moderate concentrations of K, Mg and Fe. The molecular coefficient $A = Al_2O_3-CaO-Na_2O-K_2O$ of the ratio A/K_2O is an index of alumina content [125]. It has been found that as this ratio increases from 1.5 to 2 in the enclosing rocks, the content of muscovite in pegmatitic veins increases from 100 to 600 kg m^{-3}. Ceramic pegmatites form in rocks containing $K_2O > 4$ per cent, and biotitic or amphibolitic pegmatites form in rocks rich in Fe and Mg ($\Sigma > 14$ per cent). Muscovite is either lacking, or of poor quality in these pegmatites.

(2) Alternation of rocks in the cross-section that change the composition of the solutions, i.e. their contrast range contributes to the formation of muscovite. The effect of the solutions on rocks of various composition causes the formation of unbalanced mineral associations; balanced associations are characteristic of background rocks. Quantitative estimates of the contrast of pegmatites and surrounding rocks, depending on their mica content, have been poorly investigated. The contrast range of the composition of pegmatites of the Ricolatvine type and gneisses is: SiO_2 4.6–7.2 per cent; MgO 2.83–3.25 per cent; and FeO 4.58–8.14 per cent. It seems, that an optimal degree of rock contrast is necessary for the formation of

high-grade muscovite crystals, because at very sharp differences in the composition of mica-forming solutions and feldspars of pegmatites, a high rate of crystal growth will result in the entraining of admixtures, the appearance of defects, etc. The confinement of some deposits to garnet--cyanite--biotite and cyanite--bimica gneisses, differing from similar background rocks by a stable composition, is an indirect proof of the fairness of this assumption.

(3) Elements that pegmatites are rich in, are partially evacuated by hydrothermal solutions into the substratum rocks, and, vice versa, some of the components (OH, F, Cl, etc.) are introduced from the outside. Thus, haloes of altered rocks appear near the pegmatites. An internal zone of microclinization, an intermediate zone of andesinization, and an external zone of rock hydration form at the early alkali stage. Quartz--muscovite substitution is associated with chloritization of biotite, and formation of intermediate oligoclase-microclinic and internal essentially muscovite zones. Quartz substitution in pegmatites is manifested by a halo produced by the substitution of the enclosing rocks (from the periphery to the centre): chloritization, muscovitization, and silicification. Changes occurring at different times in rocks partially or fully overlap one another, forming combined composite, superposed haloes near the majority of mica-bearing pegmatitic bodies as a result of inheriting the pathways of the solutions. When prospecting, preference should be given to anomalies that are associated with quartz--muscovite substitution. Leakage haloes,* whose shape and sizes are conditioned by those of the pegmatitic bodies, intensity of transformation, and degree of contrast in the composition of the pegmatites and surrounding rocks, fissureness and chemical activity of the latter, presence of screening horizons, etc., are of great significance. By their size, leakage haloes are usually much larger than pegmatitic veins, though it is difficult to trace the boundaries, owing to the alternation of the unaltered and intensively reworked rocks. The size of the exocontact haloes increased due to intensive fissureness, while the intensity of rock alteration decreases. Lower contrast in the composition of the enclosing rocks and pegmatites causes a reduction in the contrast of the haloes.

The structure and composition of the enclosing rocks influence the character and intensity of their transformation, and together with the underlying rocks they determine also the specific features of metasomatosis in the pegmatites. Quartz--muscovite substitution is most characteristic of pegmatites, occurring in disthene horizons, while processes of the early alkali stage are manifested most intensively in pegmatitic veins of the books rich in calcareous silicate rocks, and less intensively in rocks of quartz substitution. Metasomatically transformed rocks in the books of inter-stratified rocks are of banded structure. Biotite occurs in calcareous silicate rocks instead of muscovite at the stage of muscovitization. Muscovite forms in disthene gneisses instead of microcline at the early alkali stage, because the acidity of the solution increases under the effect of the rocks. Scapolite, epidote, and clinozoisite are widespread in carbonaceous rocks near mica-bearing pegmatites. Disthene gneisses with garnet in the direction toward pegmatite are replaced successively by disthene--garnet--biotite, garnet--biotite, muscovite--biotite

---

*Small diffusional haloes form more often near conformable non-commercial pegmatitic veins.

and sometimes purely muscovite gneisses and glimmerites, accompanied by a decrease in the amount of disthene, hornblende, rutile, garnet and sometimes biotite with a simultaneous increase in the amount of apatite, muscovite, sulphides and the appearance of microcline, tourmaline, monazite, beryl and some other minerals.

Mica-bearing pegmatites occurring within non-micaceous host rocks are mostly discordant in nature and consist predominantly of calcic plagioclases. Other types of barren pegmatites consisting mainly of K-feldspar occur within biotite schists and gneissic rocks. The mica-bearing pegmatites are concordant in nature and consist mainly of albite and bytownite with rare potash feldspar. Barren pegmatites are enriched rare earth elements with while the mica-producing types are mostly devoid of rare earth but enriched with Ba.

The basic plagioclases are replaced in exocontact haloes by acid ones with a simultaneous ordering of the crystal lattice, also characteristic of microclines. The garnet grains become larger in the direction of the pegmatite at a simultaneous decrease in the Mg content and an increase in the Mn and to a lesser extent Fe concentration, with a concomitant increase in the unit cell (by 10^{-4} cm) from 11 533 to 11 544. The concentrations of Fe, F, Li, Ba, Mn, Ga, Ca, Cu and Zr increase in biotites, while the concentrations of Mg, Ti, V and Sc decrease therein. It should be noted that the concentration of Ba, Ca, Ga, Zr and Sc in biotites near barren pegmatites is smaller than the background one, while near commercially mica-bearing pegmatites it is greater [33].

In general, the content of alkaline elements, such as SiO_2, Ba, Sr, Ga, Be, Tl, TR and Li, increases in the enclosing rocks in the direction towards the mica-bearing pegmatitic bodies, and the content of Ti, Cu and Zr decreases.

(4) Productive muscovite pegmatites differ from unproductive ones by a reduced content of Mg, Ca, Cr, Sr, V and Ba, plus reduced Rb and Cs and, at times, Pb and Be, with respect to the alkaline zones formed earlier. Generally, plagioclases, microclines, biotites, muscovites, apatites, and other minerals contain higher quantities of K, Na, Li, Fe, Mn, Pb, Cs, F, Be, Bi, Pb and other fluophyllic elements, and smaller amounts of Ba, Sr, Ca, Mg, Ti and other pyrophyllic elements [79].

Individual deviations from this trend are associated with special conditions of the formation of minerals, or with specific features of their crystalline lattice structure. The contents of Ti and Mg in quartz and quartz–muscovite zones are higher owing to the substitution of muscovite for biotite in relict recrystallized biotite; a relatively large amount of Ba, Sr and other elements, characteristic of feldspars, is contained in muscovite crystals, resulting from hydrolysis of feldspars; the content of aluminum increases in tourmalines, whereas unproductive muscovites of late alkali stages contain much Pb, Cs, Be, Ta and Zr. The content of Σ La and especially Σ TR_2O_3, Sn and Ga increases in almost all minerals, while the content of La and Ce decreases in the direction from monazite–uranic ceramic to xenotime–carburan mica-bearing pegmatites, which is associated with the accumulation of yttric lanthanoides as a result of an increase in the acidity of the medium during the period of muscovite crystallization. The values of Li/Rb, Ca/Ba, Sr/Ba, Li/K, K/Pb and Cs/Rb increase in feldspars and biotites in the zones of quartz–muscovite substitution. This increase is manifested most clearly against the background of rocks that formed at the early alkali stage, for which the values of most of these ratios change in the opposite direction as compared to unaltered rocks. The values of $Na/(Na + K)$ are minimal in muscovites from productive deposits. The index of ordering in plagioclases of commercial pegmatites is high (90–110), and the basicity is low (14–15) [33, 79]. Aluminum is in the position $T_1(O)$ in an ordered lattice; in a disordered lattice it is distributed uniformly in tetrahedrons $T_1(O)$, $T_1(m)$, $T_2(O)$ and $T_2(m)$. Many inclusions that become homogenized at temperatures below 400°C occur in quartz from commercial veins. The values of the parameter τ, the quotient of dividing the ratios of total and paleoatmospheric argon content in low and high-temperature inclusions, are < 1 in large commercial deposits, and > 1 in the secondary, non-commercial deposits. The degree of alteration of the enclosing rocks, the index of the alumina content in them, and the character of muscovitization are also used for estimating the productivity. Intensive muscovite formation in plagioclases is an indication of the amount of mica in pegmatites. Small-flake muscovite, formed from the same solutions as muscovite from pegmatites and oriented at an angle to crystalline schistosity of metamorphosed rocks, is of particular interest. The colour of the muscovite is important: the wide occurrence of light-brown 'ruby' muscovite is an indication of high productivity (unlike green-brown or late greenish-yellow muscovite) in the deposit being studied.

(5) Pegmatites that have formed at various depths differ in their content of individual elements and minerals. Pegmatites in an upright block, wherein the erosional truncation reaches maximum values, are rich in Ba and Sr but poor in Li, Rb and Cs and pegmatites, that have been concentrated in the upper part of the cross-section, are characterized by good differentiation and greater (> 500 m) extension of the vein zones to the depth at relative enrichment of the rocks and minerals in Rb, Cs, Nb, F and Sn, and depletion of Ca, Mg, Ba, Sr, V and Ti content [79]. The pressure difference during the period of pegmatite formation between the upright and trough blocks reached 100 MPa [79].

(6) Early generations of microclines and other minerals from the upper parts of the veins, as compared to their lower parts, contain more Li, Rb, Cs, Be and Pb, but less Ba, Sr and Ca. The circumstance may be used for evaluating the level of erosional truncation of these veins.

Granitic pegmatites are the source of sheet muscovite. These pegmatites are light-coloured, coarsely crystalline igneous rocks. They can be found as dykes or sills in metamorphic rocks and large granite intrusions. The mica crystals found in these deposits range from less than one inch to many feet in length. Variation in size within an individual deposit is not uncommon. Mica-bearing pegmatites have been known to exceed 70 m in thickness and 350 m in length and have been worked to depths ranging from 70 to 175 m.

Pegmatites are composed primarily of feldspar, quartz, and mica. In many geological situations accessory minerals such as garnet, tourmaline and beryl occur with the primary pegmatite constituents. The distribution of minerals in pegmatites may be even, zoned, or segregated into layers.

Prospecting for sheet mica has failed to advance beyond the trial and error method. Geological evidence is generally insufficient to supply enough information to evaluate a deposit. The most practical prospecting method for sheet mica is to sink a test pit in order to determine the percentage, size and quality of trimmed sheet that might be obtained from a deposit.

A common method of further exploration for sheet mica consists of sinking a shaft either downdip in the pegmatite body or vertically to discover the pegmatite at the desired depth. Drifts can then be driven along the strike of the deposit in both directions. Steeply dipping pegmatites can be explored by the construction of adits. Geophysical and geochemical techniques have not been applied to the exploration and evaluation of deposits containing sheet mica. The search for sheet mica remains basically a pick and shovel operation with some limited usage of mechanized equipment.

The most typical minerals of prepegmatite metasomatites are biotite after garnet, and muscovite after biotite, kyanite and plagioclase. Micas formed in altered rocks can be distinguished from those in the surrounding rocks on the basis of the colour, pleochroism, size and orientation of the mica flakes. In northern Karelia, there are increased contents of Ni, Cr, V, Li, Rb, Cs and sometimes Mg, Fe, Mn and Ba in pegmatite metasomatites and the micas from them [33]. In the Mama region in Eastern Siberia, certain garnet-biotite metasomatites are enriched with Mg, Fe and some rare elements (so-called 'orthogneisses'), but, in both regions, the muscovite metasomatites are widespread around fields and clusters of commercial muscovite pegmatites.

Synpegmatitic haloes of altered rocks and geochemical anomalies are only characteristic of mica-bearing pegmatite bodies that were transformed at the postmagmatic stage of formation [36]. It is possible to synchronize the substages of this process inside and outside pegmatite bodies [104]: 1) early alkaline, 2) increasing acidity, 3) maximum acidity. When all three stages occur in haloes, there are successive zones (progressing from unchanged rocks towards the contact) of hydration, andesinization, K-feldspathization, muscovitization and quartz replacement. Such zonation is only well developed in gneisses. In some rocks, biotite forms instead of K-feldspar, and, in marbles, phlogopite is formed instead of muscovite. In interbedded metamorphic rocks, the shape and composition of haloes are very complex.

The distribution of K, Li, Rb, Cs, Tl, Pb, Zn and Sn in samples of micas and feldspars from muscovites, rare metal muscovites and rare metal pegmatites has been studied. There is an increase in the Rb, Cs and Tl contents in all minerals as the rare metal specialization of the pegmatites increases. It is suggested that the Rb/Ba ratio in potassic minerals could be used as an indicator of the rare metal content of pegmatites and their mineral associations of different ages. Li-contents are only increased in minerals of rare metal pegmatites of the lepidolite–albite type. The levels of the Pb, and in part the Sn, Li and Cs concentrations are different in different pegmatite belts.

Results of the investigation of Li, Rb and Cs in feldspars from pegmatites from the Sayan range testify to a regular increase in rare alkalis from high-temperature non-metalliferous pegmatites to lower-temperature microcline pegmatites. Biotites from pegmatites contain on average: 7.55 per cent of K, 0.14 per cent of Na, 0.047 per cent of Li, 0.066 per cent of Rb, and 0.0049 per cent of Cs.

Biotite generations, formed in the process of the successive recrystallization and replacement of pegmatite, are enriched with lithium with a simultaneous decrease in the concentration of Rb and Cs.

Garnets from the Mama muscovite-bearing regions belong to the pyralspite group, which has a similar composition.

The manganese content of garnets from pegmatites is several times higher and the content of rare earth elements 30 times higher than in garnets from country rocks. Y, Dy, Er and Yb are predominant. When the acidity of the process increases, the role of heavy lanthanoids and yttrium grows ($\Sigma Y/\Sigma Ce$ and Y/Ce are higher in garnets from zones of quartz replacement than in garnets from zones of quartz–muscovite replacement).

In some cases, when the geochemical anomalies are not sufficiently contrasting, it is possible to use mineral from haloes, first and foremost, the so-called cutting muscovite, which is only associated with mica-bearing pegmatites. Additional information may be obtained if minerals from haloes are analysed. For instance, in the Karelia muscovite province, V. V. Gordienko (1987) has shown that biotite from haloes (up to 30 m from contacts) contains increased quantities of rare alkalis, especially Li; garnet is enriched with Mn and the Y group of REE [36].

Complex mineralogical criteria (dynamics of the variation in trace element contents in mica and feldspar during the pegmatitic process, changes in plagioclase basicity, etc.) have been developed for mica-bearing pegmatite (North Karelia). These criteria make it possible to estimate the muscovite content in a single body along a widely spaced prospecting net, and also to forecaste the mica content of vein bodies and vein series in the direction of depth, and others. The 'biotite-measure' of the enclosing rocks is the most effective method of prospecting for 'blind' muscovite pegmatitic bodies and especially for the development of large-scale pegmatite-bearing zones connected with 'superposed' metamorphism [79].

Endogenic mineral and geochemical haloes are important indications for mica–pegmatite exploration. In favourable conditions field haloes can be used to determine a little-studied field and to separate the tectonic blocks in fields. The mineralogical and geochemical study of primary haloes can be used for the exploration of blind pegmatitic deposits.

There are several examples of the use of primary haloes as tools for pegmatite exploration:

(1) wide haloes can be used to locate blind pegmatite bodies;

(2) when economic minerals are not evenly distributed in pegmatite bodies (muscovite, etc.) haloes can be used as a signal of the presence of such minerals in another part of the bodies when they are absent in visible outcrops or drill cores.

The first case is well illustrated by many examples of primary haloes around muscovite pegmatite bodies. It is necessary to use not only indicator elements (Ba, Rb, Pb for muscovite pegmatites, Li, Rb, Cs for rare metal pegmatites), but also indicator minerals in the rock discovered in samples or thin sections. For primary haloes of muscovite pegmatites the main indicator is 'crossing' muscovite. For rare metal pegmatites secondary biotite and holmquistite are the most important indicator minerals.

The haloes can be used as a signal of the presence of productive minerals. It was formerly understood that fissures along the contacts between pegmatites and wall rocks are the most accessible places for solutions. A 'generalization' of data on the mineral and element composition of pegmatites as a whole in the contact border indicates that in the contact rock Cs-rich biotite, for instance, can be seen not only near the pollucite pocket, but at any point of the contact. The same situation can be seen for Nb–Ta-rich contact minerals which are at a distance from the ore node. Well-developed zonal haloes with secondary muscovite, microcline, quartz and tourmaline occur along all the contacts of muscovite-bearing pegmatites, whereas the mica zone can be in a small part of the pegmatite body.

The presence of indicator elements and minerals in the contact border is the best sign for the discrimination of pegmatites for productive and non-productive ore bodies.

The regularities of endogenic halo-formation, their origin and the possibilities of geochemical methods in muscovite pegmatite exploration were determined in the largest pegmatite-bearing regions.

Investigations of the extreme south-western part of the mica–pegmatite belt, occurring around Chatara in South Bihar, have revealed that the emplacement of pegmatites occurred in a weaker tectonic zone, lying along the border of two Archean cratons, one to the West and the other to the East of the mica belt. The regional distribution of alkali metals leading to the enrichment of K, in the south-western part of the belt, of Na, in the central zones, and Li, in the extreme north-eastern parts of the belt is discussed [105]. The composition of the host rock exerts an active buffered influence on the development of the stages of metasomatism. Muscovite–tourmaline metasomatites are characteristic of mica schists and granites. Mineral alterations of the host rock are accompanied by the loss of Ca and Na, and the supply of K, rare alkali and volatile components in the exocontact zones. The universal and reliable element-indicators of mica–pegmatites are K, Na, Li, Rb, Cs, Ba, Sr, F, Hg, and Be, P, As, Sb and Bi. The distribution of elements in haloes is caused by their mineral peculiarities. Element haloes are wider than mineral haloes. Their sizes vary from 1–2 m to tens of metres depending on the actual conditions in different pegmatite fields.

Haloes around veins can be used successfully in the exploration of vein series and individual veins at the prospecting reconnaissance and output stages. The alkali element content and ratios of these elements in exocontact metasomatites and micas show the nature of the specialization of pegmatites. The concentrations of elements and their ratios in pegmatite minerals can be used to determine the pegmatite groups (muscovite, rare metal, miarolitic, or non-commercial) and the type of mineralization.

A high content of Ba and Sr and a low content of rare alkalis are found in the K-bearing minerals (K-feldspar, muscovite and biotite) of muscovite pegmatites. The highest Ba and Sr concentrations were discovered in K-feldspars from two-feldspar-bearing veins without mica: Ba up to 3.1 per cent by mass and Sr, 0.0002 per cent by mass, respectively [104]. The Ba/Rb ratio is as much as 172. The amounts of Ba and Sr are lower in commercial mica-bearing veins, but nevertheless they are higher than in the K-feldspars from all other pegmatites.

As can be seen in Table (2.17) there is a maximum content of Ba and a very low content of Li, Rb and Cs in K-feldspar from muscovite pegmatites in the Mama region (Russia). The concentrations of alkali elements are different in the muscovites from two commercial types of pegmatites, i.e. two-feldspar and plagioclase-bearing types.

Rare metal pegmatites have a significantly lower content of Ba and an increased content of Rb and Cs in both K-feldspar and muscovite. As a result, the Ba/Rb ratio is noticeably decreased. Lithium and especially complex rare metal pegmatites (with Li, Cs, Ta and Be ores), have very high concentrations of Rb and Cs in K-feldspars and muscovites. The geochemical anomalies confined to muscovite pegmatite haloes are not very contrasting. The contents of Rb, Ba and Pb in zones of K-feldspathization are 2 to 5 times higher than those in the host rocks. The zones of muscovitization contain positive anomalies of Li, Cs, Be,

Table 2.17.

Average Li, Rb, Cs, and Ba contents (ppm) in K-feldspar and muscovites of different miarolitic pegmatites [104]

Pegmatite group and type	Silicates	Li	Rb	Cs	Ba	Region
Muscovite group:						
Two-feldspar	K-fs	5	490	11	600	1
	Mus	130	770	33	3500	1
Plagioclase	Mus	280	370	1	5200	1
Rare metal –						
muscovite group:	K-fs	10	1500	140	70	2
	Mus	340	1000	280	14	2
Rare metal group:						
Lithium	K-fs	390	6200	560	54	1
(spodumene)	Mus	620	5100	1700	–	1
	K-fs	490	3500	150	66	3
	Mus	290	3700	120	–	3
Complex	K-fs	320	15700	2200	37	1
(spodumene)	Mus	470	16600	2000	–	1
Complex	K-fs	210	20600	3100	77	1
(petalite)	Mus	3580	15400	2300	–	1
Tantalum –	K-fs	93	3300	180	65	3
beryl	Mus	590	3800	320	–	3
Miarolitic group:						
Crystal –	K-fs	8	580	4	–	3
bearing	K-fs	2	450	6	–	4
Topaz –	K-fs	16	1400	130	–	3
beryl	K-fs	8	1500	28	–	3
Tourmaline	K-fs	210	1200	220	170	3

K-fs — K-feldspars, Mus — muscovite.

1 — Eastern Siberia, 2 — Bihar state, India, 3 — Transbaykal, 4 — Kazakhstan.

rare earth elements (REE) (1.5 to 3 times), and sometimes of B, when tourmaline is developed together with muscovite. In the muscovitization of K-feldspathized rocks, there are low contrasting positive anomalies of Ba, Rb and Pb. In the zone of quartz replacement, the concentrations of all the above-mentioned elements are minimal.

Thus, mineralogical and geochemical methods can be applied for various purposes: evaluation of the prospectiveness of the territories and the values of their erosional truncation; prospecting for concrete pegmatitic veins or their accumulations; and evaluation of the prospectiveness of the veins proper and the values of their erosional truncation using the above-listed minerals and indicator elements. The significance of the various indicators differs at different stages: elements and minerals forming broad and contrasting haloes near pegmatitic bodies are important in prospecting, whereas minerals and elements displaying clear correlation bonds with muscovite are important for evaluating productivity.

Steaming haloes form near mica-bearing pegmatites, making it possible to apply methods of decrepitation metering for locating the hydrothermally reworked rocks irrespective of the nature of the solutions. Some commercial deposits have been detected by this method.

Decreptophonic anomalies near commercial mica-bearing and barren bodies can be graded quite reliably by their shape and by number of explosions in various temperature intervals (90 per cent agreement of results). This high degree agreement has been observed at individual pegmatitic fields with clear separation of mineralization stages in time and space, but prospecting practice indicates that on the whole it is overestimated. Most clearly manifested positive anomalies are produced near pegmatitic bodies by Rb and Ba, but they are associated with the alkali stage of formation of pegmatitic bodies and therefore can be missing near mica-bearing deposits, and some of them will be false. It would seem that it is safe to use anomalies that are associated with the quartz–muscovite stage (Be, TR, B, Li, etc.), but similar, though less distinct anomalies develop sometimes in connection with the formation of muscovite at the regressive stage of rock metamorphism. Therefore it is more preferable to carry out a complex survey of anomalies within the steaming haloes of enclosing rocks, taking into consideration the differences in the background and minimal anomalous contents of various rocks. It is most advisable to use levels Rb, Cs, Ba, Tl and Li, that are two five-times greater than the background. Pyrite and pentlandite indicator minerals are also used, since their concentration increases from fractions of a per cent in unaltered rocks to 5 per cent in pegmatite. However, it is more advisable in some cases to use the reduction of mineral (disthene, garnet) and element concentrations near the pegmatitic bodies in prospecting practice. Element antagonists can be more important for prospecting than accessory elements [110]. Haloes of mica-bearing pegmatitic veins may serve as an example. These show an insignificant (two to threefold) increase in Pb and Ba content, but the reduction in V, Cr, Ni and Fe concentration is much greater (5 to 20-fold decrease) [110].

The potentiality of using indicator elements depends also on the specific features of the geochemical methods. The solubility of elements and compounds in water is of great significance in hydrogeochemical methods. Li, Rb, some-times Be, and Nb are accumulated in groundwater near muscovite pegmatites, and most interesting are Be and to a lesser extent Li, both of which are associated with quartz–muscovite substitution. Investigation of anion groups is also highly indicative. The value of SO_4^{2-}/Cl^- is increased sharply in waters near large mus-covite deposits, due to oxidation of ferric sulphides that have been concentrated at the contacts of pegmatitic bodies.

The patterns of element accumulation (dispersion) in plants are poorly investi-gated. No broad concessions has yet been reached on the object of investigation (types of plants, branches or leaves, etc.), on methods of sampling, on ways of treating the samples. Nevertheless, it should be noted that Li, Rb and Cs are accumulated in plants near pegmatitic bodies. The patterns of element and min-eral distribution in secondary haloes are also little studied, partly because of the enhancement of element dispersion in the course of hypergene processes. An increase in the concentration of typomorphic minerals should be anticipated in loose deposits. I. N. Sochevanov [110] has established that the concentration of muscovite in the 20-mm fraction of samples taken at the border of deluvial and eluvial deposits, is especially great near commercially useful pegmatitic bodies. According to some data, positive haloes near mica-bearing pegmatitic bodies are formed by Pb, Ba, Be and P (V, Cr, Ti less than background) while other data indicate that the haloes are formed by Li, Rb, and Cs, and sometimes by Be, Zr and Ga, which makes these data incomparable. Thus, when prospecting for muscovite pegmatites it is possible to recommend only lithogeochemical methods using primary haloes, hydrogeochemical methods, and investigation of minerals in secondary haloes.

Indicator minerals and elements used for evaluating the productivity of peg-matitic bodies are not always identical. For example, it is practically impossi-ble to use the index of alumina content (coefficient A) mentioned above when prospecting for deposits, but the method is good for their evaluation; acid plagio-clases with a high degree of ordering near pegmatites form only shallow haloes, but they give quite good indications of the degree of their productivity; commer-cial mica-bearing pegmatitic veins are manifested by reduced concentrations of barium or strontium, though the enclosing rocks around the pegmatitic bodies are characterized by an increased content of these elements, etc. It is advisable to use muscovite that has been extracted from the holes, the evaluate the productivity of the stripped bodies, because there is a clearly pronounced relationship between muscovite concentration in the holes and the reserves of muscovite of different grades. The coefficient of correlation for the deposits in Karelia ranges from 0.3 to 0.9.

The origin of geochemical anomalies in the mica–pegmatite fields is poly-genic. These anomalies are formed and developed during the alteration of the host rock. The alterations may be pre-, syn- and postpegmatitic. The most impor-tant prepegmatitic process is regional silica–alkaline metasomatism. It results in the formation of zones and plots of biotite metasomatites with which K, Li, Rb and Cs anomalies are connected. Weak-contrast haloes of the whole pegmatite field are formed near pegmatites in host rocks.

Haloes around pegmatite veins are connected genetically with the processes of pegmatite formation. This connection is discovered in the analogy between peculiarities in the geochemical evolution of the pegmatites and the accompanying haloes. The peculiarities of their composition are determined by the ratios of volatile components and alkali activities.

Methods of work

Geochemical exploration is carried out in a certain sequence, owing to changing tasks and objects of investigation.

The development of the regional zonality of pegmatitic fields is especially important for prospecting for pegmatites. This zonality could be expressed on geological maps of various scales in the form of vectors of the evolution of the pegmatitic process. The intensity of mineralization in single pegmatitic bodies increases in the direction of these vectors, and the probability of discovering the most productive objects also increases.

Productive enclosing rocks are contoured using the data of exploring indigenous exposures and those of metallometric survey of loose deposits during geological survey and prospecting at 1 : 200 000 – 1 : 50 000 scales. The content of Ba, Sr and other indicator elements is determined in minerals from primary pegmatitic zones, which makes it possible to evaluate the level of erosional truncation of vein series in the given tectonic block, and thereby to determine the depth of mineralization. Geochemical survey is carried out at this stage in conjunction with an examination of the metamorphism of the enclosing rocks, of structures conditioning the location of the pegmatitic fields, and of other geological objects.

Geochemical surveys at a scale of 1 : 10 000 – 1 : 5000 are aimed at prospecting for and evaluating the viability of mica-bearing pegmatitic veins while simultaneously (especially at a scale of 1 : 10 000) detailing the results of the previous stage of survey. The applicability of geochemical methods of prospecting at a scale of 1 : 10 000 depends on the size of the lithogeochemical haloes, but owing to the probability of missing commercially valuable deposits in the zones of their maximum concentration, it is advisable to prospect at a scale of 1 : 5000 – 1 : 2000. First, decrepitometric investigations are carried out, since they are cheap, highly efficient (1 g samples ground to a fraction of 0.5 – 1 mm), and precise in their location to detected anomalies (the explosiveness of the gaseous – liquid inclusions is 5 – 10 greater than background). The anomalies are graded by two indications:

(1) by shape — extended, linearly symmetrical near tectonic zones, and locally, lenticularly asymmetrical near pegmatites;

(2) by the percentage of explosions of gaseous – liquid inclusions that occur at temperatures up to 260°C compared to the total number of explosives at temperatures up to 540°C (is less than 10 per cent for barren pegmatites, less than 25 per cent for non-commercial mica-bearing pegmatites, and more than 30 per cent for commercial mica-bearing pegmatites).

The sampling grid is made denser within the boundaries of the detected anomalies and the typomorphic minerals and indicator elements are determined. When examining minerals, it is necessary to note the mineral compositions as well as the specific features of the habit, colour, crystal size and aggregate size. One is then able to judge the character of the mineral transformation. It is very important to study the ratio of the concentrations of elements accessory and antagonistic to mineralization. Application of the contrast Rb/V ratio has made it possible to establish that 75 per cent of commercial mica-bearing veins are characterized by haloes with maximum values of this ratio [110]. It is advisable to process the data in the Minsk-2 computer by the special MICA program if the number of accessory and antagonistic elements is great. Deviations in the composition, which are characteristic of commercial deposits and the enclosing rocks are studied on the basis of establishing the background concentrations of the elements and indicator minerals in the pegmatites and enclosing rocks in the given region. It is then possible to evaluate the potential productivity of deposits already discovered by analogy. Owing to the relatively low contrast in the contents of unaltered and altered (in the halo) rocks, the minimal anomalous contents of indicator elements are calculated by the formula

$$a_m = \overline{X} \pm 2\sigma,$$

where a_m is the anomalous concentration of indicator elements, $\overline{X}$ is their average concentration, and σ is the variance. The minimum values of a_m that meets the probability of entering the halo is 0.995.

Lithogeochemical haloes are studied in combination with the petrography and mineralogy of the enclosing rocks and pegmatites in order to establish the character and direction of alterations in various parts of the halo. The geochemical survey should be carried out on the basis of data of structural surveys. This contributes to a more reliable sorting out of false anomalies in the zones of the fault, etc. Structural surveys in closed regions are often based on geophysical data. Geophysical methods can be applied for direct detection and investigation of haloes of altered rocks near mica-bearing pegmatites, which is conditioned by an increased content of ore minerals. The intensity of induced polarization in the circumpegmatitic zone (whose thickness is 10 times greater than that of pegmatitic bodies) is two times greater than the background intensity; the respective values of dielectric constant are 13 and 8.4; of magnetic susceptibility 29×10^{-6} and 22×10^{-6} cm g s; and of density 2.80 and 2.76 g m^{-3}. Thus the methods of natural electric field, induced polarization, and magnetic prospecting can be applied in the search for and evaluation of haloes near mica-bearing pegmatites. Considering the increase of the quartz content at quartz–muscovite and quartz substitution, it is possible to use the piezoelectric method, electric profiling and vertical electrical sounding in pegmatites and in their vicinity when prospecting for blind deposits. This will make it possible to ascertain the strike of blind muscovite zones, that have been detected by geochemical and mineralogical haloes, and to distinguish the haloes, that have been caused by pegmatites and quartzite beds.

The possibility of tracing commercial mica-bearing zones, or the potentiality of their detection at depth in the case of non-commercial mineralization near the surface, is determined at the stage of prospecting. This problem is solved by exploring the character and intensity of rock reworking near rocks of contrasting composition. Intensive reworking of rocks, which is accompanied by redistribution of the elements and minerals, indicates commercial mica concentrations in the deep-seated deposits and enables the assessment of their viability by drilling small-diameter holes. The distribution of indicator elements in the upper and lower parts of the veins is considered at the same time. When prospecting for large single pegmatitic veins, it is advisable to combine geochemical survey with determination of the apparent resistivity. This makes it possible to define the mode of occurrence and the shape of the veins.

Muscovite has excellent basal cleavage that allows it to be split into very thin sheets exhibiting a high degree of flexibility, elasticity, and toughness. Very thin sheets of clear muscovite are transparent and colourless or almost colourless. Upon heating, muscovite begins to lose water at about 500°C. Muscovite is only decomposed by hydrofluoric acid. Mineral inclusions in mica can also limit the usefulness of sheet mica.

The location of a deposit of small-particle-size flake mica can be determined from surface geology. If the surface outcrops are of sufficient size, surface sampling may be justified, but it is more likely that the deposit would be drilled to a shallow depth to furnish samples for testing and to delineate the deposit.

Exploration for sheet mica is not amenable to most of the methodology used to search for minerals. The mica crystals in the host rock make it uneconomic to drill a deposit to any great extent to determine the mica content. Any drilling that is done can only delineate the existence of pegmatite and may reveal very little information about the mica content.

Mineralogical and geochemical methods of prospecting can be supplemented by geophysical and specific decrepitation methods (the latter being the explosion of gas-liquid inclusions in minerals). There are easily discovered haloes of 'vaporization' around mica-bearing pegmatite bodies. They occupy the same area as mineral haloes. The size of the mineral, geochemical and 'vaporization' haloes around mica-bearing pegmatites may reach several metres, sometimes up to 70 m to 80 m above veins.

Quantitative estimation of the degree of effectiveness of geochemical surveys has not practically been carried out. The expenditure on taking samples and their analysis during lithogeochemical prospecting for muscovite deposits by primary haloes are comparatively low (about the same as on drilling a 200 m deep hole) which indicates their high effectiveness. But the percentage of missed mica-bearing deposits and the percentage of false anomalies are not considered. Hence, it is possible to lose mineral resources, and to increase unproductive expenditure on verification of false anomalies. Therefore, it is untimely to speak about a quantitative estimation of the degree of effectiveness of geochemical work in the region of mica-bearing deposits before carrying out special economic studies. However, it is possible to outline measures that contribute to improving the quality and effectiveness of the work:

(1) More careful sorting of the territory (exclusion of regions with rocks weakly contrasting in composition, as well as structures that are not prospective, etc.).

(2) Integration of exploration with the exception of those survey methods (bio-chemistry, secondary geochemical haloes) that are still of low effectiveness.

(3) Observance of sequence of geochemical exploration.

(4) Introduction of new prospective, but rarely applied methods (hydrogeo-chemistry, etc.).

(5) Wider use of the anion group elements for detection of anomalies, which is necessitated by decrepitation-metering and hydrogeochemical studies, and the great influence of the pH of solutions of the character of the processes occurring in the anion part of the solutions relatively independently of the composition of surrounding rocks.

(6) More intensive investigation of the distribution of elements in individual minerals (better through ones), but not in rocks. The content of the ex-amined minerals in the enclosing rocks has a great effect on the potential for extracting elements from mineral-forming solutions and the degree of their accumulation in the minerals. It was considered that an increase in the halo thickness and in the content of indicator elements in minerals was determined by the intensity of the mica-forming solutions. However, in-vestigations carried out earlier on the distribution of fluorine in biotites from granites, indicate that at a concentration of fluorine in granite that is approximately constant, its concentration in biotites decreases, with a concentration increase in biotite content.

(7) Increase of sensitivity and accuracy of analyses that contribute to the de-tection of new indicator elements.

(8) Improvement of the methods of sample analysis, allowing determination of the forms of occurrence of mercury and other elements, the use of multi-plicative haloes, as well as data not only on the content, but also on the patterns of distribution (variance), of elements, etc.

2.9. MICA—PHLOGOPITE

Phlogopite has a pearly to submetallic lustre and varies from translucent to trans-parent in thin sheets. It has been identified in colours ranging from brownish red through yellowish brown, greenish brown, and dark pearl grey to an almost colourless pale green. The colour intensifies as the iron content of the mineral increases. Phlogopite can contain fluorine which is usually most prominent in reddish brown samples and least in greenish specimens.

Sheet phlogopite is found in areas of metamorphosed sedimentary rocks into which pegmatite-rich granite rocks have intruded.

Deposits of phlogopite mica that can economically provide large quantities of sheet mica are available in only a few places and occur in regions of metamor-phosed sedimentary rocks that have been intruded be pegmatite-rich granite rocks. Phlogopite is classified into vein, pocket and contact deposits. Vein deposits, which are generally narrow and are enclosed in fine- to medium-grained pyroxenite, are the major source of phlogopite sheet mica. Pocket deposits have been found to be irregular in shape, size, course and persistence. The pyroxenite surrounding this type of deposit is usually more coarsely crystalline and open textured than that found with vein deposits. The phlogopite crystals may be distributed irregularly throughout the pocket along with crystals of other minerals such as pyroxene and calcite, or may occur as very large solid masses of phlogopite. The Precambrian phlogopite deposits connected with ultra-abyssal granite–gneiss complexes belong to the magnesian skarn formation. They occur in intensively granitized and migmatized metamorphic rocks, including dolomitic and magnesitic marbles.

Industry uses large perfect crystals of phlogopite with a low iron content and an increased concentration of fluorine. These form deposits, associated with highly metamorphosed Mg–Ca rocks of the Archean and Proterozoic, or with younger complexes of ultrabasic, alkaline rocks and carbonatites. The general prospecting criteria of phlogopite deposits are as follows:

(1) most ancient geosynclinic regions of occurrence of Mg–Ca rocks with superimposed metamorphism of the amphibolite stage on the rocks of the granulitic facies;

(2) the presence of granitization in the regions of the areals with sites of weakly altered productive rocks within which intrusions of granites and pegmatites are detected;

(3) metasomatic (diopside) rocks at sites of manifestation of isoclinic folding among pyroxene gneisses, slates and amphibolites along tectonic dislocations;

(4) deep faults in shields, at the edges of platforms, and in intrageosynclinic uplifts and medium massifs;

(5) zonal intrusions of ring structure, wherein bodies of pyroxenized or melanitized olivinites are most advantageous; and

(6) combination of ultrabasites with bodies of ijolite–melteigites, nepheline and alkali syenites, carbonatites, and sites of increased fissureness in these rocks.

Conditions for the application of geochemical methods

Favourable conditions include combinations of contiguous magmatic and metamorphic rocks that are contrasting in composition and rich in Mg and K. The greater the difference between the concentration of these two elements in the contiguous rocks, the greater the probability of deposit formation under the effect of mineral-forming solutions. The commercial value of phlogopite occurrences is determined by the intensity of K–Mg metasomatosis that causes the formation of zonal bodies (phlogopite, diopside, diopside–phlogopite, diopside–amphibole, diopside–orthoclase) with a central nucleus in the form of a phlogopite zone [74]. Commercial phlogopite veins, containing most valuable phlogopite crystals in the bulges, sometimes occupy a secant position in relation to metasomatic banding.

The mineral composition of phlogopite zones and veins is quite diverse: diopside, amphibole, fluorite, fluorine–apatite, calcite, etc. (metamorphogenetic deposits). Deposits that are associated with complexes of ultrabasic and alkaline rocks and carbonatites (see Apatites), are still more diverse.

The indicator elements of phlogopite mineralization are Mg, K, F, the OH group, Sr, Ba, Li, Rb and Cs, from the evidence of data from surveying individual regions. Owing to the redistribution of elements in the zones of rock contact, the composition of the haloes around metasomatites is diverse. It has been established that replacement of granite gneisses by diopside bodies associating with phlogopite mineralization has resulted in the removal of K and Rb and the introduction of Mg, Li and Cs. Potassium and rubidium were accumulated most intensively in exocontacts of diopside bodies: in the diopside–feldspar zone on the side of the gneisses and in the calciphyre zone on the marble side [74]. The content of Mg, derived from the marbles, decreases successively from marbles (12.4 per cent), through calciphyres (11.2 per cent), to diopside (10.2 per cent) and diopside–feldspar (1.3 per cent) rocks, and to granite gneisses and gneisses with diopside (0.4 per cent). Lithium and caesium have been introduced from the outside, and their concentration has diminished on both sides of diopside bodies containing phlogopite, i.e. toward granite gneisses and marbles. A correlation dependence exists between Li, Cs and Mg, which is manifested in a gradual increase in the value Mg/Li from granite gneisses to marbles, that indicates equalization of their chemical potential with respect to magnesium.

Fluorine and chlorine are major indicator elements, and their presence in mineral-forming solutions is confirmed by parageneses of phlogopite with scapolite and gummite, as well as by high concentration of these elements in gaseous–liquid inclusions and in the phlogopites proper. Magnesium and fluorine are bonded in phlogopite more strongly than iron and fluorine; therefore an increase of the chemical activity of fluorine contributes to the crystallization of less ferruginous and more qualitative phlogopites. This is also favoured by a reduced iron content in the surrounding rocks.

Low-ferruginous phlogopites with a high fluorine content have been found in marbles and calciphyres of the Pamirs, Aldan and some other regions in the capacity of rock-forming minerals. However, these phlogopites have no economic value owing to their small size and low content of crystals. Small amounts of recrystallized large standard crystals of slightly more ferruginous phlogopite occur in circumskarn calciphyres. Commercial concentrations form sometimes in magnesian-skarn bodies, but the basic mass of phlogopite is confined to veins and nests in circumjoint magnesian skarns, and is distinguished by an even higher iron content. An increase in the iron content in the phlogopite in this case is conditioned, to our mind, by a reduction of the pressure in the process of fissure opening that contributed to the formation of larger crystals. The results of experimental geochemical surveys indicate the possible use of fluorine as an effective indicator of the presence of phlogopite. Differences have been established in the amount of fluorine in phlogopites, diopside rocks (0.007–0.26 per cent), and enclosing granites (0.009 per cent), as well as in crystalline schists and gneisses. The amount of fluorine in biotite, phlogopite, and amphibole decreases by 30–40 per cent

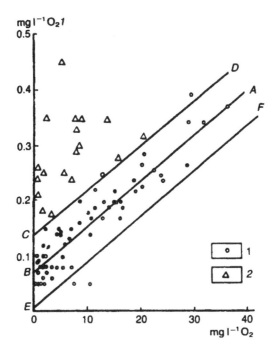

Figure 2.40. Content of fluorine in water, mg/1 [28].   1 — on the territory outside phlogopite deposits; 2 — the territory associated with phlogopite deposits.

through weathering. The degree of fluorine removal increases with the reduction of the particle size. Maximum concentrations of fluorine have been established in surface swamp water in landscapes of the acid gley class, wherein there is much organic matter (permanganate oxidizability 25–30 mg per litre O_2). The amount of fluorine and organic matter is low (permanganate oxidizability 1 mg per litre O_2) in superpermafrost ground water in landscapes of the acid class. The content of fluorine and organic matter in brooks recharged by swamp water and superpermafrost groundwater correlates with the fluorine content in plants, i.e. fluorine passes first through the biogenic cycle. When investigating the distribution of fluorine in water, it is necessary to consider the value of permanganate oxidizability (indicating indirectly the content of organic matter) in accordance with the experimentally established linear dependence $y = 0.066 + 0.0087x$. The straight line BA in Fig. (2.40) satisfies this equation, and the straight lines CD and EF limit the background values of fluorine in waters outside the effective zone of phlogopite deposits, depending on the value of permanganate oxidizability at a confidence level of 0.90. The minimal anomalous value of the coefficient F, calculated from the equation of the straight line CD ($y = 0.66 + 0.067 + 0.087x$), is 0.009. The anomalous values of the coefficient F are traced 150 m down the slope from the surveyed phlogopite show, which is considerably greater than the halo of anomalous values of Cl, S and Mg. Permanent streams in the area of the drainage system, wherein phlogopite shows and deposits are concentrated, did not possess anomalous values of the coefficient F.

Increased fluorine concentrations are observed in primary lithogeochemical haloes. Fluorine anomalies outline clearly phlogopite deposits that have formed at the contacts of olivinites with alkaline rocks. The content of fluorine in ijolite melteigites of the external contact ranges from 0.02 to 0.08 per cent, and in the zone of occurrence of phlogopite rocks that replace altered olivinites it increases to 0.4 per cent [60]. Minerals rich in F and Cl are of prospecting significance.

Much Sr (0.14 per cent) and Ba (0.024 per cent) is found in enlarged grains of calcite from marbles near phlogopite–calcite veins. The external alterations in the marbles are not determined at some distance from the veins, but the content of barium in calcites (0.017 per cent) is approximately five times greater than in unaltered marbles. The content of Sr and Ba increases in calcites with the development of regressive metamorphism from dolomitic to calcitic marbles, carbonaceous metasomatites, and calcite–sulphide veinlets. Particular deviations are established on the background of general sequence toward a sharp increase of the barium content, which is conditioned by a drop of pressure near tectonic dislocations, a temperature rise near magmatic bodies, or sharp changes in the composition of the solutions at the interface of various rocks. The sharp rise of the barium content in calcites from productive veins is associated with a sharp drop of the pressure at the sites of vein formation and with some increase in the alkalinity of the solutions. However, the established patterns are based on a limited number of samples taken within one deposit, making it impossible to recommend Ba and Sr for the time being as indicator elements for wide introduction into practice.

Commercially valuable phlogopites display high concentrations of hydrogen among high-pressure adsorbed gaseous admixtures that are confined to defective parts of the phlogopite crystals. This makes it possible to hope for the application of this correlation in preliminary evaluation of crystal quality from small fragments and sludge.

Methods of work

Geochemical exploration is carried out during prospecting surveys at a scale of 1 : 50 000 and less with the aim of detecting rock complexes, productive for phlogopite deposits, on the basis of such indicator elements as K, Mg and Ca (Ca is most important in turf-covered areas). Hydrogeochemical sampling for fluorine can be recommended on sites that have been detected by geological surveys at a scale of 1 : 50 000–1 : 10 000. Sources of superpermafrost groundwaters and surface swamp waters should be sampled. The water of permanent streams is not prospective for sampling. Fluorine dispersion haloes reach maximum sizes (150–200 m) in excessively humid landscapes of the acid gley class. The discovery of secondary dispersion haloes of fluorine has made it possible to reduce greatly the areas that are promising for detailed prospecting [28]. Similar problems are solved in prospecting at a scale of 1 : 10 000–1 : 2000 on the basis of lithogeochemical haloes. In this case, neutronactivation geophysical investigations have successfully been employed for determining F and Cl content. Electric prospecting, magnetic prospecting, and γ-survey are worthy of application in complex with geochemical methods. Contacts of various rocks and tectonic dislocations are located and mapped using radio control, measuring instruments and magnetic prospecting. Phlogopite bodies are detected by methods of dielectric constant and ground γ-surveying [125].

There are various zoning types of skarn bodies. The analysis of mineral parageneses shows that their diversity is determined by fluctuations in the chemical potential of K, Na, F and Fe in solutions and the magnesian carbonate composition of the rock [70].

Phlogopite deposits are connected with the post-magmatic stage of skarn formation. They are formed at the earlier alkaline stage of skarning and to wards the beginning of the subsequent acid stage. The most valuable low-iron and thermoproof phlogopites are formed with increased activity of fluorine in the solutions and a small content of iron in alumosilicate rock involved in the skarning process. A phlogopite-bearing province on the Aldan shield is composed of polymetamorphic Precambrian rock complexes.

Regional forecasting is of definite significance. In the region in question it is based on outlining the areas where essentially ferro-magnesian rocks are found. The presence of phlogopite was prospected for on the basis of a generalized discriminatory analysis of factorized initial data on all the samples carried out by an computer [97]. The best results are obtained by using the typomorphic association: Ni, Co, Cr, V, Mn, Ti, Y, Yb, Nb, Pb, Zn. By summarizing the points mentioned, it is possible to carry out a local forecast of potential ore-bearing areas, zones and regions.

It is found that phlogopite-bearing diopside bodies have a higher Li- and Cs-content than the country rocks, whereas Rb and K are concentrated in the exocontact zones. Balance calculations of matter indicate the necessity of the addition of Li and the subtraction of Rb in the formation of diopside rocks. A higher Li- and Cs-content in a diopside body can be used to detect the genetic relationship between phlogopite mineral deposits and zones of secondary schistosity.

A local forecast of potential phlogopite-bearing areas was also carried out on the basis of an analysis of data on the elements in the above typomorphic association in all the samples studied using the technique of chief components. This analysis revealed the superimposed nature of the typomorphic association and made it possible to subdivide it into two geochemical parageneses, connecting them with the products of silicic-alkali (Nb, Pb, Zr, Y, Yb), and magnesian–ferro–calcic (Ni, Co, Cr, V, Mn, Ti) metasomatism.

An increase in the Rb, Li and F content is observed towards the end of the process of rock formation in the complex.

The existence of a positive correlation between the Rb and F contents in biotites indicates the common nature of the forms of occurrence of Rb and F in the mineralization solution.

Different radiometric methods of determining the K line showed an abnormal content of the radioactive isotope ^{40}K in mica. The method which excludes the influence of U–Th mineralization by measuring the γ-spectrum intensity in three bands, is recommended as universal.

The detected deposits are evaluated by the degree of their productivity, proceeding from the scale and intensity of K—Mg metasomatism and respective zonality, the extreme member of which is phlogopite–diopside rock. Phlogopitization is sometimes accompanied by ferruginous metasomatism with the formation of amphibole, pargasite or magnetite, and sometimes, conversely, by the removal of iron. The presence of minerals with a high concentration of F and Cl in the zones (fluorite, fluorapatite, phlogopite, scapolite, etc.) indicates commercially useful mineralization, which is often associated with phlogopite veins occupying a secant position in relation to primary metasomatic banding. The quality of the raw material is evaluated by the content of Fe and F in the phlogopites and accessory minerals: the smaller the iron content and the higher the fluorine content, the better the phlogopite quality. Low-ferruginous phlogopites can be anticipated in regions with moderate tectonic activity (wherein the formation of large deposits is highly improbable), where composition of the enclosing rocks is favourable and the concentration of fluorine in solution is high.

The differences in the content of elements in the surrounding minerals can be used in prospecting. In particular, productive phlogopite bodies differ from the surrounding rocks by negative values of γ-activity, which favours γ-spectrometric investigations aimed at specifying the outlines of productive micatization by the concentration of isotope ^{40}K. These outlines have been confirmed when sinking workings.

2.10. ROCK CRYSTAL

Colourless and flawless crystals of low-temperature quartz relate to rock crystal. Two types of deposits are known, (a) pegmatitic, and (b) hydrothermal. Hydrothermal rock-crystal-bearing veins, forming as a result of circulation of mineral-forming solutions in the upper zones of the earth's crust, are of major significance. Crystals of rock crystal occur more often in the flat wall of cavities that are older than those containing vein quartz. Two successive stages are characteristic of the process of piezoquartz formation: the formation of quartz veins proper, and the formation of cavities with crystals of rock crystal. In the first case, the vein bodies formed during a relatively short period of time, under conditions of sharp equalization of thermodynamic potentials. High-silicic solutions separated rather quickly from their source and their acid—alkali evolution also occurred at a high rate. The composition of the veins is almost independent of the character of the enclosing rocks. Circumvein alterations are manifested poorly. Rock-crystal-bearing nests were formed during the second stage of evolution, under the effect of specialized, solutions unsaturated in alumina, on quartz veins. At this time, conditions favoured the longer existence of various thermodynamic levels that facilitates the mobilization of components from the enclosing rocks, their redistribution, and the creation of equilibrium systems. Cl, F, Hg, CO_2 and H_2O were introduced from deep sources. The basic part of the nest components

— Si, Na, Ba, Ti, Ca and Ni — was derived from the enclosing rocks. Minerals parageneses of nests depend on the composition of the enclosing rocks.

The prospecting indications of hydrothermal deposits of rock crystal are:

(1) Manifestation of quartz rock blocks (framed by faults), and anticlinal structures of the first order (meganticlinoria) in regions with jointed heterogenous structures.

(2) Occurrence of roughly layered and coarse-grained quartzites, quartz–chlorite schists, enclosing quartz veins; the presence of screening slate horizons; and the manifestation of pre-ore progressive regional metamorphism of rocks with the predominance of green slate facies.

(3) Occurrence of brachy-anticlines of asymmetric structure that are complicated at the sides by small folding and faults, as well as linear-block structures.

(4) Large tectonic faults and accompanying fan-like dislocations; small-radius seam contortions with rupture dislocations.

(5) Presence of secant or concordant quartz veins with indications of recrystallization, and a coarse and giant-grained columnar structure.

Conditions for the application of geochemical methods in prospecting

The mineral composition of hydrothermal rock-crystal-bearing zones is diverse. The minerals are divided into three groups by the degree of abundance and extent of segregation:

(1) main minerals (quartz, muscovite, chlorite, ankerite, dolomite, calcite, albite and barite);

(2) secondary minerals (haematite, pyrite, tourmaline and halite); and

(3) rare and accessory minerals (cinnabar, rutile, anatase, brookite, xenotime, monazite, apatite, fluorite, sphalerite, chalcopyrite, arsenopyrite, ilmenite, jarosite, anhydrite, scheelite, zircon, sphene, tourmaline, kaolinite, palygorskite, scapolite, etc.).

Rock-crystal-bearing quartz veins are characterized by a more diverse composition of genetic types and a high mineral content, greater than in barren veins. The presence of sericite, chlorite, ankerite, dolomite, calcite, barite, haematite, rutile, anatase and monazite in quartz veins is a most advantageous indication in prospecting for rock crystal. The initial compositions of the rocks enclosing quartz veins changes essentially under the effect of rock-crystal-forming solutions, resulting in disturbance of the previous quantitative relationships of the rock-forming components. This is manifested in the neogenesis of minerals containing water as of hydroxyl groups and as alkalis, calcium, and carbon dioxide (sericite, chlorite, epidote, ankerite and dolomite). The altered rocks near the nests with crystals of rock crystal are characterized by an increased content of pyrite, apatite, monazite and rutile, due to the redistribution of elements under the impact of rock-crystal-forming solutions. When approaching productive bodies, the amount of ilmenite and magnetite in the enclosing rocks decreases, but the content of apatite, monazite, pyrite, zircon, rutile and haematite increases. Cinnabar, barite, xenotime, chalcopyrite and rutile are found in the altered rocks.

Investigation of the mineral composition of quartz veins, altered rocks, and the nest-filling material aids in understanding the geochemical specificity of the rock-crystal-bearing zones containing minerals of chlorine (halite, sylvite, apatite, scapolite); fluorine (fluorite, apatite); titanium (sphene, anatase, brookite, rutile); potassium, sodium, rubidium, and lithium (feldspars, adularia, sericite); copper (chalcopyrite, arsenopyrite); lead, zinc, arsenic, and mercury (cinnabar, calcite, barite, pyrite); barium, and strontium (barite); and rare elements (apatite, monazite, xenotime). Changes in the genetic types and amount of minerals around the nests and rock-crystal-bearing veins indicate redistribution and migration of the admixture elements near the nests.

Elements participating directly in the processes of forming quartz crystals are indicators of rock-crystal-bearing deposits. Sodium, potassium, magnesium, calcium, carbon dioxide, chlorine and fluorine occur in the gaseous–liquid inclusions. Chlorine, sodium and carbon dioxide play the key role. The gaseous–liquid inclusions of quartz crystals contain increased concentrations of chlorine ($450–800$ g kg^{-1} H_2O). Multiphase inclusions, containing jointly liquid carbon dioxide and mineral-prisoners (halite, sylvite), have been detected. Bromine is associated with chlorine. The mobility of titanium in the process of rock crystal formation can be explained by the high concentration of chlorine in the solutions. This is confirmed by an exceptionally wide occurrence of rutile, brookite, anatase and sphene in the crystals, enclosing rocks and vein quartz. During formation, titanium migrated in the form of complex halogen compounds $(TiCl_6)^{2-}$ (the excess of chlorine over sodium contributed to the transition of titanium into a solution). The presence of chlorine in rock-crystal-forming solutions conditions the increased concentration of mercury near the nests of rock-crystal-bearing quartz veins. It has been proved experimentally that in addition to the sulphide form, mercury can be transported in the form of chlorine complexes.

The presence of fluorine in rock-crystal-forming solutions is proved by the paragenesis of quartz with fluorite, tourmaline and apatite, as well as by direct detection of fluorine (up to 0.3 g l^{-1}) in the liquid phase of the inclusions in the quartz crystals, and by the analyses of rocks in the rock-crystal-bearing zones. Barium relates to indicators of rock-crystal-bearing veins, and barite occurs often in quartz veins and rocks around the nest. The content of barium in the quartzites is 10^{-3} per cent, and 200×10^{-3} per cent near nests in the same rocks. Increased barium concentration (as compared to the local geochemical background) have been found in altered rocks around the nest. The content of arsenic and antimony is an average 0.3 and 0.4×10^{-3} per cent in unaltered quartzites, 2.1 and 1.5×10^{-3} per cent in altered quartzites, respectively. Thus, fluorine, chlorine, mercury, alkali elements, titanium and barium, participating directly in the processes of quartz crystal formation, are typomorphic indicator elements of rock-crystal-bearing mineralization. Hydrothermal quartz-forming solutions that are characterized by aggressiveness and high penetrating power cause migration and redistribution of elements in rock-crystal-bearing zones. The enclosing rocks, that occur beyond the boundaries of rock-crystal-bearing deposits contain minor concentrations of chlorine (0.002 per cent) in all cases, irrespective of their lithological composition. A considerably greater concentration chlorine (up to 0.02 per cent) has been detected in the zone of altered rocks (near the nests). The high chlorine content in these zones is conditioned by overlapping rock-crystal-forming processes that caused an increase in the number of gaseous–liquid inclusions containing chlorine, and in the number of chlorine-bearing minerals: apatite, epidote, amphibole and scapolite. A greater content of chlorine (38×10^{-3} per cent) is found in rock-crystal-bearing veins as compared to veins lacking rock crystal (2×10^{-3} per cent). Various elements can be indicators, depending on the specific features of the geological structure of the deposits, and on the composition of the enclosing rocks. Altered quartzites and amphibolites near rock-crystal-bearing zones contain considerably greater concentrations of bromine as compared to unaltered varieties ($n \times 10^{-4}$ per cent): 1.8 (variance 0.28) and 5.4 (0.40); 7.6 (0.22) and 94.5 (0.36), respectively. Rock-crystal-bearing vein quartz differs from vein quartz that contains no rock crystal by an increased content of bromine, whose concentration increases hundreds of times at sites with highly-productive rock-crystal-bearing nests. The content of mercury in unaltered or slightly altered rocks of the external rock-crystal-bearing zone ranges from 1×10^{-6} per cent to 4×10^{-6} per cent, and it increases greatly (from 10 to 25 times) in the zone of intensively altered rocks (0.5 m from the nests). A small amount of this element is found in quartz veins (25×10^{-7} per cent). An increase in the mercury content is observed in rock-crystal-bearing vein quartz (up to 6×10^{-5} per cent). The content of mercury in quartz reaches maximum values near the nests (61×10^{-6} per cent). Fluorine is distributed irregularly on rocks enclosing quartz veins. The average content of this element in slightly altered rocks of the external zone is 2×10^{-2} per cent. The amount of this element increases sharply in the zone of intensively altered rocks (5×10^{-2}–12×10^{-2} per cent), which is associated with an intensive occurrence of sericite and apatite therein.

Alkaline elements and titanium undergo a certain amount of redistribution in rock-crystal-bearing zones. The content of sodium and titanium decreases directly near rock-crystal-bearing quartz veins (0.5 m therefrom). The content of sodium and lithium is considerably greater in commercial rock-crystal-bearing quartz veins that in veins containing no rock crystal (0.05 (Na) and 0.001 (Li); 0.1 and 0.003 per cent, respectively). The content of these elements in the quartz increases close to the nests.

Thus, three groups of admixture elements are detected in rock-crystal-bearing zones. The first group is characterized by endogenic introduction of mercury, potassium, fluorine and chlorine to all zones; the second by migration of titanium, barium and sodium in the zones, often with enrichment of the intermediate and depletion of the internal zones; and the third by a successive removal of the ferric group elements that is more marked in the internal than in the intermediate zone. Hence, geochemical methods can be applied when prospecting for rock-crystal-bearing quartz veins.

Decreptophonic trains and dispersion haloes have been identified around rock-crystal-bearing veins. This permits a logical conclusion on the possible formation of secondary haloes of indicator elements around rock-crystal-bearing veins. Dispersion trains form according to the following pattern: rock-crystal-bearing vein body → endogenic halo → dispersion train. When surveying decreptophonic haloes, it is possible to distinguish zones (areas) of considerable size at the level of the common background of decrepitation activity. This method contributes to the expression and rapid evaluation of potential rock-crystal-bearing zones at the stage of preliminary prospecting.

Secondary haloes result from the exposure of veins at the surface, and denudation of rock-crystal-bearing zones and their endogenic anomalies. The morphological and other specific features of secondary dispersion haloes are closely associated with the structure of the endogenic anomalies, differing from the latter by the smoothness of the forms and downslope displacement. Blind rock-crystal-bearing bodies, as well as those exposed on the surface, are located by using secondary haloes of the stated elements, when the endogenic anomalies are exposed to the present-day surface [86].

Anomalous mercury concentrations have been detected at a distance of 15 to 20 m from the contact of rock-crystal-bearing bodies with the enclosing rocks. Negative anomalies of this element are observed often at the contact of the vein bodies. The contrast of the secondary haloes is three. Displacement of the mercury dispersion haloes downslope is insignificant. Secondary haloes of sodium display slightly greater displacement, and the negative sodium anomaly is near the outcrop of the quartz vein [86]. The data supplied prove the principal possibility of applying geochemical methods of prospecting to rock-crystal-bearing zones, using secondary haloes. Hydrothermal rock-crystal-bearing deposits are accompanied by primary haloes of chlorine, mercury, potassium, sodium and fluorine, whose sizes exceed those of the vein bodies. The content of chlorine in rocks enclosing quartz veins is characterized by the normal law of distribution. This element is characterized by a high background content, and the fluctuations of the values of chlorine concentration are insignificant in various deposits. Positive geochemical anomalies are observed at a distance of three to four metres from rock-crystal-bearing nests. The effective half-width of the haloes is 5.0 m, the detected half-width is 10 m, with a contrast of 15. Broader endogenic haloes occur near thick and rich nests.

Mercury forms broader haloes than chlorine around rock-crystal-bearing zones. Positive geochemical anomalies of mercury are detected at a distance of 20 to 30 m from rock-crystal-bearing nests. A sharply reduced content of mercury, reaching sometimes the level of probable negative anomalies, is observed at the very contact and in the quartz vein. Anomalously increased concentrations of mercury develop asymmetrically in some cases, i.e. along one of the contacts of the vein zone with the enclosing rocks. In the case of closely converging veins and nests, the mercury haloes join together, forming broad anomalies with a sharply increased content of this element (up to 6×10^{-5} per cent). The effective half-width of the halo reaches 8 m, the detected half-width is 25 m, the contrast 5. The temperature ranges of mercury sublimation in the rocks are diverse. The following mineral forms have been established (°C): (1) low temperature 80–100; (2) first medium temperature 170–220; (3) second medium temperature 220–300; (4) third medium temperature 300–400; (5) high temperature > 400. The low-temperature form corresponds to metallic mercury (native mercury). The first medium-temperature form corresponds to its compounds with halogens (mercurous chloride), the second partially to calomel and metacinnabarite, the third to cinnabar. The high-temperature form corresponds to isomorphic admixtures of mercury and partially mercurous oxide and sulphate. Thermal forms of mercury sublimation in unaltered rocks occurring far away from the mineralized zones, are observed in the range 200–220°C. The second type of thermal sublimation curve is associated with altered quartzites, occurring near rock-crystal-bearing nests, that are characterized by the following temperature forms: 100, 220, 300, 400 and 450°C.

Haloes plotted by the forms of mercury at one of the deposits are illustrated in Fig. (2.41). Haloes of the low-temperature forms of mercury make up two anomalous sites, one of which is an isolated zone in the south-western part of the region and extends along the zone of dislocation. The second site is in the

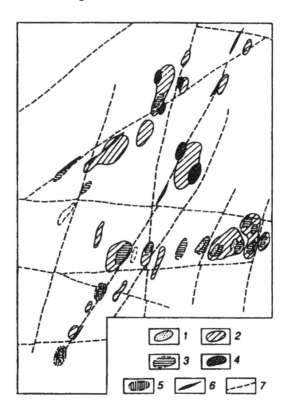

Figure 2.41. Primary haloes of temperature forms of mercury at a deposit in the Urals. 1 — low-temperature form; 2 — first medium-temperature form (area of the curve of thermal sublimation up to 100 conditional units); 3 — second medium-temperature form (area of the curve of thermal subli-mation more than 100 conditional units); 4 — third medium-temperature form; 5 — high-temperature form; 6 — quartz veins; 7 — tectonic dislocations.

eastern part of the surveyed area. The width of the haloes of the low-temperature form of mercury is relatively small (10–15 m). Haloes of the first medium-temperature form of mercury (200–300°C) are distinguished by maximum sizes (up to 50 m) and intensity; these haloes outline anomalies of low-temperature forms of mercury and correlate closely with the latter. Haloes of the second medium-temperature form of mercury (300–400°C) consist of two anomalous narrow bands that are arranged at the periphery (north-eastern part of the area) of the common composite halo. They outline the peripheral part of the halo of the first medium-temperature form of mercury. The high-temperature form of mercury is distributed locally along the zones of dislocations. Its haloes coincide with those of the first medium-temperature form with the maxima of sublimation below 300°C. Close correlation of the low-temperature form of mercury with the first medium-temperature one (200–300°C) and the high-temperature form with the medium-temperature one is observed at most prospective areas, that are detected by the haloes of sodium, chlorine, and fluorine. This association indicates the potential existence of native (metallic) mercury and its halides with cinnabar in the haloes, and medium-temperature mercury with its isomorphic admixtures in the sulphides. As a rule, three temperature forms of mercury occur around rock-crystal-bearing nests: the first and second medium-temperature forms, and high-temperature forms. The presence of several temperature forms of mercury in the rocks is a major prospecting indication of rock-crystal-bearing veins.

The distribution of sodium and potassium in rocks enclosing quartz veins is approximated by the log-normal law. The boundaries of the haloes are quite distinct. Generally, the shape of the sodium halo is more irregular than that of the potassium one. Vein zones of irregular shape are characterized by narrow and contrasting haloes of sodium with a sodium content of up to 3 per cent. The effective half-width of the halo above vein bodies of linear shape is 5–6 m, the detected half-width is 10–12 m, the content of the element in the halo is 0.5 per cent, the halo contrast is 10–12. Endogenic haloes of potassium near linear vein bodies are broader (effective half-width 6–10 m, detected half-width 10–20 m), than near zones of irregular shape (effective half-width 2.5 m, detected half-width 3 m). The contrast of the haloes near the veins of the first type is 6–10 and 3–4 near stockworks.

Haloes of potassium and sodium are screened by various structural elements that control the localization of rock-crystal-bearing vein bodies. A sharp decrease in the content of sodium down to the level of negative anomalies is observed near the nests. The width of the endogenic haloes of fluorine usually exceeds by two to four times the thickness of the veins. The haloes of fluorine around commercial rock-crystal-bearing veins are longer than near small nests. The distribution of fluorine in the haloes is irregular and ranges from 0.02 to 0.8 per cent. The effective half-width of the haloes is not more than 10 m, the detected half-width is 20 m, and the contrast of the haloes is 6. Haloes of barium, antimony, zinc, lead, arsenic, bismuth, copper and nickel, distinguished by their linear character and distribution parallel to the contacts with the veins, have also been detected in rocks enclosing rock-crystal-bearing bodies. The specific features of halo formation around rock-crystal-bearing bodies are illustrated

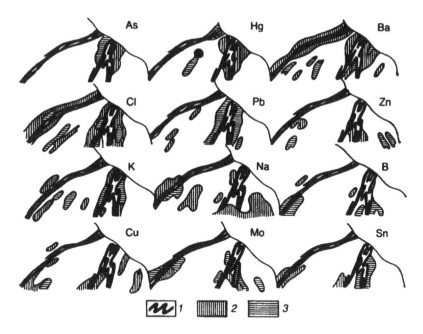

Figure 2.42. Primary haloes around rock-crystal-bearing veins in unaltered and altered quartzites. 1 — rock-crystal-bearing veins; 2–3 — limits of element contents, per cent (2 — As 4×10^{-3}, Hg $(400–800) \times 10^{-7}$, Ba $(3–6) \times 10^{-1}$, Cl $(15–20) \times 10^{-3}$, Pb $(0.2–0.3) \times 10^{-3}$, Zn $(0.4–0.5) \times 10^{-2}$, K 2–3, Na 2–3, B $(25–30) \times 10^3$, Cu $(2–3) \times 10^{-3}$, Mo $(4–5) \times 10^4$, Sn $(4–5) \times 10^{-3}$; 3 — As $(5–7) \times 10^{-3}$, Hg $> 800 \times 10^{-7}$, Ba $> 6 \times 10^{-2}$, Cl 40×10^{-3}, Mo $(6–25) \times 10^{-4}$, Sn 5×10^{-3}, Pb $(0.3–1) \times 10^{-3}$, Zn 0.15×10^{-2}, K, Na 3–5, B $(30–100) \times 10^{-3}$, Cu 3×10^{-3}, Mo $(6–25) \times 10^{-4}$, Sn 5×10^{-3}).

in Fig. (2.42). Primary haloes surround rock-crystal-bearing quartz veins irrespective of their origin and age. The haloes are characterized on the whole by a similar element composition and common forms of the chemical elements. The shape of the primary haloes is quite diverse and depends on the conditions of occurrence, geological structure, configuration, sizes of the vein bodies, and composition of the enclosing rocks. The configuration and sizes of the haloes are closely associated with the specific features of the lithological composition: porosity, permeability of the rocks enclosing the quartz veins. An increased content of elements is observed in quartzites possessing high permeability at a greater distance from the veins than in slates. High contents of mercury and fluorine are observed in quartzite horizons that are characterized by increased effective porosity and permeability. Screening of endogenic haloes is detected at the contact of quartzites and aleurolites. Hydrothermal rock-crystal-forming solutions, penetrating from the forming nests with quartz updip and sideways into the enclosing rocks, develop syngenetic primary steaming haloes exceeding by size the anomalies of the indicator elements. Within the steaming zones, the solutions caused regeneration of minerals and healing of tiny fissures in them. Self-conservation of the solutions occurred in the form of various gaseous–liquid inclusions. Primary steaming haloes near vein bodies are detected by determining the number of inclusions in a unit of mineral volume. Vertical zonality is characteristic of rock crystal deposits; it is accompanied by changes in the geochemically specific features of the vein bodies, and qualitative and quantitative characteristics of the haloes in space. Two groups of elements are distinguished:

(a) those elements with a positive geochemical gradient, whose content in the vertical direction increases upwards (mercury, barium, strontium, antimony and silver); and

(b) those elements with a negative gradient, whose content decreases in the upwards direction (sodium, boron and elements of the ferric group).

The zonality of the primary haloes is an integral reflection of geochemical zonality. Primary haloes around rock-crystal-bearing zones are presented in Fig. (2.42). A quite clear differentiation is observed in the distribution of the elements along the vertical line: the haloes of Ba, As, Sb, Hg, Sr and Ga have maximum width in the upper part of the cross-section and wedge downward, while the haloes of Ni, Co, V, Cr, Sn, Cu, Mo, W, Na and B expand sharply with depth and reach maximum width in the sub-ore part of the circumore space. The haloes of Ba, Cl and Br occur symmetrically in relation to the vein bodies. An alteration in the phase relationship of mercury is a reliable criterion for discriminating supra-ore haloes from sub-ore ones. The native (low-temperature) form of mercury is detected in the supra-ore zones, where it is characterized by a large number of forms and is sublimated in a wider thermal spectrum than in the sub-ore zones. The geochemical series of zonality characteristic of supra-ore zones is as follows, with consideration of the mercury forms (from the bottom upwards):

$$Hg(> 450°C)-Ba-Ga-Hg(> 330°C);$$
$$Ag-Sb-Hg(< 330°C)-Br-Hg(< 260°C)-Hg(100°C).$$

Figure (2.43) illustrates an example of predicting a blind vein body within a deposit by means of primary haloes. Primary haloes of Ba, Sb, As and Pb, that are indicators of the supra-ore sections of the haloes, were discovered in the upper horizons of the haloes. The value of the indicator ratios has made it possible to consider the detected anomaly as a supra-ore one and to recommend it for evaluation. A rock-crystal-bearing body has been stripped.

Figure (2.44) illustrates the change with depth in the ratios of linear productivities of haloes of individual elements to the sum of all the elements. This change demonstrates clearly the vertical zonality of the haloes. The values of the multiplicative index diminishes monotonously and contrastingly with depth; this index changes 5000-fold over a distance of 800 m. It is possible to solve one of the most important problems of geochemical prospecting by means of zoning: to determine the position of the detected geochemical anomalies with respect to the vein bodies. When the anomalies are at the surface, it is possible to determine the level of their erosional truncation in relation to the veins by means of the haloes.

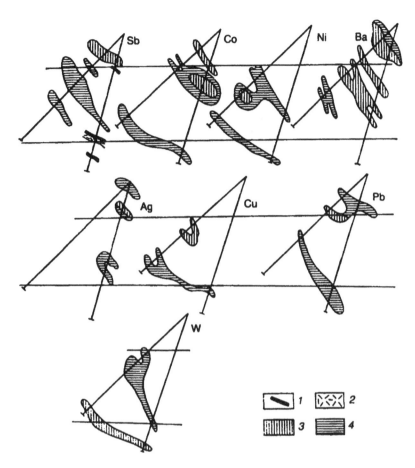

Figure 2.43. Primary haloes around rock-crystal-bearing veins in quartzites. 1 — quartz veins; 2 — hydrothermally altered quartzites; 3 — content of elements in the interval $(0.5-10) \times 10^{-3}$ per cent; 4 — content of elements exceeding 10×10^{-3} per cent.

Halo zonality can be used for discriminating supra-ore haloes from sub-ore ones when prospecting for hidden mineralization.

Evaluation of the deposit depends on the quality of the quartz crystals, which is determined by the effect of radiation on the mineral phases. Interaction of the fast particles and hard quanta with quartz is accompanied by ionizing effects and structural changes in the lattice of the crystals. The following types of disturbance are distinguished:

(a) non-equilibrium, free and localized carriers of a charge (electrons and holes); and

(b) defects of the 'vacancy-introduction' type, and their complexes.

Ordinarily, natural quartz is colourless and transparent. Colourless crystals gain some colour as a result of γ-irradiation, and the colour in initially coloured crys-

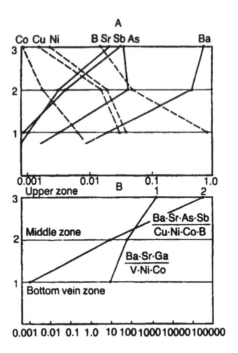

Figure 2.44. Zonality indices of elements (A) and graphs showing changes in the multiplicative index of haloes along the vertical line in bottom, middle, and upper rock-crystal-bearing zones of a deposit (B).

tals turns richer under γ-irradiation. Quartz may have a smoky, smoky-citrine, citrine, or amethyst colour. The smoky colour is conditioned by the formation of hole-type paramagnetic centres under the effect of ionizing radiation at substitution of aluminum for silicon. The missing charge is usually compensated by sodium. The smoky colour of quartz is formed at mineralization of the latter in aluminum-bearing solutions. The colour turns richer with a rise in temperature on increased alkalinity of the solutions. When a proton is present in the capacity of a compensator, centres of smoky colouration are not formed, owing to the high energy of ionization. Unlike alkaline ions, hydrogen is strongly bonded with oxygen in the quartz lattice, and there is continuous recombination of stationary holes with free electrons, and aluminum centres of smoky colouration are not form. Citrine quartz crystals form in less alkaline systems in the presence of lithium admixtures. Citrine crystals contain greater concentrations of hydrogen and lithium admixtures as compared to smoky quartz. There are OH(Al–Li) defects with hydrogen in contrast to substitution only with sodium. After irradiation, citrine crystals gain a yellow colour, and the richness of the colouration is considerably less than in smoky quartz.

Complex investigation of crystals by γ-irradiation and physical methods is most effective. The data obtained by investigating the infra-red spectra of synthetic quartz, grown under different conditions (different media, temperatures, growth rates), have made it possible to establish certain patterns in the change of the relative intensity of the bands of various OH-defects. The concentration of OH(Al–Li) and OH(Al) defects changes with a rise in the solidification temperature, which is evaluated by the relationship of the areas of respective bands. The relative intensity of the diffusive band increases proportionally to the growth rate, which can be estimated by the ratio of the diffusive band area to the total areas of the OH defects. In turn, the degree of acidity–alkalinity can be characterized by the ratio of the areas of OH(Si) bands to the total area of the bands of OH defects. It has been established that there is a correlation between the change of the homogenizing temperature of the inclusions in the quartz of various deposits and the variation of the ratio of the areas of the bands OH(Al–Li) and OH(Al) defects. γ-irradiation makes it possible to detect high-grade optical crystals. Optical quartz is generally homogeneous and is characterized by a slightly smoky or citrine colouration without zonality, sectors, or spottiness, due to uniform distribution of admixtures in the volume of the crystals.

It has been established by the γ-irradiation of vein quartz that productive rock-crystal-bearing quartz acquires a smoky-citrine colour with an irregular, spotted colouration. Quartz containing no rock crystal is coloured uniformly, whereas zones of metasomatic reworking in productive vein quartz, which are characterized by the appearance of citrine colouration on a background of a slightly smoky or smoky colouration of various intensity, can be detected by irradiation.

Certain patterns have been established in the variation of the intensity and character of colouration in the vertical section of the vein zones. The smoky colouration changes in the quartz up through the cross-section to a smoky-citrine and citrine colour. The number of hydrogen-bearing defects increases in the crystals of rock crystal in the upper horizons of the vein bodies. The formation of optically radiation-stable crystals is observed under conditions close to neutral (upper horizons). The quality of the crystals improves in the upper horizons of the veins.

Quartz crystals are divided into four types, in correspondence to the stages of their growth: type I — smoky colouration; type II — smoky-citrine colouration; type III — uniform citrine colouration; and type IV — no colouration under the effect of ionizing radiation. It has been established that the direction of the variation of the colouration from type I to type IV does not change. A clear variation of the radiation colouration of quartz is observed in all the cases in the vertical section of the vein zones: a change of the smoky colouration to a smoky-citrine and citrine colouration. This indicates a decrease of the relative alkalinity of the solutions in the vertical direction. As the hydrothermal solutions moved to the upper horizons, their alkalinity changed as a result of the lowering of the temperature of mineral formation. On the background of a common wave of acid–alkaline evolution, associated with gradual cooling of the solutions, the local changes of the pressure during the formation of tectonic dislocations caused fluctuation of the acidity regime, and a repeated increase of alkalinity, which is manifested in the presence of zonality in the quartz. γ-irradiation makes it possible to carry out objective estimation of the depth of formation of vein bodies, and to separate them by the degree of the erosional truncation.

Methods of work

Lithogeochemical methods of prospecting by secondary dispersion haloes consist in exploring the distribution of indicator elements in loose formations, overlapping rock-crystal-bearing quartz veins, by means of systematic sampling. It is advisable to conduct prospecting lithogeochemical surveys by secondary haloes at 1 : 50 000 and 1 : 25 000 scales, and detailed surveys at 1 : 10 000, 1 : 5000 and 1 : 2000 scales in selected areas, which *a fortiori* have shows of rock-crystal-bearing mineralization, which have been detected in the process of some other work. Lithogeochemical prospecting along secondary dispersion haloes is carried out by sampling loose deposits on a uniform grid at the depth of the representative horizon. The grid of the routes and the density of the sampling points depend on the scale (see Table (2.5)), the shape of the vein bodies, and the size of the haloes. The profiles should be oriented across the strike of the structures and veins. The distance between the profiles of the main grid should be a maximum of 0.9 of the length of the halo, and the distance between the sampling points should be a maximum of half the halo width, to ensure reliable detection of the anticipated dispersion haloes of the vein zones. The most rational sampling grid for detailed prospecting is 50×10 and 25×10 m. The chosen sampling grid can be made denser or sparser, depending on the geological conditions and the results obtained. Additional sampling is carried out in zones that are of special interest for detecting rock-crystal-bearing bodies. The required intensification of the grid for co-ordinating the results of work in adjacent profiles and for reliable selection of sites for making workings is ensured by reducing the distances between the profiles and the spacing between the sampling points in the profile. The scale of the geochemical exploration should be successively increased in selected prospective areas. A fine (> 0.5 mm) sandy-argillaceous fraction of eluvial–deluvial formations should be sampled from the representative horizon during lithogeochemical prospecting; the mass of the sample should be 50 g. The initial mass of the sample is increased to 300 g in some cases, when using a coarse fraction (< 0.5 mm). Lithochemical survey is efficient in the Urals and Siberia with shallow sampling (20–50 cm) from beneath the humic horizon of the soil, or from the bottom part of the humics. A 2 g fraction (1 mm) is separated from the samples when surveying secondary decreptophonic haloes within prospective zones. The decrepitation activity is determined as a sum of pulses (explosions) at heating this material up to 750°C.

The correctness of sampling should be verified by repeated sampling at an amount of 3 per cent of the total number of samples. Control sampling is carried out: (a) in selected profiles, that are arranged uniformly on the area; and (b) in individual points or profiles, that disturb the regular geological pattern of the dispersion field. The samples should be rubbed to powder in mechanical grinders, or manually in agate mortars, to 150–200 mesh (powder) and analysed for chemical elements: Cl, F, Br and Hg (by special methods), K, Na, Li, Rb and Cs (by flame photometry), as well as Be, Ti, V, Cr, Co, Ni, Cu, Zn, Y, Zr, Nb, Mo, Ag, Sn, Ba, La, W, Pb and Bi. The mathematical law of distribution of decrepitation activity and its parameters within prospective areas is determined by plotting an integral graph of frequency distribution according to a probable pattern. Ordinarily, graphs of decrepitation activity are characterized by great variation, and they are smoothed out by the 'sliding' window (moving average) technique. The values and lower level of probable anomalies of decrepitation activity are determined on the basis of calculated and graphic data. The hypotheses of the equality of parameters of decrepitation activity distribution are verified by means of Fisher's variance ratio and the double Student t-test. This latter method is applied to determine whether the difference between the functions of distribution of the regional fund of decrepitation activity and at the prospective areas are essential or accidental.

Evaluation of the migratory capacity (comparative mobility) of the indicator elements and their coefficient of dispersion uses the formula:

$$\sigma = \frac{M}{2.5} C_{\max},$$

where M is the linear productivity of the halo (%) and $C_{\max}$ is the maximum content of the element in the given section of the halo [46].

Owing to log-normal distribution of errors during geochemical surveys the systematic error is determined by the results of control work [46].

The lithogeochemical method of prospecting by primary haloes consists in exploring the distribution of indicator elements in the bedrocks. The surveys are divided into searching, detailed, and prospecting ones. The first is aimed at detecting the primary haloes of endogenic deposits exposed at the level of erosional truncation. The bedrocks are sampled in the exposures, workings, and shallow boreholes. The surveys are conducted at scales of 1 : 50000 and 1 : 25000 in areas devoid of loose formations; in some cases they are carried out in closed territories if the data demonstrate that the primary haloes will not be sufficiently distinct in the loose cover owing to their impoverishment (at a thickness of the loose formations in excess of 5 m). Detailed geochemical surveys are conducted along primary haloes at scales of 1 : 10000 and 1 : 5000 on the sites with detected rock-crystal-bearing occurrences, with the aim of contouring the haloes on the surface, and the potential mineralization at depth. Sampling is carried out at the surface, in workings, and in boreholes. Geochemical sampling of test wells is performed when prospecting in closed areas. Lithogeochemical surveys using primary haloes (sampling of boreholes and workings in the course of sinking) are also conducted in the process of exploration of deposits. The purpose of this stage is to prospect for blind vein bodies at the flanks and in the deep horizons of the deposits, and evaluation of the erosional truncation of the zones.

To this end, it is most advisable to sample workings and boreholes that cross the rock-crystal-bearing body at different levels, which makes it possible to build vertical geochemical sections. The intervals between the boreholes and workings should go beyond the zones of hydrothermally altered rocks.

Primary geochemical haloes are explored by sampling ore-bearing bedrocks in a series of cross-sections or profiles, usually oriented across the strike of the rock-crystal-bearing zones. The sampling grid is selected in such a manner as to cross the assumed geochemical anomaly minimum by two profiles. The profiles should go beyond the zones of hydrothermally altered rocks in different directions.

The sampling grid should be 500×50 and 250×20 m when prospecting at a scale of $1:50000-1:25000$ scales in well-exposed regions. When sampling in areas with irregular exposure, routes should run along the sites with maximum exposure. It is permissible to deviate in this case up to 30 per cent from the given grid [46].

The sampling grid for detailed lithogeochemical prospecting by primary haloes is 50×10 m. Geochemical samples are taken, with consideration of all the specific geological features of the rocks, along the lines of the profiles (bedrocks on the surface, walls of workings, core from boreholes) at 5-m intervals. Sampling is conducted continuously by the 'dotted-line furrow method' by taking 5–6 small rock chips, sized 3–4 cm^2 across, from the sampling interval and uniting them into one sample. The sampling interval is reduced to 1 m near the assumed vein bodies. Veins occurring in the profile and zones of faults are sampled individually. At a change of rocks, sampling is conducted in such a way that each sample contains fragments of only one rock variety. This can be ensured by changing the sampling interval. Five to ten small fragments sized 3–4 cm^2 each, which are distributed uniformly every 10–20 cm, are taken from these intervals. The drilling mud is sampled when sampling boreholes in the intervals of poor core recovery. Rock samples taken for analysis, must be representative (corresponding by composition to the sampled rock), i.e. similar multiples of the mineralogical cell. An empirical rule is applied in practice implying that the real sample should exceed the elementary mineralogical cell by at least 10 times. The necessary and sufficient geochemical sample should have a mass of 100 g for homogeneous fine-grained rock, and up to 200 g for coarse-grained rock. It is enough to take 30–40 samples (with a relative error of maximum ±20 per cent) to characterize confidently the elementary composition of the geological formations (determination of background content and variance). The samples are processed in stages:

(a) grinding in jaw crushers to 0.5 mm (a 5 g sample is taken from this fraction for decrepitation);

(b) crushing in a roll crusher to 0.1 mm;

(c) mixing by the ring and cone method, and reduction of the samples by quartering to 50 g each; and

(d) mechanical grinding of the quartered batches to 200 mesh (powder).

One half of each sample is stored as a duplicate, and the other half is sent for analysis. It is necessary to note that the low content-contracts of elements in geochemical anomalies of piezoquartz deposits demands most thorough processing of the geochemical samples, eliminating their contamination by foreign material.

It is not possible to process geochemical samples in units that are used for processing ore samples. It is recommended that differential processing of samples is conducted. Samples from hydrothermally altered rocks, quartz veins, and veinlets are selected, based on results of recording bedrocks along the sampling profiles, and processed after processing background samples. The crushing units should be cleaned before processing each batch. The chemical heterogeneity of the samples is manifested when determining small concentrations of elements in the samples of rock-crystal-bearing deposits.

The dependence of reproducibility of the analysis on the size of sample is expressed by the formula:

$$S = \sqrt{\frac{k}{m}} \times 100 \text{ per cent,}$$

where S is the relative standard deviation of the results of analysis, conditioned by heterogeneity of the sample; k is the sampling constant; and m is the mass of sample used. The mass of the sample should be increased to ensure reliable determination of the total element concentration. The element whose determination requires the maximum volume of the sample should be the reference one when determining the volume of the sample.

The high sublimating capacity of mercury should be considered when storing samples. Mercury should be determined soon after sampling. All samples taken during geochemical exploration should be subjected to rapid spectral analysis for a wide range of elements. Alkali elements are determined by flame photometry, or by the neutron-activation method. Chlorine is determined by the nephelometric method, and mercury in a mercury atomic-absorption photometer (RAF-I). Fluorine is determined by the rapid emission spectrometric method. The forms of mercury are determined using the FLUR-I instrument.

In order to reduce costs the following sequence of analytical work is advisable: rapid spectral analysis and determination of chlorine in all the samples; contouring of geochemical anomalies; and analysis only of samples within the contours of geochemical anomalies by more expensive methods.

It is necessary to consider the capacity and sensitivity of the methods used when analysing the samples. For confident determination of anomalous element concentrations, the sensitivity of determination should be three times lower than the background levels of the elements. The quality of sampling during geochemical exploration by primary haloes is controlled by repeated sampling up to 3 per cent of the volume in areas where the detected anomalies are hardly probable by the geological data. The error of the work is determined by the data of primary and control sampling, and it is considered when determining the value of the geochemical background and detecting geochemical anomalies.

Geochemical anomalies are contoured by the values of the minimal anomalous contents of elements, calculated for each of the various rocks that are typical of the region being explored. The values of the minimal anomalous concentrations of elements are determined, proceeding from the values of the basic parameters of indicator distribution: average concentration (arithmetic mean), standard distribution of concentrations assuming a normal distribution, and, in the case of a log-normal distribution (most typical for trace elements), the average and standard distribution of the logarithm of the element concentration.

It is advisable to contour anomalies by the value of the minimal anomalous concentration, calculated at a 5 per cent level of significance (unilateral 2.5 per cent).

The parameters of explored sites are compared with the parameters of the background distribution of the elements. The following methods are applied when processing analytical data:

(a) the method of variational curves;

(b) the method of probable pattern, or method of straightened (cumulative) graph of accumulated frequencies; and

(c) the calculation method. Calculation methods are more preferable. Indexes exceeding the background levels by two standard deviations are adopted as the lower limit of anomalous values. This method is applied to detect weak anomalies, which are characteristic of rock crystal deposits. The probability increases in this case of detecting false anomalies, conditioned by errors of sampling, processing, sample analysis, and by variation in background values.

However, such large numbers of anomalies are not usually detected within rock-crystal-bearing zones as within the limits of ore deposits. The functions of element distribution in the rocks within rock-crystal-bearing deposits differ from one another and are conditioned by fluctuations of the parameters of distribution caused by heterogeneity of the sites and errors in the methods applied. But they can be conditioned by concrete geological geochemical factors. Fisher's variance ratio and the double Student t-test are used to determine these differences by comparing the character of element distribution within the vein zones. It is established whether the functions of element distribution at these sites are considerably different, or whether the differences are accidental.

Positive anomalies are detected that reflect the accumulation of elements owing to their introduction by hydrothermal solutions (mercury, fluorine, chlorine and potassium). Negative anomalies, conditioned by the removal of such elements as titanium, sodium and barium from this area are also detected. Special attention is given to sites where positive and negative anomalies meet, where the probability of discovering concealed vein bodies increases greatly. For mercury, it is advisable to explore the forms of its occurrence, alongside determination of the gross concentration, for objective judgement of the presence of commercial mineralization. A comparison of haloes by the determined forms of mercury makes it possible to interpret anomalies, detect those most prospective for evaluation, and to judge the level of erosional truncation of the deposits.

The conditional coefficient of productivity, K, reflecting on the graphs of the areas the ratio of the primary form of mercury (chloridic) to its superposed forms (metallic, oxide, calomel, sulphidic and isomorphic), is determined. This coefficient is equal to 2.8–3.0 in mineralized zones and fluctuates around unity at sites remote from rock-crystal-bearing nests. The coefficient is 1.5, on average, near tectonic zones. An increased gross content of mercury and a variety of forms of its thermal sublimation, with predominance of the chloridic form, are characteristic of large rock-crystal-bearing zones.

It is advisable to construct multiplicative haloes, using the indicators of deposits known in the region. Additive haloes are constructed only for elements that form haloes due to their introduction. Elements forming haloes characterized by their removal should be excluded from the number of indicators to avoid weakening of the effect of anomaly intensification. Multiplicative anomalies are delineated by the minimal anomalous value of the product of the content of indicator elements, calculated by the results of sampling background sites, and rocks which are distributed in the region.

The following methods are applied to calculate the parameters of background distribution of elements:

(1) the results of geochemical sampling of rocks are used on carefully chosen background sites remote from the vein bodies and deposits, and bearing no traces of rock crystal mineralization;

(2) maps are draw of the distribution of values of indicator element concentration products in isolines (gradation in decimal scale) within the explored area, with subsequent identification of most probable background sites, characterized by minimal values of element concentration products, as well as uniform distribution of these values; and

(3) calculation of minimal anomalous values of indicator element concentration products in the chosen background sites (at a 5 per cent level of significance), and contouring of geochemical anomalies by these values.

The results of geochemical prospecting by primary haloes are illustrated in the form of graphs of concentrations. A coordinate field is used when plotting these graphs. The values of the geochemical background or the percentage of the content of elements are units for measuring the content. The results of areal sampling of the bedrocks are presented graphically in the form of geochemical maps compiled on a geological basis. The concentration products for each element are given in the maps in isolines, and geochemical anomalies are marked out on this basis. The gradation of the values within the anomalies depends on their intensity and is presented usually in the decimal scale. It is advisable to draw maps for each anomaly on a scale of 1 : 5000 (the grid for spacing the results of analysis is 20×2 mm) when sampling the bedrocks at a scale of 1:10 000.

The zonality of the haloes is used as a criterion for evaluating the level of erosional truncation of the geochemical anomalies. The ratio of the mean concentrations and the half productivity ratio are determined for a pair of elements, as well as the ratios of the parameters of particular composite haloes. The parameter of the supra-ore haloes is given in the numerator, and that of the sub-ore ones is stated in the denominator. The level of the erosional truncation of the anomalies is evaluated by comparing the values of the parameters stated above with those of haloes of known vein bodies. It is most advisable to use the parameters of multiplicative haloes $(Ba \times Sr \times Ga)/(V \times Ni \times Co)$, because their zonality is of greater contrast.

Besides graphical material depicting the distribution of the absolute concentrations of elements, it is necessary to compile tables and draw graphs characterizing the changes in the ratios between elements in a vertical direction (upper horizons, middle section, roots of quartz veins above the rock-crystal-bearing body), and in the horizontal direction (centre and edges of the ore body). The ratio of linear productivities of particular composite anomalies is used as a criterion for grading the zones of dispersed mineralization. The detected anomaly can be referred to the category of zones of dispersed mineralization when the value of this ratio corresponds to the primary haloes occurring at the level of the lower parts of commercial mineralization. The values of the indicator ratio are compared in this case with the values tabulated for the deposits of the graphs of changes of indicator ratios along the vertical line.

The problem of enhancement of weak haloes is solved by various methods: rational sampling (method of continuous taking of spot samples), analysis of minerals and heavy fractions of geochemical samples (pyrite, calcite, sericite, chlorite); construction of multiplicative haloes $(Ba \times Sr)$, $(Y \times Ge \times La)$, $(Zn \times Pb)$, calculation of the ratios of polar pairs of elements (accompanying element — antagonist); application of highly sensitive methods of analysis; determination of different forms of mercury; and detection of geochemical anomalies by averaging methods (sliding window).

The following indications are considered when grading anomalies:

(1) location of site in concrete geological space; anomalies are evaluated positively, whose element composition corresponds to direct geochemical indicators (Cl, F, Na, K);

(2) contrast of geochemical anomalies with respect to geochemical background; high contrast of the anomalies is a positive factor for their evaluation;

(3) spatial coincidence of geochemical and mineralogical indications; haloes that are confirmed by mineralogical data, merit primary positive evaluation.

Anomalies are divided into true and false ones. Depending on their prospectiveness, the former are divided into the following classes:

- quite probable occurrence of rock crystal;
- probable detection of rock-crystal-bearing mineralization;
- unlikely occurrence of rock-crystal-bearing mineralization; and
- probable occurrence of rock crystal at depth.

The indications of true anomalies are as follows:

(a) Sharply irregular distribution of indicator elements in the zones of modified rocks with considerable deviations in concentration levels.

(b) Presence of the following elements in haloes, participating directly in the processes of rock crystal formation: chlorine, fluorine, potassium, sodium and mercury.

(c) Maximum superposition in space of haloes of diverse indicator elements.

(d) Coincidence of positive and negative anomalies, confidently indicating the epicenters of rock-crystal-bearing mineralization.

(e) Wide thermal spectrum of mercury sublimation in altered rocks in the presence of low-, medium- and high-temperature forms.

(f) Presence of supra-ore indicator elements in the haloes (Ba, Sr, Ag, As, Sb, Hg, low-temperature forms of mercury sublimation) that are characteristic of weakly eroded vein bodies.

When evaluating anomalies, the sequence is determined by the stage of prospecting – exploration work. General (review) evaluation of the anomalies is conducted at the stage of prospecting at a scale of 1 : 50 000, as well as their grading and identification of primary anomalies for a more detailed exploration. Examination of the most prospective anomalies by means of workings is conducted at the stage of detailed prospecting at a scale of 1 : 10 000. The anomalies are evaluated at the exploration stage by the method of analogy with the known ore bodies. Boreholes and workings for examining anomalies are subjected to geochemical sampling with the aim of obtaining data on the distribution of elements in a vertical direction. This is important for interpreting the detected anomalies. The volume of drilling and mining operations for stripping the primary halo depends on the complexity of the geological structure and the dimensions of the halo; it should be minimal and ensure only stripping of rock-crystal-bearing bodies, with preliminary evaluation of the quality of the quartz crystals, using the method of γ-irradiation. The character of radiation colouration of the quartz, arising under the effect of radioactive sources of ^{60}Co, is determined.

Disintegrated blocks containing quartz crystals piled up on mountain slopes are sampled when conducting prospecting work. The crystals are sampled in workings when rock-crystal-bearing veins are stripped in the indigenous occurrence. When exploring, samples are taken from the nests. The number of crystals, if possible, should range from 10 to 12. Two plates (3 mm thick) are cut out of the crystals and are polished on both sides. One of the plates is kept as a reference, and the other is covered with aluminum foil and is subjected to γ-irradiation by a source of ^{60}Co (the dose of irradiation is 0.7 A kg^{-1}). The optical density of the arising colouration is measured in a spectrophotometer within a wavelength range of 200 to 650 μm. The percentage transmittence is reduced to a common thickness of the preparation equal to 1 cm. Optical absorption spectra are recorded in locally pure zones of crystals, lacking inclusions and fractures. The saturation dose is determined. The density of the smoky colouration is expressed by a five-point system (the thickness of the samples is equated to 1 cm): one point, 0.0–0.2; two points, 0.2–0.5; three points, 0.5–1.0; four points, 1.0–2.0; five points, > 2.0.

Quartz is divided into four groups, depending on the colouration (smoky, smoky-citrine, citrine and irradiation-stable). The frequency of encountering quartz of these types is calculated. Zones that are prospective for optically irradiation-stable quartz (uniform colouration and low density) and citrine, are determined by the type and density of the colouration. Qualitative evaluation of the level of erosional truncation of the vein bodies is conducted by the predominance of citrine and smoky crystals if the veins are localized in quarzites, or by the density of the smoky colouration when mineralization occurs in shales and amphibolites. The data obtained are plotted on plans and cross-sections. Graphs are plotted reflecting the gradient of variance of colouration density. A map of isolines reflecting the density of colouration in a five-point system in the plane of the ore bodies is drawn.

The method of infra-red spectroscopy, consisting in measuring the areas of the main characteristic absorption bands, is applied to determine the quantitative criteria of evaluating the degree of the erosional truncation of the vein bodies. Two indices are used: the ratio of the areas of absorption bands OH(Al–Li) to OH(Al). The geochemical gradient is determined, which is the variation of the stated ratios per 100 m drop within the known deposit.

The depth of the erosional truncation is calculated using the formula

$$h = \frac{a - b}{Q} \times 100,$$

where h is the difference in the depth of the erosional truncation, a is value of maximum content of the ratio OH(Al–Li)/OH(Al) in the reference deposit, b is the same, but minimal in the new zone, and Q is the geochemical gradient.

A scale of lithium and hydrogen concentrations is drawn, proceeding from the ratio of the absorption bands and the percentage of crystals of types 1 to 4 in the case of complex character of mineralization, or when evaluating vein zones in new regions where no data are available on reference deposits. The percentage of crystals of type 1, characterizing the bottom part of the ore bodies and formed in the early period, is taken with a negative sign, and the number of crystals of types 3–4, formed in later periods, is taken with a positive sign. The content of crystals of type 2 is considered as neutral and does not participate in the calculations. The percentage of quartz of the early and late stages is added up algebraically. The relationship of the crystal types will set in proportions from +100 to −100 per cent. Zonality is expressed by the gradient of quartz changeability in points.

The choice of one or the other method is determined by the character of the problems to be solved, and, in the general case, by the degree of exploration of the given territory. It is possible to apply in the Urals and Siberia the heavy concentrate–explosion method of prospecting by secondary dispersion haloes.

Primary attention in geochemical prospecting should be given to shows whose outcrops present no practical interest, but which may at depth pass into commercial deposits. γ-irradiation can be used to grade quartz veins. The specific features of the distribution of indicator elements in the primary haloes are specified by the results of geochemical sampling, and this basis can be used to evaluate the prospects of rock crystal occurrence in the deep horizons of the sites being explored. The use of geochemical criteria makes it possible to perform substantiated delineation of commercial blocks and detection of productive zones for exploration (using data from sampling the core from boreholes) and to reduce expenditure on expensive sinking of workings. Sampling of bedrocks at the edges and in areas adjacent to the deposit being explored is most effective for detecting prospective geochemical anomalies and for expanding the front of exploration work on this basis.

The presence of larger multiplicative haloes around vein bodies makes it possible to recommend drilling of individual holes within prospective areas and more intensive exploration work at those sites where haloes of indicator elements have been detected as a result of geochemical sampling of holes of the first stage.

2.11. ICELAND SPAR

Transparent calcite, or Iceland spar ($CaCO_3$) rarely occurs in nature and is a raw material in short supply. Crystals of Iceland spar form at the low-temperature hydrothermal stage. Five types of deposits are known, characterized by constancy of the main mineral associations: chalcedony–zeolite–calcitic, calcitic, calcite–quartz, quartz–sulphide–calcitic, and microcline–calcite–morionic. The first and most commercially interesting type of deposit is associated with volcanic and sub-volcanic rocks of basic and moderately basic composition: basalts, dolerites, andesites and their tuffs. Deposits of Iceland spar are associated in Russia with trappean magmatism of the Siberian platform within the occurrence of laval sheets and tufogenic deposits.

The distribution of deposits of Iceland spar of various orders in a lava field is controlled by negative depression structures. The deposits are associated closely with lenses of spherulitic lava, exposing at the base of lava sheets. Nests with Iceland spar are concentrated mainly at sites of sheet warping, and are localized in the spherulitic lavas proper and in the more ancient underlying sheets, forming productive horizons that are several metres thick.

Deposits of Iceland spar occur in a tuff field, either inside sub-volcanic dolerite bodies (dykes, stocks), or in agglomerate tuff in the crater zones of volcanoes, or near these zones. They are controlled quite sufficiently by tectonic dislocations. Large nests with crystals of Iceland spar are localized in zones of large block disintegration of rocks.

Most intensive calcitic mineralization is characteristic of the periods of successive volcanic eruptions (final stages of tuff outburst and initial stages of lava outflow). The primary composition of the hydrothermal juvenile-wad solutions, entering the 'productive' rocks, was essentially chloride–sodium, but changed to essentially sodium calcium bicarbonate chloride, owing to interaction with the substratum rocks. Iceland spar crystallized out of these low-temperature solutions.

The common prospecting indications of Iceland spar deposits are as follows:

(1) presence of lava and tufogenic rocks of the trappean formation;

(2) intensive manifestation of tectonic ruptures in predominantly negative intraeffusive structures (troughs, calderas, etc.);

(3) wide occurrence of textural varieties of rocks of the spherulitic lava type, sites of large block crushing in the rocks of the tuff field, etc., that contributed to the penetration of large portions of hydrothermal solutions and to crystallization in large cavities under the screens;

(4) low-temperature hydrothermal reworking of rocks in the presence of chalcedony–zeolite–calcitic mineral association.

Conditions for the application of geochemical methods in prospecting for and evaluation of deposits

Crystals of Iceland spar formed in cavities. Hydrothermal solutions, entering from the outside, were the source of CO_2, and the substratum rocks were the source of Ca. The degree and character of this interaction depended on the composition of the solutions and the enclosing rocks, their structural–textural features, and degree of disintegration.

An indispensable condition for the formation of calcite is the presence of the HCO_3^- ion in the hydrothermal solutions, that conditions the principal possibility for applying the carbon metering method for prospecting purposes. The concentration of HCO_3^- near deposits of Iceland spar is so great that calcite is sharply predominant in the composition of the amygdules of the basaltic sheets in associations with paragonite, saponite, and chalcedony, which are more widespread than productive horizons with Iceland spar. When mowing away from the spar-bearing nodes, the composition of the amygdules changes, owing to a sharp increase in the content of analcite, thomsonite, hydromica, prehnite, and quartz at a reduction of the amount of calcite.

The relict parts of the solutions are located in the gaseous–liquid inclusions in the calcite and associated minerals in the enclosing rocks that makes it possible to anticipate the application of decreptophonic methods of exploration for prospecting purposes. All the rocks enclosing deposits of Iceland spar are characterized by increased losses on ignition.

Gaseous–liquid inclusions in crystals of Iceland spar and associated minerals are distinguished by quite a high concentration of chlorine ions, which are sometimes predominant over HCO_3^- ions. The content of chlorine in dolerites from spar-bearing dykes is by one order of magnitude greater as compared to dolerites from dykes containing no spar. The effect of chlorine is manifested also in the change of the initial concentration and distribution of elements closely associated with it (Hg, Cu, Pb, Zn, Ag, Mn, Cr, Ti), that is established clearly when comparing their contents and variance of contents in unaltered and hydrothermally reworked rocks. Particularly significant among these elements are Hg and Ti.

Four forms of mercury of different temperatures have been established: native (100–150°C), chloride (140–260°C), sulphide (220–350°C), and isomorphic.

The chloride form of mercury found in all 76 analysed samples which produces maximum peaks in fluorescent charts is the most widespread one. The sublimation temperature of this form of mercury corresponds approximately to the homogenizing temperature of gaseous–liquid inclusions (80–220°C) in crystals of Iceland spar, displacing slightly to the region of higher temperatures.

Low-temperature native mercury, which is absent in the majority of hydrothermal deposits, occurs quite often. A combination of two low-temperature forms of mercury occurs sometimes in regions with nests of Iceland spar crystals. The sulphide form of mercury is encountered more rarely, and isomorphic mercury (434°C) has been detected only in one sample.

The given data indicate the association of the basic amount of mercury with the stage of productive zeolite–calcitic mineralization. It is confirmed by a sharp increase in the content of mercury (2×10^{-6}–7×10^{-6} per cent) in the productive parts of the mineralized zone, while the content of mercury in altered rocks of higher temperatures (skarned with magnetite and without it, quartz-carbonate rocks, etc.) is usually lower ($< 1 \times 10^{-6}$ per cent). As a rule, the content of the chloride forms of mercury increases with an increase of productivity of one or the other part of the spar-bearing zone. Thus, mercury is a sensitive indicator of spar-bearing mineralization. Cinnabar has been found in small amounts at some deposits in the lava field.

The addition of titanium trichloride into an aqueous–chloride–carbonate solution decreases the capacity of this solution for spontaneous crystallization and contributes to the formation of large and pure crystals of calcite. The content of titanium is usually increased in high-grade crystals of Iceland spar. Undoubtedly, the enclosing rocks are the source of titanium, which is indicated by sharp variation in the dispersion of titanium in the proximity of spar-bearing nests. The effect of titanium on the quality and sizes of the Iceland spar crystals is sometime so great that it is possible even to distinguish commercial and non-commercial blocks within the mineralized zones of the deposits in some regions by the degree of titanaugite concentration in the dolerites. Apparently, a high content of alkalis, especially potassium, in the enclosing rocks is also advantageous for the formation of Iceland spar. Therefore, productive basalts and dolerites differ from unproductive ones by the high values of the coefficients $CaO/(\Sigma FeO + MgO)$, $K_2O/(K_2O + Na_2O)$, as well as Fe_2O_3/FeO, which is associated, evidently, with oxidation of the rocks under the effect of volatiles. There are other differences in the initial composition of the enclosing rocks that are reflected, for example, in Table (2.18).

Deposits of Iceland spar form in volcanic regions, where, owing to inheriting the ways of solution flows, there occurs often (though it is not always so) superposition of low-temperature hydrothermal reworking of rocks on high-temperature one, i.e. telescoping. Directional alteration of rocks during the entire hydrothermal process is most favourable. Thus, accumulation of calcium in the dykes of the Tungussky–Munkambinsky fault occurred at the skarn stage of mineralization, during the formation of siliceous–carbonaceous rocks, as well as lower temperature calcitic veins with nests of Iceland spar. This directivity of the process contributes to the formation of high contents of calcium in relatively easily soluble compounds in the proximity of mineralized zones with Iceland spar. The deposit of Iceland spar is confined to that part of the mineralized zone in dykes of dolerites, wherein high-temperature siliceous–carbonaceous rocks of hydrothermal genesis are most widespread, and these rocks are intensively leached in the proximity of the deposit and accompanying manifestations.

Some of the more high-temperature hydrothermal minerals, for example, magnetite, garnet, tourmaline, and fluorite, and their respective elements (F, Li), can be used when prospecting for spar-bearing zones proper, owing to the frequent spatial superposition of high- and low-temperature mineralization.

The degree of mineralization of spherulitic lavas usually decreases with the distance from the volcanic centres — courses of hydrothermal solutions. Similar

Table 2.18.

Chemical composition of dolerites

Components	Dolerites enclosing deposits of Iceland spar					Dolerites with manifestation of Iceland spar		Dolerites containing no calcite		
	1	2	3	4	5	6	7	8	9	10
SiO_2	44.7	45.2	42.6	45.7	45.15	46.45	45.85	47.6	47.2	47.7
TiO_2	1.88	1.89	1.84	1.98	2.16	1.88	1.87	1.14	1.7	1.49
Al_2O_3	14.33	15.0	14.35	14.56	15.07	14.23	14.79	15.0	16.58	14.64
Fe_2O_3	8.37	8.49	9.29	8.04	7.72	8.1	8.61	2.56	11.1*	13.4*
FeO	4.71	4.96	4.56	4.56	4.93	4.81	4.68	8.35		
MnO	0.16	0.16	0.18	0.18	0.15	0.28	0.1	0.26	0.24	0.2
MgO	7.15	6.44	5.39	5.93	5.51	6.14	5.73	7.14	4.64	5.95
CaO	10.65	10.22	11.8	11.08	11.08	9.92	9.82	11.72	11.07	11.58
Na_2O	2.43	2.7	2.6	2.7	2.89	2.94	2.85	1.97	2.04	2.1
K_2O	0.6	0.48	0.46	0.46	0.56	0.4	0.43	0.23	0.29	0.29
Ignition losses	4.87	4.25	6.45	4.6	4.3	4.43	4.71	3.57	4.7	2.2
Sum	99.55	99.79	99.52	99.79	99.52	99.58	99.44	99.54	99.56	99.55

All the dolerites have neither amygdules, nor visible hydrothermal reworking, which has been confirmed by analysis of thin sections. 1–10 — sample numbers.

* — total iron.

phenomena have also been established for deposits confined to zones of tuff crushing. The proximity of volcanic centres is indicated by such minerals as garnet, disthene, staurolite, tourmaline, fluorite, etc., and sometimes by slightly increased basicity of the rocks proper, lesser sorting of the material in the tuffs, and by other factors that can be used to evaluate potential productivity of the mineralized zones or productive horizons in the lenses of spherulitic lavas and in the proximity of the latter; the probability of discovering larger deposits is greater in the proximity of volcanic centres.

Finally, the majority of elements in the proximity of nests are redistributed as a result of interaction of the hydrothermal solutions and enclosing rocks, which is expressed in a sharp increase of their irregular distribution, dispersions, that can also be used to evaluate the productivity of the zones proper.

Mineral-forming solutions change their composition in time and space, owing to which mineralogical–geochemical zonality develops in deposits of Iceland spar that is conditioned by differential distribution of elements in space. Mordenite–calcitic association with a relatively low content of chalcedony is characteristic of spherulitic lavas, and the roof of the underlying covers is characterized by a chalcedony-quartz-calcitic association. Palagonite, saponite, chalcedony, calcite and zeolites (mordenite) prevail in amygdules of the underlying cover roof, and sites with chalcedony and calcite coincide with the occurrence of Iceland spar. Tourmaline and fluorite are associated with the subnest, underlying parts of the mineralized zone. The distribution of zeolites and calcites in mineralized zones is quite significant. Analcite and natrolite are replaced from the bottom upward in pyroxene zeolite rocks by sodic calcic and potash zeolites; large nests with crystals of Iceland spar are confined to the upper parts of mineralized zones. The ratios of calcite and zeolite contents in the productive parts of the mineralized zones (in dykes of dolerites explored by us) change from 10 to 40 and less in unproductive parts, wherein zeolites sometimes are predominant. Local variations in the acidity and alkalinity of the solutions that are identified by the variation in the character of zeolitic–calcitic mineralization in productive zones, also have an effect on the character of reworking of the enclosing rocks. The directivity of this process is somewhat different in occurrences of sodium zeolites as compared to sites with potash zeolites and Iceland spar. The content of silica is increased (50 per cent) in the areas with nests of large crystals of Iceland spar (which is sometimes accompanied by mordenite and heulandite), while, on the contrary, it is decreased (43–44 per cent) in regions with comprehensive accumulation of analcite and poor calcitic mineralization.

Analcite and zeolites poor in silica are in equilibrium with solutions that are undersaturated with silica (as a result, the latter should be derived from the enclosing rocks), and silica-rich zeolites (mordenite, etc.) are in equilibrium with solutions that are oversaturated with silica (i.e. silica should deposit in the rocks and minerals, for example, chalcedony, and quartz). The content of Ca, K, volatiles, and some other elements usually increases in the rocks around the nests. The degree of iron oxidation increases too. The clearly expressed correlation ($r = 0.75$) existing in basalts between K and Na, disappears during hydrothermal reworking of rocks ($r = 0.08$), but a correlation that was absent earlier appears between Zn and Cu ($r = 0.45$). This correlation is associated with their redistribution and subsequent concentration in the same minerals.

At the destruction of carbonates and other associated minerals, containing inclusions with HCO_3^- ions, the latter dissolve in water, forming hydrochemical haloes that are partially assimilated by plants (biochemical anomalies). Carbonate ions are detected rather rapidly and easily in rocks, plants, and solutions by chemical methods. Increased concentrations of these ions indicate the potential presence of Iceland spar.

The possibility of using primary haloes in the search for and evaluation of showings and deposits of Iceland spar in lava and tufogenic fields is not the same.

Spar-bearing mineralization is associated in a lava field with sufficiently large lenses of spherulitic lava. Prospecting for these lenses by structural indications (confinement to troughs, etc.) is quite reliable in combination with geophysical exploration of lava sheets and deluvial dispersion haloes of indicator minerals. Geochemical exploration at this stage is of indirect significance and consists in exploring the usually well-exposed upper parts of the basaltic sheets for volatiles (alkalis, Fe^{2+}/Fe^{3+}, K/Na, and chalcophilic elements). Increased concentrations of volatiles (H_2O, Cl) and chalcophilic elements at high values of Fe^{3+}/Fe^{2+} and $K/(K + Na)$ indicate that the sheets considered are productive and contain large lenses of spherulitic lavas. Geochemical exploration of hydrothermally reworked rocks within the lenses of spherulitic lava proper and adjoining productive horizons is conducted to locate the potential occurrence of nests with crystals of Iceland spar in their individual parts, i.e. in order to determine potential productivity. To this end, the character of rock variation in the bottom parts of lenses of spherulitic lavas is studied, proceeding from the presence here of calcic zeolites, a sharp decrease of the magnetite content, and an increase of the silica and potassium content in the rocks, and in individual minerals. This exploration is necessary when analysing the core from boreholes, thereby making it possible to avoid sinking large workings in, *a fortiori*, unpromising sites of mineralized zones.

The following problems can be solved when exploring deposits in the tuff field of the Siberian platform:

(1) prospecting for mineralized zones;

(2) evaluation of their prospects; and

(3) evaluation of prospects of individual parts of these zones.

Anomalies of indicator elements encompass not only the dykes of dolerites, but also the enclosing tuffs. The elements are characterized by varying mobility, forming leakage-type haloes. A decreased concentration of certain elements (for example lead) sometimes corresponds to an increased content of others (e.g. copper), but there is also a simultaneous increase (or decrease) in the content of all the chalcophilic elements. These specific features of the distribution of the concentrations of certain elements make it possible to distinguish unaltered tuffs from hydrothermally reworked ones. The thickness of the haloes in the tuffs (in the proximity of mineralized zones in dolerites, in individual cases in the proximity of large deposits) attains 70 m, and individual elements form haloes of various dimensions. Thick and sufficiently contrasting haloes, indicating the presence of mineralized zones with spar-bearing mineralization, are formed by Cl, Hg and H_2O. The thickness of haloes of the reworked rocks with anomalous contents of the above-stated minerals and indicator elements in spar-bearing fields, associated with zones of large block crushing in the tuffs proper, is not in excess of 20 m, i.e. it is several times less.

Secondary dispersion haloes are poorly explored. It is possible to apply carbonometric and chlorometric surveys (in the absence of carbonaceous cement in the enclosing rocks) in eluvial–deluvial deposits to prospect for mineralized zones with crystals of Iceland spar.

Method of work

Dispersion haloes of silica, zeolites and analcites are explored in the case of small-scale (1 : 200 000–1 : 25 000) work. It is simpler to discover these minerals, and they occur more often than local manifestations of Iceland spar. It is advisable to use these minerals when prospecting for mineralized zones overlain with drifts. Spatial coincidence of earlier ore (especially magnetitic, zeolite–calcitic) mineralization makes it possible to apply heavy concentrate sampling in prospecting work. The prospects of the basaltic sheets and dykes of dolerites are evaluated in lava and tuff fields by the specific features of the mineral and chemical composition, and the main attention is given to the contents of Ti, Cl, and H_2O, the degree of iron oxidation, and the value of $K/(K+Na)$; also to the abundance of amygdaloidal rocks, and to the composition of the amygdules. The predominance of calcite in the composition of the amygdules is a positive indication. It is possible to apply carbonometric survey in a tuff field, but the presence of carbonaceous material in the composition of tuff cement produces false anomalies, whose size depends on the occurrence of carbonized tuffs. In principle it is possible to grade anomalies by differences in the isotopic composition of the carbon, but it is hardly advisable to conduct such complex analysis on a mass scale at this stage.

When prospecting at a scale of 1 : 10000 and larger, sampling is conducted along topographically controlled profiles across the strike of spar-bearing zones. It is more advantageous to orient the profiles in the north-west direction in the tuff field of the Siberian platform, owing to frequent confinement of the mineralized zones to north-easterly faults. The profiles can be oriented in any direction on the slope in a tuff field wherein the basaltic sheets occur sub-horizontally. The spacing between the profiles is 100–50 m at a maximum furrow length of 10 m. The length of the dotted furrow should be chosen so that the primary non-contrasting haloes are characterized by at least nine samples. An increase of the sampling interval smooths out the curve of concentrations. This smoothing out is permitted only to a value at which it is possible to detect a concrete anomaly (i.e. a maximum of 10 m in the majority of cases). The task of prospecting work at this scale is to identify practically all the lenses of spherulitic lavas and mineralized zones in the most prospective parts of the lava and tuff fields. Attention is given in a lava field at this stage of work to increased concentrations of garnet, disthene, staurolite, tourmaline, and fluorite, indicating the closeness of incurrent canals, in the proximity of which the most valuable lenses of spherulitic lavas are often concentrated.

The task of geochemical work at 1 : 5000 to 1 : 2000 scales is to evaluate the prospects of the discovered mineralized zones and their most productive parts. When evaluating the prospects of the discovered mineralized zones, it is possible to proceed from two basic criteria: patterns of distribution, and concentrations of elements that indicate spar-bearing mineralization. More intensive redistribution (with introduction or leaching) of the primary components of the rocks points ordinarily to the intensity of hydrothermal reworking of rocks and, in the final analysis, to the potentiality of commercial concentrations of Iceland spar.

In a spar-bearing field wherein several mineralized zones were surveyed, the more productive zones were characterized by a sharp increase in the variance in the concentration of the majority of elements, first and foremost of alkalis, Ca, Cu, Pb, S, Mn and Ag (Cl and Hg were not explored in these zones). The productive zones are characterized by higher contents of Ca, Mg, Pb, Co, Cu and Ag. Productive mineralized zones in dykes of dolerites are also characterized by increased variance in the concentration of the majority of analyzed components. Detailed sampling has indicated that it is possible to identify sites of rocks within these zones that are characterized by increased concentration of Ti (and respective minerals) in combination with chlorine, as well as Hg, Cu and Ag, and the temperature of mercury sublimation is slightly higher (by 20–30°C) in relation to the most widespread values in the mineralized zone. All forms of mercury, including the low-temperature one, are present in spar-bearing deposits.

Evaluation of the prospects of mineralized zones by associations of minerals depends greatly on the type of deposit. Chalcedony–quartz–calcitic parageneses without zeolite in effusive traps indicate the presence of high-grade crystals of Iceland spar. The presence of zeolitic (desmine–heulanditic) mineralization in zones of large block crushing in the tuffs at a sharp decrease of analcite is a sufficiently reliable indication of potential commercial mineralization. A generally more diverse composition and high content of accessory minerals are favourable indications of spar in mineralized zones. Attention should be given to garnets, disthene, minerals of titanium, magnetite and sulphides (especially chalcopyrite, galenite, etc.), and rather rarely occurring tourmaline and fluorite.

Investigation of the specific features of the chemical composition of the more widespread minerals also contributes to correct evaluation of the prospects of various parts of the mineralized zones. The content of calcite and alkalis in desmines from dykes of dolerites determines to a great extent the productivity of the mineralized zones in Iceland spar. Productive sites are characterized by increased concentrations of calcium and especially potassium with a concomitant decreased concentration of sodium. The greatest content of K_2O (0.55 per cent) will be found in desmine from a commercial block of a deposit with a particularly high content of optical calcite; desmine from a commercial part of the zone with a moderate content of optical calcite will have K_2O content of 0.36 per cent; desmine from a non-commercial part of the mineralized zone the K_2O content will be 0.28 per cent; in desmines from siliceous–carbonaceous rocks it will be 0.18, and in desmine from a site wherein calcitic mineralization is limited 0.13 per cent.

Other mineralogical indications are also used when evaluating the potential productivity of the mineralized zones; the presence of Iceland spar is a favourable factor. The shape and habitus of these crystals enable to form an opinion on the type and, consequently, on the potential prospects of the identified spar-bearing manifestation.

In general, it is necessary to take at least 30 samples in fine-grained rocks and 50 samples in coarse-grained rocks to characterize mineralized zones in general. It is desirable to distribute the profiles in space so as to ensure uniform sampling within the volume of the entire spar-bearing body, i.e. the samples in a tuff field should be taken in profiles at various horizons, and in a lava field in the depth of a slope at some distance from the erosional surfaces, otherwise a part of the spar-bearing zone will be assessed that adjoins the surface along which the profiles are located. The spacing between the profiles should be 50–20 m, in correspondence with the commonly adopted systems of workings and boreholes. The length of the furrow should be 1–2 m. The length of the furrow (and the spacing between the profiles if possible) should be decreased by a factor of two in the direction of maximum variation of the rocks.

The zonality of the primary haloes is revealed by comparing the absolute concentrations of components at various levels and distances from the centre, calculating the ratio of metre per cents to the entire width of the halo, the value of the multiplicative index, the specific features of distribution and content of elements by means of the rank coefficient of Spearman, and the probability coefficient of the number of jumps, etc. It is necessary to bear in mind the possibility of halo distortion owing to the multiphase character of the pulsating process of mineral formation, which is manifested most vividly in some deposits in the tuff field.

The task of geochemical exploration is to identify the position of spar-bearing nests in space. This can be done by detailed sampling of quarries and other workings in the region of conjectural spar-bearing nests, because specific changes occur in the composition of the surrounding rocks in the proximity of the nests.

It has been established that two zones can be clearly identified in the productive horizon of the lava field in the region of spar-bearing nests:

- A zone of intensive leaching of rocks at a large distance from the nests, encompassing the sites of gruss from the nearby lenses, or sites of spherulitic lava; the hydrothermal solutions are saturated in this zone with calcium.

- A zone of partial leaching of some components and simultaneous deposition of others (in particular, calcium); this zone adjoins the nest.

The distribution of calcitic and potash zeolites, analcite, and sodic zeolites are explored at the stage of detailed prospecting–exploration work. Sites are most favourable for the concentration of commercial mineralization where calcitic and calcic zeolites are predominant. An increased content of titanium minerals, a high content of chlorine, and the presence of low-temperature forms of mercury also point to the potentiality of discovering calcite in this part of the mineralized zone.

When conducting all types of field work, geochemical samples should be taken from bedrocks that have not been affected (or very little affected) by weathering, because the hypergene processes can greatly distort the primary content of components being analysed. The mass of the samples should be within the commonly adopted standards for each type of analysis.

The basic mass of each sample is subjected to spectral analysis for a wide range of elements when conducting lithogeochemical prospecting at scales of 1:50 000 to 1:10 000. It is necessary to analyse a limited number of samples from reference profiles for a wide range of elements when conducting detailed work; samples are also taken from these profiles for mineralogical and chemical analyses.

Mobile chlorine was identified according to the following procedure. Easily soluble chlorides were extracted from 2–3 g lots by a weak solution of nitric acid (1:30) with subsequent nephelometric determination of chlorine in AgCl. In the case of such processing, halite, sylvite, carnallite, chlorapatite and partially sodalite go into the solution.

When conducting lithogeochemical prospecting at a scale of 1 : 200 000– 1 : 100 000, the basic mass of the samples is analysed for chlorine and a wide range of elements by the method of approximate quantitative spectral analysis. The purpose of this method is to determine the most 'working' minerals, and indicator elements that are not quite identical in various regions owing to the differences in the composition of the original rocks and their preceding high-temperature reworking. When conducting more detailed work, it is necessary to analyse a limited number of samples taken from the reference profiles for a wide range of elements. Samples are also taken for mineralogical and chemical analysis. The rest of the samples are used for quantitative determination of the indicator elements already identified in the rocks and (with a greater degree of confidence) in individual minerals. The work conducted at individual deposits makes it possible to identify commercial blocks by such elements as Ti and Hg in combination with Cl.

To locate weak anomalies in the proximity of spar-bearing bodies it is necessary to use the criterion of 'two standard deviations', that increases considerably (up to 2.28 per cent against 0.14 per cent) the possibility of detecting false anomalies. However, it is advisable to lower successively the level of a Ca depending on the m-number, in accordance with the expressions $Ca = C_f + 3S/\sqrt{m}$ (normal law) or $Ca = C_f \times \varepsilon^{3/\sqrt{m}}$ (log-normal law), owing to the usual correlativity of the anomalous contents of elements in several m-points on the profile and several contiguous profiles. The number of false anomalies does not increase above their probable number, which is 0.14 per cent. Therefore, the criteria $3S$ or $2S$ (ε^3 or ε^2) are used to identify isolated point anomalies, the majority of which are of no interest.

The results of comparative exploration of Iceland spar deposits indicate that primary haloes of rather small size and intensity are observed around them in the majority of cases. Therefore the necessity arises of 'enhancing' the contrast of weak anomalies by adding up the deviations of the absolute (or relative) values of various indicator components from the background values, with subsequent analysis of respective summed or additive haloes, thereby making it possible to get rid of false anomalies, wherein variations in the content of one component are not conditioned frequently by respective variations in the content of other components. It is possible to use also multiplicative haloes (content, variance of contents, frequency of occurrence, degree of association between correlated values) on the basis of resemblance criteria that are used in mathematical statistics. In principle, it is advisable to use the method of multiple correlation, considering the entire totality of investigated minerals and elements.

It is necessary to check first the anomalies with a maximum set of positive geological–structural, petrographical, mineralogical and geochemical indications. The following sequence can be recommended when evaluating prospective anomalies:

(1) detailed exploration of the geological situation in the region of the anomaly, and, if necessary, application of mineralogical–geochemical methods for this purpose;

(2) geophysical exploration with additional geological and mineralogical–geochemical survey of the deposit in shallow workings;

(3) preliminary evaluation of the identified anomaly; and

(4) stripping of the mineralized zone (spar-bearing body) by workings and boreholes with the aim of evaluating its prospects.

The data of sampling are evaluated on the basis of dividing the deposit into groups. When interpreting these data, it is necessary to consider the specific features of the geological–structural position of the identified anomaly, rock alteration around the ore, mineral composition (presence or absence of Iceland spar crystals) and other geological indications, as well as the ordinary considerable excess of the halo sizes not by the basic component, but by the minerals and accompanying elements. The objects are usually divided by the complex of all the indications into two antagonistic groups in a certain succession: unaltered rocks or mineralized zones, spar-bearing and barren mineralized zones, productive sites (to a various degree), and unproductive sites in spar-bearing and barren objects by all the main parameters.

The formation of vein Iceland spar occurs with the hydrothermal metamorphism of environmental beds, in which there are geochemical haloes around the vein bodies and the deposits. The fluid compositions changed regularly with increasing proximity to the paleosurface, and in this connection the ratios of oxidized to reduced forms of gases (H_2O/H_2, $CO_2/(CO+CH_4)$) in mineral inclusions increase from the lower to the upper levels of the ore bodies, and the ratios of various nitrogen compounds change. The ratios of oxidized to reduced forms of gases in mineral inclusions can be used, along with other data, as criteria for determining the relative level of the erosional truncation of the deposits.

Analysis of the work carried out demonstrates the potentiality of combining geochemical and geophysical surveys.

1. Airborne magnetic surveys in a tuff field make it possible to identify the basic rupture structures that can control spar-bearing mineralization; exploration of these structures under field conditions by geochemical methods make it possible to separate mineralized zones with zeolitic–calcitic mineralization from non-mineralized ones.

2. Identification of large prospective spar-bearing structures is conducted in a lava field on the basis of gravimetric, magnetic, and electrical surveys. It is advisable to evaluate at this stage the productivity of the sheets by the petrochemical and mineralogical criteria stated above.

3. Identification of favourable structures enclosing a group of lenses or individual lenses of spherulitic lava can be achieved with a sufficiently high degree of reliability (80 per cent) by a combination of magnetometry and electrical prospecting. Mineralogical criteria can be used with the same aim. Exploration of the character of mineralization, as well as the degree and character of rock reworking within the lenses makes it possible to judge the productivity of the identified lenses with greater confidence.

4. Favourable sites within mineralized zones (sites traversing fracture zones) are located by electrical methods, and the adjoining parts of the mineralized zones are analysed by means of indicator minerals and elements.

5. The locality of the cavities within the exploration quarries can be determined by microelectrical surveys (the cavities are identified at a distance of 10 m). Exploration of the character of rock reworking in the proximity of the cavities by their mineralogical–geochemical haloes makes it possible to determine the directivity and character of hydrothermal reworking of the rocks within approximately the same limits, and thereby the prospects of the detected geophysical anomaly can be evaluated. This combination of geophysical and geochemical methods will contribute apparently to a rise of labour productivity and a decrease of losses during exploitation of deposits.

2.12. MALACHITE

Malachite, which relates to the group of gem-quality stones, occurs in zones of oxidation of sulphide deposits. Where aggregates of malachite are extracted alongside with other minerals. Mineralization occurs in zones of oxidation of:

(a) sulphide copper and copper–iron ore skarn formations;

(b) sulphide copper and cobalt–copper stratiformed manifestations.

Primary ore mineralization is associated in the deposits of the first sub-group with metasomatic processes that occurred at the contact of granitoids or effusives with carbonaceous rocks, accompanied by the formation of sulphides (pyrite, chalcopyrite and pyrrhotite) in skarns. Paragenesis of the minerals in these formations is quite diverse: pyroxene, garnet, tetrahedrite, molybdenite, sphalerite, galenite, barite, native copper, chalcocite, cuprite, tenorite, lampadite, azurite, haematite, martite, atacamite, turgite, copper phosphates, etc. In the deposits of the second group malachite is concentrated in karst carbonaceous formations, occurring in the zone of oxidation of stratiformed Co–Cu deposits. Accumulations of malachite are among redeposited ore minerals and clays. Copper and iron sulphides were the initial material for the formation of malachite.

The basic minerals of the zone of oxidation are: malachite, chrysocolla, an earthy mixture of copper oxides (tenorite and melaconite), cobalt oxides (heterogenite, linneite, and korrolite), oxides of iron and manganese, azurite, cuprite, native copper, dioptase, carnotite and libethenite. Malachite occurs mainly in the planes of schistosity and in fractures, as well as in karst caverns in limestones.

Prospecting for malachite is conducted in regions of copper mineralization with the occurrence of ancient crusts of weathering and zones of oxidation of primary sulphide ores. Finds of malachite in the form of films, incrustations, and earthy masses in carbonaceous formations, which are the sources of calcium needed for the formation of malachite and creating conditions for its formation, are prospecting indications. Geochemical survey aimed at the identification of copper anomalies can be carried out when prospecting for malachite in closed regions. The activities should be concentrated in regions of prospecting and prospecting–exploration work for the major components (copper, iron, cobalt). When grading anomalies, prospective ones are those that are located among carbonaceous formations in zones of oxidation of skarn and stratiformed manifestations. Gas–mercury surveying is applied to trace dislocations with a break in continuity and fracture zones.

Integrated geochemical and geophysical investigations (method of natural field, vertical electrical sounding, gravimetry, seismic exploration) are conducted to identify karst caverns, which that are then checked by core drilling.

Hydrochemical research was carried out using the method of the preliminary concentration of elements on cadmium sulphides. The preliminary dosed reactives, cadmium chloride (0.85 g) and sodium–sulphide (0.75 g), were introduced into a sample of 1 litre and it was shaken for 1–2 minutes to guarantee a complete reaction. As a result, a colloidal sediment of cadmium sulphide was formed, on which elements of the chalcophile group (Cu, Co, Zn, Pb, Ag) contained in water precipitated. The sediment was filtered out on an ashless filter, dried and subjected to spectral analysis.

The study described here concerns the application of selective chemical extraction in malachite deposits in Kazakhstan. In order to obtain the Cu-bearing phases for a kinetic study in soil, the samples were submitted to a sequential extraction procedure using the following reagents: NH_4 acetate, hydroxylamine hydrochloride, NH_4 oxalate under γ-radiation and finally strong acids.

The use of NH_4 oxalate under γ-radiation in partial extraction, compared with the hot mixed-acid extraction for copper, shows the anomalous geochemical area better and contrasts areas.

According to our study, the following evidence gives grounds for optimism, and the geochemical operation can be recommended for prospecting purposes:

(1) occurrences of geochemical copper anomalies (considering the regions along copper-mineralized peripheries);

(2) the presence of chalcopyrite and hypergene copper mineralization in the areas where copper and molybdenum ores occur;

(3) higher copper contents in lumps of ores;

(4) an abundance of polymetallic mineralization (very often in association with silver) in the peripheral parts and near stockwork zones;

(5) the zoning of metasomatites, and the intensity and scale of metasomatic alterations.

2.13. CHRYSOPRASE

Chrysoprase is a green-coloured, microfibrous and fine-grained variety of silica that is used in jewellery. Its mineralization is associated with serpentinized and silicified nickel-rich ultrabasites whereon crusts of weathering occur (of linear and mixed types). In chrysoprase formation nickel and magnesium derived from the ultrabasites, went into the solution and migrated with the silica into the deep horizons, wherein independent mineral phases (nickel and nickel–magnesium hydrous silicates) formed at pH of about 6–7.

Chrysoprase deposits occur relatively rarely (Australia, Poland, Russia, Brazil). They are mainly associated with silicification zones of serpentinites formed by the metamorphism of ultramafic and mafic rocks. Chrysoprases are excavated from the nickel-bearing weathering covers of serpentinites. Chrysoprase-bearing serpentinitic massifs are cut by numerous dykes and veins of gabbros: dolerites, granites, aplites, pegmatites and quartz. Besides chrysoprase, the above-mentioned massifs contain numerous veins of magnesite (to 4 per cent Ni) opal and chalcedony.

The common prospecting indications of chrysoprase mineralization are:

(a) occurrence of nickel-rich ultrabasites belonging to the gabbro-peridotitic formation, within which altered rocks occur of the silicified (ocherous–siliceous) profile of the crust of weathering;

(b) spatial association of ultrabasic rocks with intrusions of granitoids; and

(c) presence of sites of silicification and cementation with chalcedonic material in the zones of rock brecciation.

Quartz, chalcedony, opal and prasopal, which are resistant to weathering and are preserved in alluvial formations, as well as tourmaline, talc, asbestos, garnierite, cacholong and sapphirine, are associated with chrysoprase. Mineralogical criteria of the presence of chrysoprase include occurrences of mineral associations of opal + prasopal + chalcedony + magnesite + kerolith, alloyed iron ores, and nickel silicate minerals. Chrysoprase is rarely of green colour in natural occurrences. It is often covered with a brownish crust of ferric hydroxide, and is transformed intensively into a loose mass (marshallite). Massifs of ultrabasites containing chrysoprase have been subjected to the median stage of serpentinization; they are composed of chrysotilelizarditic or essentially lizarditic serpentinites with wide manifestation of chloritization and transformation into talc. Magnetite, magnesite, chlorite, limonite, saponite and talc occur in altered rocks.

The most prospective areas for chrysoprase are the least deep intrusive ultrabasitic massifs, characterized by high values (0.16–0.21) of the facies coefficient $K_f = Fe_2O_3/MgO$, and low (7.8–9.0) indices of the magnesium/iron ratio [114].

A favourable indication for applying geochemical methods when prospecting for chrysoprase is the enrichment of the ultrabasic rocks in nickel (up to 3.3 per cent). Favourable conditions are created in this case for the formation of chrysoprase under hypergene conditions. According to geochemical data, serpentinites of chrysoprase-bearing massifs are also enriched in iron. A relatively stable (about 40 per cent) silica content, which increases sharply (up to 68–84 per cent) in talc–limonite rocks and silicified serpentinites, has been determined in serpentinized rocks in Kazakhstan; the content of magnesium oxide (about 30 per cent), on the contrary, decreases in silicified serpentinites (to 2.5 per cent).

Maximum concentrations of calcium oxide were detected in carbonatized serpentinites, while in silicified serpentinites calcium oxide content is minimal, but the summary content of potassium and sodium is increased. The zones of silicified serpentinites are enriched (in per cent): in nickel to 0.37, in cobalt to 0.03, in chlorine to 0.41, and in copper to 0.006.

Chrysoprase, like the chalcedony and opal associated with it, contains, 0.2–3.3 per cent nickel, 0.02 per cent cobalt, 0.03 per cent chromium, 0.01 per cent zinc, 0.03 per cent copper; 0.3 per cent sodium, and 0.3 per cent potassium.

Methods of work

Chrysoprase deposits are rarely prospected for as an independent stage of the geological exploration process. It is ordinarily found in passing, when conducting prospecting work for other minerals (nickel, cobalt and magnesite). Areas that are favourable for chrysoprase mineralization are identified in the course of route geological explorations at a scale of 1:50000. To this end, all the records on the prospecting work and on the rocks are reviewed. The main attention is given to the colour of the rocks and minerals. Zones favourable for chrysoprase mineralization can be identified by mercury anomalies, owing to the confinement of chrysoprase bodies to sites with occurrences of silicified serpentinites and developed quartz–chalcedonic veins along faults. These zones are identified clearly on the surface of weathering crusts by disintegrated blocks and fragments of brown and yellow-brown serpentinites with ochre limonitic incrustations, and by the presence of fragments of opal, chalcedony and quartz.

It is necessary to increase the density of the grid of routes to trace the zones of silicified serpentinites at sites where direct indications of chrysoprase have been identified.

It is most effective to conduct (on prospective sites) areal geochemical sampling of bottom sediments and eluvial–deluvial formations at the stage of prospecting in semi-closed regions. Sampling scales are 1 : 25 000 and 1 : 10 000. It is most advantageous under mountain relief with a developed drainage system to sample fine silty-argillaceous fractions of alluvial deposits (bottom sediments) in small rivers and dry ravines, as well as deluvial and colluvial fans. When exploring secondary dispersion haloes, samples are taken from the upper horizon of the eluvial–deluvial formations where the thickness is not more than several metres.

A detailed complex of geophysical and geochemical operations is used to eval-uate the nature of the detected anomalies. The most reliable results have been obtained with a lithogeochemical survey of the sorbed and water-soluble forms of elements, as part of the detailed research complex.

In the zone of chrysoprase mineralization, an increase in nickel and copper concentrations, with a predominance of nickel depending on depth, was noted in the composition of the superimposed sorptive saline subsurface aureoles.

Efficient use was made of the method of extraction with subsequent colorimetry on paper for natural water at the sampling points with the determination of nickel and copper (up to 2–3 mkg/l).

In order to increase the effectiveness of hydrogeochemical prospecting for chrysoprase deposits within overburden terrains the following operations are rec-ommended:

(1) On-the-spot filtration of water samples through micron-size porous media and immediate CO-precipitation of the microelements from the filtrates.

(2) The investigation of the composition of thin suspensate microelements as a sampling medium.

(3) On-the-spot analytical determination of certain components in filtrate and in natural water.

Samples are taken from a depth of 15–20 cm in dry arid regions with alka-line and neutral soils, and from a depth of 0.4–0.8 m in areas with a humid climate and podsolic, grey, brown, and highly alkaline soils. It should be noted that when determining the method of areal geochemical sampling, it is necessary to use to the maximum information on the metallogenic features of the region being explored, gained as a result of geological surveying and geochemical ex-ploration at the preceding stage: e.g. typical ore deposits (nickel and cobalt) and structural conditions of their localization. The prospecting grid is laid out with tools when prospecting at the 1 : 25 000 and 1 : 10 000 scales. The prospecting lines are oriented in the direction of maximum variation in the distribution of indicator elements (transverse the strike of conjectural structures). The samples

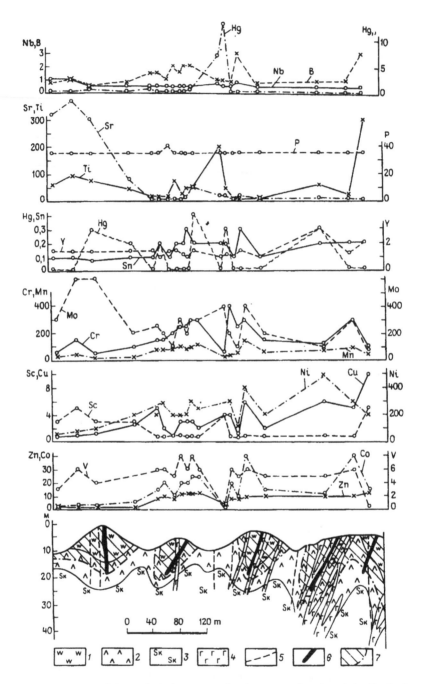

Figure 2.45. Diagram of the geological structure of a chrysoprase deposit and the distribution of elements across the strike of a chrysoprase zone (Kazakhstan). 1 — silicified nickel-rich rock, 2 — ocherous-siliceous zones, 3 — serpentinized ultrabasites, 4 — gabbro veins, 5 — zones of tectonic dislocations, 6 — highly productive bodies with chrysoprase, 7 — low productive bodies.

are taken from boreholes when there is a thick weathering crust many metres in depth, owing to the fact that the dispersion haloes are located at a considerable depth. The drilling of boreholes can cause an increase of expenditure, so loose deposits are usually sampled in test wells drilled when prospecting for nickel or cobalt deposits. Samples are taken from bedrocks in the form of small fragments of rock with a size of 1×2 cm at the most prospective sites of silicification zones; the mass of an individual sample is 100–150 g. All the samples taken during geochemical exploration are subjected to rapid spectral analysis for nickel, cobalt, copper, zinc, tungsten, bismuth, boron, beryllium, barium and lithium.

Experience of investigating primary haloes in the process of prospecting for and exploring nickel and cobalt deposits makes it possible to recommend delineation of geochemical anomalies by the value of minimal anomalous concentrations, calculated with a 5 per cent level of significance. Special attention is given to the epicentres of positive and negative anomalies, because high-quality chrysoprase gravitates toward local accumulations of nickel silicates with a high content of nickel against a background of impoverished high-silica ore (Fig. (2.45)).

The potential nickel content in the ultrabasites and the iron content are determined when interpreting the results of geochemical surveys of primary dispersion haloes. Ultrabasites, containing from 2 to 3.3 per cent nickel and having a ferruginosity factor F of 0.12, are the most favourable ones. The increased ferruginosity of the massifs is manifested in a wide occurrence of accessory minerals: haematite, pyrite and andradite. One of the characteristic features of chrysoprase-bearing massifs is their enrichment in admixture elements that are typical of rocks of the granitoid series — Li, Zn, Be, P, W, Nb, Bi, Ba and B — accompanied by the formation of such mineral phases as orthite, monazite, and tourmaline. Geophysical methods include magnetic prospecting to identify serpentinites.

Bulk samples are analysed in a laboratory to determine the quality of the chrysoprase. The colour, translucency, and size of conditioned zones are the basic indices of chrysoprase quality.

2.14. TURQUOISE

Turquoise belongs to the group of hydrous phosphates of copper. The turquoise showings of exogenic (or hydrothermal–metasomatic) genesis are divided into three groups, depending on the composition of the enclosing rocks:

(a) deposits in magmatic rocks of porphyritic formation;

(b) shows in sedimentary–metamorphic formations, including those occurring in the zones of contact with intrusive rocks; and

(c) shows in zones of oxidation and secondary sulphide enrichment of copper-porphyritic and polymetallic deposits. Mineralization is confined to the weathering crusts of rocks rich in phosphate, copper, and organic remains, and to hydrothermal quartz veins.

The common prerequisites of prospecting for deposits are as follows:

(1) occurrence of quartz veins in granitoids, quartz–feldspar porphyries, andesite porphyrites, as well as in sandy–clayey rocks, and argillite rich in phosphorus and copper that are necessary for the formation of turquoise;

(2) manifestation of zones of crushing, brecciation, limonitization and increased fissureness;

(3) occurrence of low-temperature facies of secondary quartzites that replace intrusive rocks; manifestation of metasomatites, represented by keratinized, silicified, sericitized and pyritized rocks.

Conditions for the application of geochemical methods in prospecting for turquoise

The following minerals are detected at the manifestations of turquoise: alunite, halloysite, jarosite, pyrite, haematite, kaolinite, wavellite, chalcosiderite, barite and calcite. The associations of minerals that are paragenetic for turquoise are diverse, depending on the character of the enclosing rocks. If the rocks of the substratum, whereon weathering crusts occur, are represented by rhyolites, trachytes, monzonites, porphyrites and granites with dispersed ore mineralization, the associated minerals of turquoise are: sphalerite, dolomite, anhydrite, zeolites, limonite, halloysite, jarosite, sericite and kaolinite. Chalcosite, malachite, and azurite are detected in the case of intensive ore mineralization in the zones of sulphide concentration. When turquoise mineralization occurs in weathering crusts with ore mineralization, limonite, malachite, azurite, chrysocolla, wavellite, halloysite, sericite and kaolinite replacing schists, sandstones and aleurolites enriched in phosphorus are widespread. Feldspar, barite and calcite are located when turquoise occurs in quartz veinlets. Jarosite is characteristic of zones rich in chalcopyrite, chalcosite, pyrite and marcasite. Hydromica, kaolinite, quartz, jarosite, lepidocrocite, goethite, halloysite, chalcosiderite, svanbergite, woodhouseite, barite and natrojarosite are detected in zones of sulphuric acid leaching.

The indicator elements of turquoise-bearing zones are as follows: phosphorus, titanium, zinc, copper, lead, nickel, vanadium, chromium, cobalt, strontium, barium, zirconium and molybdenum, whose content depends on the character of the enclosing rocks and their alteration. Some of the stated elements are characteristic admixtures in turquoise proper: phosphorus, copper and sulphur, while iron, aluminum and silica are of considerable significance; besides that, admixtures of chromium, titanium, zinc, barium, beryllium, nickel, molybdenum, vanadium and zirconium are identified. Migration and redistribution of the elements occur at sites of tectonic dislocations and in turquoise-bearing zones as a result of the effect of solutions on the rocks. Phosphorus, iron, copper and carbon dioxide are introduced into the metasomatites. The ratios of the contents of antagonistic elements, i.e. phosphorus to hafnium, barium to boron, and zirconium to hafnium, are also indicators of turquoise mineralization. The stated specific features of turquoise manifestation condition the potentiality of applying geochemical methods when prospecting for this mineral. Dispersion trains of copper, barium, strontium, zinc and phosphorus form in the clay and silty fractions of bottom river bed formations of permanent and temporary natural streams near shows of turquoise. Turquoise-bearing bodies are identified by hydrochemical haloes (by increased contents of copper, sulphate- and phosphate-ion).

It has been established experimentally that it is possible to apply lithochemical methods of prospecting for turquoise by secondary haloes and dispersion trains, as well as by primary haloes. Haloes of copper, phosphorus, molybdenum and vanadium occur around turquoise mineralization when the deposits are located in sedimentary–metamorphic rocks represented by coaly–carbonaceous–siliceous formations.

Turquoise-bearing zones in sedimentary–metamorphic rocks are spatially associated with carbonaceous shales and quartzites, surrounded by haloes of phosphorus, copper, vanadium and molybdenum. Only positive endogenic haloes of phosphorus and copper are detected in hydrothermal altered rocks without turquoise mineralization, whereas sharp fluctuations in the contents of these elements are observed in productive zones. Vertical zonality (from the bottom upward) is characteristic of turquoise deposits:

(1) a zone of primary carbonaceous–micaceous–quartz schists with concretions of phosphorites and pyrites;

(2) a zone of 'lacy' turquoise, occurring below the groundwater table; and

(3) a zone of hydromicaceous rocks with turquoise mineralization.

Geochemical zonality reflects the mineral one and is closely associated with the latter. As was to be expected, considerably higher concentrations of Ba, Hg, Ag, Sb and As occur in the upper ore zones; on the contrary, the content of Cu, B, Ni, Co and Cr increases in the lower horizons.

Methods of work

It is advisable to conduct detailed hydrochemical exploration at a 1:10 000 scale in areas prospective for turquoise mineralization. The hydrochemical method is based on the determination of the degree of mineralization and the ionic composition of the ground and surface waters by taking samples and analysing them for Cu, SO_4^{2-}, Fe, Cl, P and Ba. In this case, it is possible to identify indications of the presence of turquoise-bearing bodies occurring at some depth within the area of the drainage system. All the sources of groundwater and aquifers are sampled. The spacing between the sampling points should be not more than 1 cm on a map of the same scale as that of the work conducted. When sampling surface-stream flows, preference should be given to streams with a small discharge.

Secondary dispersion haloes are explored in dry and arid regions. The effectiveness of this method is determined by intensive distribution of leaching and weathering zones at turquoise deposits. The density of the sampling grid at scales of exploration of 1:50 000 and 1:10 000 is given in Table (2.1). The samples are taken at a depth of 25–40 cm. It is most rational to determine the sampling depth experimentally under various conditions when conducting detailed prospecting. Samples of eluvial–deluvial formations are taken from haloes, pits, and trenches.

A fine fraction (< 1 mm) with a mass of 50 g is sampled. The samples are sieved through a screen with 1–0.5 mm meshes. A fine fraction with a mass of 20 g is sent to the laboratory, where it is further ground to a powdered state.

When conducting lithochemical survey to identify secondary dispersion haloes, special attention is given to the presence of positive anomalies of molybdenum (with a 0.001 per cent content), barium and strontium, as well as to the presence of haloes of introduction and removal of copper. Positive anomalies of copper (0.01 per cent) are located usually at the periphery of the mineralized zones. It should be taken into account that the epicentres of mineralization are confined to the zones of contact of positive and negative anomalies of copper and phosphorus. When grading anomalies, prospective ones are those demonstrating spatial coincidence of sulphide mineralization with haloes of various elements. Prospective anomalies are those containing deluvial fragments of quartz with inclusions of turquoise. Zones of crushing can be identified by the gas–mercury method, also by the anomalies of mercury in the rocks by means of high-sensitivity analysis.

Primary haloes are very important when prospecting and evaluating if the ore zones occur at a depth and are characterized by a zonal structure, which makes it possible to judge spatial location of the productive bodies.

The bedrocks are sampled to identify haloes of phosphorus, copper, barium, strontium, mercury, molybdenum and other elements when prospecting by primary haloes. Special attention is paid to anomalies of phosphorus that usually encompass turquoise mineralization and accompanying low-temperature metasomatites. Samples are taken by the dotted furrow method. From five to six rock fragments, taken at the normal sampling (3–5 m, at equal spacing), are combined into one 200 g sample. The veins and zones of faults are sampled individually.

2.15. CONCLUSIONS

1. Accessory elements occurring around productive bodies, are used as indicators when conducting geochemical surveys at deposits of the first group. This is associated with the wide distribution in the earth's crust of the basic petrogenetic elements that are contained in the minerals of the deposits under consideration: C, Si, Al, Mg and Ca. Geochemical methods of prospecting are based on direct and inverse correlation dependences between the contents of the sought for minerals and anomalies.

2. The composition of the environment, and rock enrichment in the components needed for the formation of minerals, was of great significance for the formation of many deposits of the first group. Deposits formed ordinarily in two stages. A complex of geological formations (rocks and veins) developed during the first stage as a substratum containing chemical elements needed for the formation of commercial minerals: ore-free quartz veins (for rock crystal), magnesian-silicate metamorphic rocks (for phlogopite), intrusions of ultrabasites of the dunite–harzburgitic formation (for asbestos), and trappean intrusions rich in calcium (for Iceland spar). High thermobaric parameters of formation were characteristic in the majority of cases for mineral associations of products of the first stage: magmatic, ultrabasic

and basic rocks (asbestos and Iceland spar), metamorphic magnesian formations (phlogopite), and veins of high-temperature quartz (rock crystal).

Deposits of mineral resources, associated with the effect of hydrothermal solutions circulating at a retarded rate and causing redistribution of admixture elements in circumore zones, formed during the second, productive stage proper. A limited number of components were introduced from a depth (CO_2, F, K, H_2O and Cl), whereas the bulk of the components were derived from the enclosing rocks.

This common course of the process of formation of many deposits of the first group conditions the use of two groups of indicator elements when applying geochemical methods of prospecting:

(a) identification of haloes of introduction of mainly anion-forming elements (Cl, F and Hg); and

(b) identification of positive and negative anomalies, and their points of contact as most reliably indicating the epicentres of mineralization. The role of the geological environment is considered in this case, which is favourable for the formation of deposits and conditions the use of petrochemical criteria of prospecting for their evaluation.

3. The specific conditions of formation of some deposits stipulate the peculiarities of prospecting work. Taking into account the xenogenic character of diamonds, it is necessary to apply primarily mineralogical methods in prospecting for and evaluation of deposits. Lithogeochemical methods of prospecting by secondary haloes and trains are advisable at the first stage when detecting kimberlite pipes proper under conditions of poor exposure and turf cover. It is necessary to consider in this case that kimberlites differ from the enclosing sedimentary carbonaceous formations, traps and ultrabasic rocks by the content of such elements as nickel, cobalt, vanadium, chromium, boron, copper, zinc, lithium, niobium, rubidium and radioactive elements. The main criterion of diamond content in kimberlites is the presence therein of associations of mantle (extra-kimberlitic) minerals of the diamond–pyropic facies: garnet, chromespinellids and ilmenites rich in chromium.

4. Combined application of mineralogical and geochemical methods is advisable in the search for topaz, beryl, garnet and corundum. Bank contents of representative minerals and their groups that can be used to draw mineralogical and geochemical maps are identified by the data of processing the results of mineralogical and geochemical surveys by methods of multiple variation and correlation analysis.

5. Mineralogical criteria are most effective also for minerals, wherein the useful component is a product of melt protocrystallization (chrysolite, sapphire, etc.).

6. The dimensions of the endogenic haloes around productive bodies in the group of deposits being considered exceed appreciably the parameters of the productive bodies and are identified more easily than the ore bodies themselves. The dimensions and shape of the endogenic haloes depend on the concrete structural conditions of localization of mineralization, chemical, and mineral composition, and physico-mechanical properties of the enclosing rocks. Endogenic haloes are absent at some deposits (kimberlites). It is advisable to construct multiplicative

haloes and forms of mercury occurrence to enhance haloes of weak contrast. A wide thermal spectrum of mercury sublimation in altered rocks in the presence of sorbed (100°C), chloride (220°C), sulphide (320°C), and isomorphic (420°C) forms is a favourable indication for discovering blind bodies.

7. It is necessary to consider the quality of the mineral, size of crystals, and absence of growth flaws when evaluating deposits of the first group. To this end, it is advisable to apply γ-irradiation and infra-red spectroscopy.

CHAPTER 3

Deposits of the second group

Geochemical methods of prospecting for major types of mining chemical raw materials, such as apatite, phosphorites, boron, fluorite, native sulphur, and salts, are discussed in this section. The chemical composition of a mineral, i.e. the content of phosphorus, fluorine, boron, potassium, sulphur, bromide, and so on, is most important for the deposits of this group. Many of these elements are extracted from fossil and recent deposits, and from groundwater.

Geochemical prospecting for such deposits is aimed at detecting those elements that are the major component of ores that is the case in the search for metalliferous minerals. Deposits of chemical raw materials form in rather diverse conditions. They are related to endogenic, metamorphogenetic and exogenic genetic types. The first comprise the largest deposits of apatite, volcanogenic sulphur, boron, and fluorite. Apatites may also be of metamorphogenetic origin. Deposits of phosphorites, native sulphur and potassium salts are exogenic ones. The specific features of prospecting for deposits of mining chemical raw materials depend on the diversity of their genetic types. The application of different methods of geochemical prospecting for these deposits and their assessment is conditioned also by a non-uniform distribution of the reserves of chemical raw materials over the territory of Russia and Ukraine.

The last decade was characterized by the most intensive consumption of mining chemical raw materials, primarily, for the development of agriculture. Apatites, phosphorites, and potassium salts, used in the production of mineral fertilizers, are of tremendous importance for the development of the national economy. The boron-containing raw materials and sulphur are very important for the development of the chemical industry. Owing to an increasing volume of prospecting for these raw materials coupled with the assessment of the potentialities of deep horizons in the traditional mining regions within the confines of the deposits being explored, geochemical methods may prove to be effective for practical geologists. These methods may be very useful in searching for both the outcropping deposits and those that have not been exposed at the present level of erosion.

The geochemical methods employed in the search for various minerals are described below in different detail depending on the degree of exploration of the respective raw material.

3.1. APATITE

Apatite belongs to the group of calcium-phosphate salts. The following types of its deposits are known: (1) magmatogene; (2) carbonatitic (magmatogene and metasomatic); (3) pegmatitic; (4) contact-metasomatic; (5) hydrothermal; (6) volcanogenic-sedimentary; and (7) metamorphogenetic. The first two are commercially most important. Magmatogene deposits of apatite are associated with alkaline rocks: agpaitic nepheline syenites. Apatite–nepheline and apatite–nepheline–rare metal formations of apatite ores are distinguished.

The majority of researchers advocate the magmatogene origin of apatite deposits of this group.

Apatite deposits in calcitic and dolomitic carbonatites are associated with complicated intrusive complexes of ultrabasic alkaline composition. The first stage of their formation was characterized by the injection of ultrabasic magma followed by the emplacement of ultrabasic alkaline formations alternating with alkaline intrusions. This resulted eventually in the formation of carbonatites proper. The apatite mineralization is usually assumed to occur at the early and late calcitic stages. There are two extreme points of view on the origin of carbonatites: magmatogene and metasomatic (hydrothermal).

The prospecting indications of apatite deposits are given below.

1. Magmatogene deposits

(a) Occurrence in the region of complex, multiphase ring-shaped intrusions of alkaline rocks of the central type composed of nepheline syenites, syenites, and ijolite–urtites.

(b) Distinct differentiation of rock masses, conditioning the formation of rock varieties with intensive layering.

(c) Increased tectonic dislocation of alkaline rocks; presence of zonal ring-shaped faults with developed rocks of the ijolite–urtite complex.

2. Carbonatitic deposits

(a) Occurrence in the region of endogenic accumulations of calcite and dolomite associated with complex multiphase intrusives of ultrabasic alkaline composition and with a zonal ring-shaped structure. Distinct differentiation in the ultrabasic-alkaline rocks.

(b) Presence of local crushing zones and extensive fissuring

(c) Development of nepheline–pyroxene and pyroxene–amphibole accumulations in ultrabasic rocks.

Appreciable contributions to the development of mineralogo-geochemical methods of prospecting for apatite deposits have been made A. A. Kukharenko, Ye. I. Semenov, V. I. Gerasimovsky, and O. B. Dubkin.

Indicator minerals and elements

Apatite deposits are notable for a diverse and complicated set of minerals. The apatite-bearing alkaline rocks contain such minerals as sodium, iron, titanium, zirconium, rare earth elements, and strontium, as well as mineral phases having phosphorus, fluorine, chlorine, sulphur, and lithium. Titanium-bearing minerals are: ramsayite, lamprophyllite, murmanite, lomonosovite, loparite (agpaitic nepheline syenites), sphene, ilmenite, and titanomagnetite (miaskitic varieties of rock). Minerals characterized by an increased quantity of lithium include lipidomelane, biotite, and spodiophyllite. Pyrochlore, loparite, and murmonite are rich in niobium, while apatite and lomonosovite are characterized by raised phosphorus. Zirconium levels are increased in eudialyte, lovozerite, and zircon, while rinkolite, steenstrupite, nordite, apatite, pyrochlore, loparite, Be-chkalovite and tungunite have raised thorium levels. Minerals rich in fluorine are villiaumite, arfvedsonite, fluorite, and apatite; chlorine and sulphur-minerals are sodolite and cancrinite, while eudialyte, lamprophyllite, loparite, belovite, nordite, apatite, and sphene have raised levels of Sr−Na association. Many rare elements (Nb, Ta, Zr, TR) form not only independent phases but also enter morphologically the structure of many minerals.

The agpaitic varieties of nepheline-bearing rocks contain minerals in which Na is associated with Ti and Nb. Such rocks are characterized by increased amounts of rare lithophilic and radioactive elements including Zr, Be, TR, Li, Hf, Nb, Ta, U, Cl, F and S [34]. The miaskitic rocks bear typomorphic minerals containing Ca in association with Ti and Nb, or phases having no Na and Ca. The mineral composition of the above-mentioned rocks reflects geochemical conditions of alkaline rock formation. The nepheline–syenite intrusions of the Khibiny Massif, whose agpaite ratio equals unity, contain minerals of both agpaitic and miaskitic rocks (eudialyte + sphene + ilmenite + apatite) [34].

Garnet and sphene are closely associated with apatite. Sphene is characteristic of the upper contact zone of apatite deposits, and of tectonic dislocations.

More than 40 elements have been identified in alkaline rocks. According to the degree of their abundance and content, these elements can be integrated into the following four groups (Figs (3.1) and (3.2)):

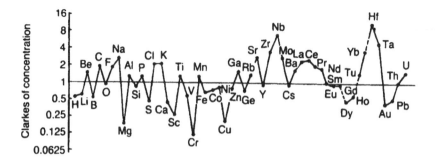

Figure 3.1. Relative abundance of indicator elements in the Khibiny Massif [65].

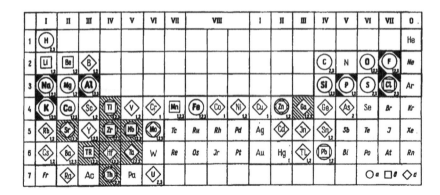

Figure 3.2. Geochemical table the nepheline–syenite intrusions of the Khibiny Massif [65]. Abundance of major and accessory (bold type), trace (light type) elements. Main stages of formation: a — protocrystallization; b — residual crystallization; c — post-magmatic processes. Typomorphic specific feature of elements: characteristic (black angles), typical (hatching).

(a) Major (petrogenetic) elements Si, Al, Fe, Na, K and Ca, contained in the main rock-forming minerals as significant constituents in the amounts of no less than 1 per cent.

(b) Petrogenetic elements of secondary importance (less than 1 per cent) such as Ti, Mg, P, Zr, F, Cl, S, Sr and Mn. These elements are contained in rock-forming accessory minerals that do not influence the species characteristics of rocks.

(c) Rare elements whose average amount is commensurable with or somewhat exceeds the clarke for the lithosphere. The majority of these elements form independent accessory minerals or are present as appreciable admixtures in the rock-forming and accessory minerals (Ba, Rb, TR, Nb, Ta, Ga, Th, Be, Li, Zn, Pb, Mo, U and V).

(d) Very rare elements occurring as isomorphic or atomically dispersed admixtures in different minerals in amounts of less than one clarke (Cr, Ni, Cu, Co, Y, Cs, B, Sc, Ge, Zn, Cd, Sn, W, Ag, Au, Hg, Tl, As and Se).

An analysis of the data presented in Figs (3.1) and (3.2) shows that (a) alkaline rocks contain an exceptionally great range of various elements, and (b) alkaline rock masses are characterized by a generally anomalous content of phosphorus and titanium as well as by high concentrations of volatiles (Cl, F, S, H_2O) and radioactive rare lithophilic elements (TR, U, Th, Zr, Be, Li, Hf, Hb) which makes it possible to apply geochemical methods and radiometry in the search for apatite ores. The greatest number of compounds contain Na, TR, Nb, Li, Ti, P and S. For apatite–carbonate and apatite–quartz ores there is a particularly increased content of Y, Yb, Ce, La, Ba, Sr, Pb, Ti, Zr, Ni, Co and B. Of these it is convenient to use Ce to indicate apatite ore, since the background value in crystal line rocks makes up 4–20 g/t and in apatite more than 80 g/t.

Express X-ray radiometric analysis gives the contour of ore bodies based on the secondary dispersion haloes. In this case there is a contrast of 1–2 orders of magnitude with a Ce content of 100 g/t in loose deposits with apatite. Therefore, the use of analytical data from the survey is realistic. In order to obtain promising areas on the basis of these data, the factor and discriminating analyses should be processed by IBM. The spectrum of indicator elements for apatite−carbonate ores is: Ce, La, Y, V, Mn, Ti and P; for apatite−quartz: Ce, La, Pb, Be, As, Y, B, Cr, Bi, Mn, Cu and P; and for apatite-bearing loose deposits: Ce, La, Y, Sr, Ba, Yb, Zr, Be and Nb.

The following petrogeochemical indications [23] have been established for alkaline rocks enclosing apatite ore bodies:

(a) a high content of alkali (Na > K), Na/K = 1.28;

(b) low content of calcium;

(c) agpaite ratio of 0.99;

(d) high content of titanium and ferrum (Fe^{3+} > Fe^{2+});

(e) predominance of titanium over zirconium and of cerium over yttrium;

(f) high concentrations of fluorine, chlorine and sulphur;

(g) distinct correlation between phosphorus and calcium.

Apatite deposits are usually confined to ijoltie urtites that in many respects are close to apatite bodies by the specific features of their mineral and chemical composition (agpaite ratio, content of volatiles) [23]. The leading role in agpaite nepheline syenites is played by Cl (0.16 per cent) and F (0.14 per cent), whereas phosphorus is characterized by polarity (antagonism) in respect to magnesium and potassium. Correlation analysis has proven the strongest positive bond of P_2O_5 with MnO in rocks and the significant effect of the latter on the phosphorus bonds with SO_2, TiO_2, Al_2O_3, FeO and CaO. The bond of phosphorus with Mn, Ti and Fe is conditioned by the association of apatite with magnetite, sphene, ilmenite and amphibole. The main minerals of carbonatites are carbonates (80−99 per cent of the rock volume) represented by calcites, dolomites, ankerites, and rarely by siderites. Other minerals are present in minor concentrations and, therefore, may be related to the group of secondary and accessory minerals. The varieties of carbonatitic minerals are listed below.

- *Silicates*: diopside, augite, aegirine, riebeckite, arfvedsonite, cataphorite, phlogopite, biotite, vermiculite, chlorite, serpentine, gieseckite, chrysotile, serpophite, muscovite, bastiorthoclase, albite, nepheline, forsterite, monticellite, chondrodite, sphene, garnet, zircon, cerite, epidote, zoisite, melilite, vesuvianite and lamprophylite.

- *Haloids*: fluorite and sellaite.

- *Phosphates*: apatite, monazite, florencite and isokite.

- *Sulphates*: barite and celestine.

- *Oxides*: quartz, beddelite, rutile, anatase brookite, thorianite, haematite, ilmenite, perovskite, dysanalyte, pyrochlore, hatchettolite, samarskite, columbite, zirkelite, titanomagnetite and magnetite.

- *Sulphides*: galenite, sphalerite, pyrrhotite, chalcopyrite, pyrite, marcasite, chalcosine, molybdenite, tetrahedrite and valleriite.

- *Carbonates*: siderite, rhodochrosite, bastnäsite, parisite, synchisite, lanthanite, strotianite, sahamalite, roentgenite, burbankite, ancylite and calkinsit.

The typomorphic minerals carbonatites include barium pyrochlorepandaite, phlogopite, and apatite as well as the less abundant: (a) baddeleyite ZrO_2; (b) perovskite–knopite (Ca, Ce)(Ti, Fe^{3+}, Nb)O_3; and (c) rare earth carbonates. The typomorphic admixture elements of carbonatites are Ca, F, P, Sr, Ba, TR, Ti, Zr, Nb and Pb, whose behaviour is determined by the semblance of their geochemical properties with those of calcium.

Apatite is usually formed at an early stage of the carbonatitic process and is associated with the calcite and dolomite varieties. Potential occurrence of apatite in ultrabasic alkaline rocks and carbonatites depends on the degree of their magmatic differentiation and depth of the present level of erosion. The platform complexes with extensive apatite mineralization are notable for the predominance of sodium over other alkaline elements, whereas carbonatitic complexes dominated by potassium are usually devoid of apatite mineralization of economic value μ [30]. Increased contents of F, Cu, Zr, Ta and Ti are characteristic of complexes prospective for apatite. Apatite mineralization occurred at the stage of accumulation of volatile components. The specific behaviour and distribution of phosphorus in alkaline ultrabasic rocks and carbonatites are determined by high concentrations of this elements as well as Ca, F, Cl, Na and K.

Basicity indices have been calculated for the main rock types of the Khibiny Massif and alumina-, titano- and zirconosilicates distributed in the rocks and pegmatites. It has been shown that the basicity indices of the rock-forming and accessory minerals tend to increase to words rocks with higher basicity. The types of mineral associations based on mineral basicity may be used as indicators of minerogenesis of medium basicity.

In the deposits under examination the apatites differ substantially in their content of fluorine, chlorine, iron and strontium.

Geochemical methods of apatite prospecting

The use of geochemical investigations for predicting, prospecting, and exploring apatite deposits is based on the differences in the chemical composition of apatite ores and enclosing rocks. The hypergenesis of ore bodies results in the formation of lithochemical and aqueous haloes. Destruction of the bedrocks leads to the development of loose deposits of different thickness which overlie an apatite mineralization. Haloes of phosphorus, fluorine, and other elements develop in loose formations around apatite bodies in the zone of weathering. There are examples of successful application of geochemical methods in the search for apatite.

Secondary dispersion haloes of elements have been detected in loose sediments of apatite deposits associated with alkaline rocks (anomalies of P, F, Sr, Ti, Cl, Na and Zr), carbonatites (haloes of P, F, Ce, Y, La, Ba, Nb and Ta), basic rocks (haloes of P, F, Ba, Sr, Ti and other elements), and metamorphogenetic formations (haloes of P, Be, B, Pb and Ce).

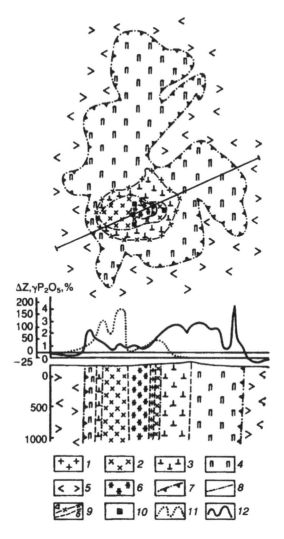

Figure 3.3. Diagram of the geological structure of carbonatitic massif in the Kola Peninsula and graphs showing the distribution of phosphorus in and geophysical anomalies above this massif [53]. 1 — carbonatites; 2 — amphibole carbonatites; 3 — apatite-bearing pyroxenites; 4 — peridotites; 5 — effusive sedimentary rocks of the Ismandra–Varzuga series; 6 — urtite (nepheline carbonatites); 7–9 — contacts: 7 — massif, 8 — carbonatitic core, 9 — rocks (a, identified; b, assumed); 10 — pits that stripped the crust of weathering of carbonatites; 11 — content of P_2O_5; 12 — magnetic anomaly (Δz).

The fluorometric survey has turned out to be effective in locating apatite bodies in carbonatites. Poorly exposed (owing to peat formation) apatite–carbonate ore bodies associated with carbonatites are contoured by the secondary haloes of phosphorus (Fig. (3.3)). It should be noted that phosphorus anomalies have not been detected in the peat beds overlying the ultrabasic rocks in such sites [53].

Apatite mineralization is identified by secondary dispersion haloes of phosphorus whose content exceeds the background value (0.6 per cent P_2O_5) 1.5–4 times. Phosphorus content also exceeds Ce contents (0.01 per cent), La content (0.01 per cent), and Y content (0.002 per cent), which elements exceed background concentrations by several times. The displacement of haloes downslope (10–12° gradient) reaches 150 m. The content of P_2O_5 and TR in the secondary haloes decreases in a thick (up to 6 m) overburden. Sites with a maximum P_2O_5 content coincide almost completely with the haloes of Y, Ce, and to a lesser extent, of La [26]. These examples indicate the potential use of lithochemical survey in the search for apatite mineralization. The applicability of the hydrochemical method in prospecting for apatite mineralization has been proven both experimentally and theoretically. This method is based on exploring the variations in the chemical composition of the subsurface waters and associated primary haloes in the surrounding rocks under the effect of apatite bodies. When prospecting for magmatogene apatite deposits, elements are used whose abundance in natural waters is explained by their easy solubility and stability in solution. These elements are fluorine and HPO_4 ion. Concentration of the latter in water occurring in the zones of ore bodies, reaches anomalous values (0.5 mg/l, compared to 0.02 mg/l outside apatite deposits) [102]. Figure (3.4) shows the content of fluorine in the waters of apatite deposits. Maximum concentrations of fluorine are characteristic of waters in bedrocks (1.01 mg/l) due to the high content of this element in apatite–nepheline deposits (0.75–2.58 per cent). Local contrasting fluorine anomalies in waters are, in the majority of cases, conditioned by the dissolution of NaF that is present in villaumite ore bodies.

Other indicator elements of apatites are lithium, rubidium, chlorine, and rare earth elements, characterized by high migrational ability and mobility and an ability to form distinct aqueous haloes in alkaline rocks. Sodium and calcium are less mobile.

The chemical composition of fissure water in apatite deposits depends on the composition of the enclosing rocks and their constituent minerals, from which various elements are leached out by aggressive solutions. The aggressiveness of water is conditioned by the presence of fluorine, sodium (leached from apatite), lomonosovite, villaumite, and nepheline. Owing to the high alkalinity of the enclosing rocks, the waters in the zones of agpaitic nepheline syenites are characterized by an appreciably sodic composition (Na–HCO and Na–CO_3) and a high content of fluorine. The contents of the elements are related as below (in mmole/l):

$$Na > K \lessgtr Ca > Mg > TR; \quad HCO_3 \lessgtr CO_3 > SO_4 \approx Cl \approx F > HPO_4.$$

The haloes of Be, Zr, Nb, and Ti that have various migrational mobility (Ti > Zr > Be > Nb) are detected in the groundwaters of the Khibiny alkaline rock mass. Thus, increased alkalinity of the subsurface waters and the presence of phosphate ion, fluorine, lithium, and rubidium are the hydrochemical prospecting indications of apatite mineralization.

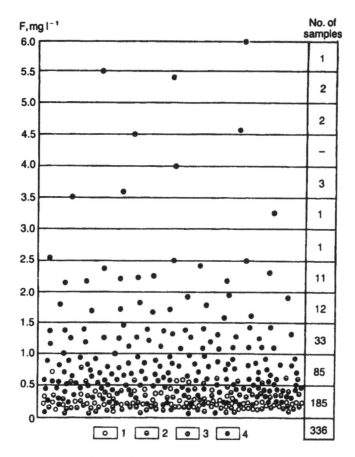

Figure 3.4. Fluorine content in natural waters of apatite deposits [102]. 1 — surface river and lake water; 2 — water in Quaternary deposits; 3 — tap water; 4 — fissure water in crystalline rocks.

The apatite ore bodies associated with basic and ultrabasic rocks are indicated by Cu, Zn and Ba haloes in water. The Cu and Zn contents (mg/l) in the water near an apatite mineralization amount to 5.6 and 5.1 versus their background values of 0.01 and 0.002, respectively.

Nepheline syenites are highly radioactive due to the presence of uranium, radium, thorium, and potassium in tantalum–niobium and zircon minerals, and in orthoclase. Rocks of apatite–nepheline ore bodies and ijolite urtites are least radioactive. However, their radioactivity increases in the transition from high- to low-grade ores. Carbonatites with apatite are also characterized by increased radioactivity due to the presence of pyrochlore, monozite, and partly of zircon, baddeleyite, perovskite, sphene, and xenotime.

Ultrabasic rocks are less radioactive. The high radioactivity of apatite deposits aids in their detection by radiometric method. Figure (3.5) shows the results of prospecting for the apatite-bearing carbonatites by the airborne γ-ray spectrometric

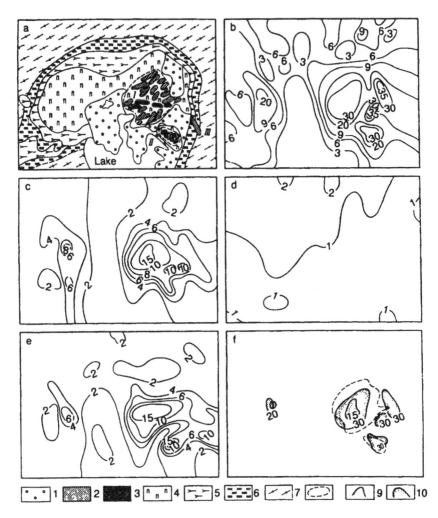

Figure 3.5. Recording pattern of a carbonatitic rare metal deposit [1]. Maps: a — geological; b — of thorium contents, 10^{-4} per cent; c — of uranium contents, 10^{-4} per cent; d — of potassium contents, per cent; e — of gamma field, 15^{15} A/kg; f — interpretation. 1 — Quaternary deposits; 2 — carbonatites; 3 — apatite–forsterite–magnetite rocks; 4 — pyroxenites; 5 — ijolites and melteigites; 6 — fenitized gneisses; 7 — granite-gneisses; 8 — deposit contours. Sites with a maximum content of 10^{-4} per cent; 9 — uranium; 10 — thorium.

method. Several uranium and thorium anomalies, coinciding with the zones of development of carbonatites with apatite and rare metal mineralization, manifest themselves against a low radioactive level that is typical of the enclosing rocks (pyroxenites and ijolite melteigites).

At the stage of reconnaissance prospecting, geochemical methods can be applied to detect and outline the apatite ore bodies and to predict the prospects of ore occurrence at depth, making use of the close correlation between apatite and nepheline. The close functional dependence between the contents of P_2O_5, F and TR (due to the presence of these elements in apatite) can be used to locate the outcrops of apatite bodies by fluorometric survey. Rocks are also explored by activation methods (integral and spectrometric variants of neutron-activation logging; sensitivity: 1.5 per cent of P_2O_5, error up to 6 per cent) that make it possible to delineate the apatite-bearing formations, to assess quantitatively their phosphorus content, to determine the quality of apatite–nepheline bodies, and to specify the boundaries between balanced and low-balanced ores. Spectrometric activation logging for fluorine permits the application of remote sensing analysis of ores and rocks directly in boreholes. This is especially important in case of poor core sampling. The method is rapid and highly efficient.

The content of isomorphic admixtures has been shown to decrease in nepheline syenites, as the content of apatite, and the deposit's dimensions, increase. The apatite–sphene–nepheline formations are characterized by an inconsistent composition of apatite: if an apatite–nepheline body contains an interbed with prismatic sphene, the content of admixtures in the apatite increases and the concentration dispersions are observed in this interlayer [23]. Therefore, a drastically non-uniform content of admixtures down the cross-section may be indicative of the epicentres of apatite bodies. Sphene mineralization in alkaline rocks is a prospecting indication of deep-seated apatite mineralization.

The content of admixture-elements in the vertical series of nepheline syenites and apatite rocks changes on a regular basis. There are two groups of admixture elements:

(a) Those with a positive geochemical gradient increasing upward in the vertical direction (Hg, Sr, Ba, TR, F, Cl, CO_2 and Ca). The content of phosphorus increases in the closing members of the series, which one enriched in nepheline parallel to its accumulation.

(b) Those with a nepheline gradient decreasing in the following direction: B, Na, and elements of the ferric group (Cu, Ni, Co and V).

Mercury and its forms in apatites are a sensitive indicator in the supra-ore zones. We have found that these zones in apatites are characterized by a broad thermal spectrum of mercury sublimation and by the presence of up to three or four forms of mercury (native, 120 °C; chloride, 220–240 °C; sulphide, 270–320 °C; and isomorphic, 420 °C). In contrast, the sub-ore apatite bodies contain only two forms of mercury: sulphide (280–300 °C) and isomorphic (420 °C).

Zonation by dip has been discovered for apatite deposits; it is expressed as changes in the morphology of the bodies, the rock sequence, and the structure and composition of the rocks and minerals with depth.

The middle levels contain the most ore. Levels which are insufficiently deep bear less apatite ore. The sub-apical levels are considered to be quite promising, but the picture is complicated by the branching of lodes and interstratal post-apatite tectonics. The apical and supra-apical levels and, especially, those levels at a distance from the apex are poor in ores. Vertical zonation is manifested

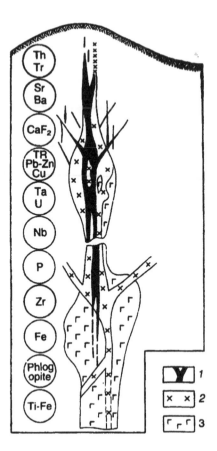

Figure 3.6. Diagram of vertical zonality, composition, structure and distribution of mineralization in the massifs of ultrabasic alkaline rocks and carbonatites [30]. 1 — carbonatites; 2 — ultrabasic alkaline rocks; 3 — ultrabasites.

as increased concentrations of phosphorus and titanium down the dip of the ore bodies over a distance of at least 200 m from the frontal zones.

Apatite deposits associated with carbonatites are also characterized by the vertical zonality stipulated by regular occurrence of magmatic and metasomatic processes in time and space. The alteration of different types of deposits in the vertical section of ultrabasic alkaline rocks and carbonatites is shown in Fig. (3.6). Commercial accumulations of various types of mineralization occupy certain positions in the vertical section. Medium-deep intrusives of complicated structure and great petrographic diversity are most prospective for apatite. Apatite-rich bodies extend downwards to a considerable depth. It is significant that in some cases carbonatites are replaced by apatite ores at depth. The mineral zonality of carbonatites is characterized by an increase in the concentrations of phologopite, amphiboles, magnetite, and apatite with depth. The occurrence of barite, fluorite, and rare earth minerals is typical in the upper horizons. Apatite forms commercial accumulations within an appreciable range of the vertical section which is some-what displaced with depth relative to pyrochloric carbonatites that have not been detected as commercial ores in deeply eroded rock masses (Kovdor, Phalaborwa).

Being closely interrelated, the geochemical zonality mirrors the mineralogical one. The contents of copper, nickel, vanadium, and zinc in carbonatites increase downwards, whereas the opposite trend is characteristic of barium, strontium, mercury, fluorine, and chlorine. Some deposits in the USA and Uganda are known to contain at a certain depth carbonatites containing apatite that has an increased concentration of copper. Apatite is also characterized by a direct change in the TR composition, i.e. by the predominance of average TR in the early (deep-seated), high-temperature generations of these minerals, and by the prevalence of heavy lanthanoids in the near-surface apatites. Magnetites from carbonatites are, under near-surface conditions, characterized by high concentrations of magnesium, and low concentrations of titanium and manganese. These characteristics are conditioned by a decrease in temperature and by the growth of oxygen activity, which is responsible for the deficiency of F^{2+} in the mineral-forming system.

The presence of a deep-seated apatite mineralization is based on the following criteria: predominantly carbonate–magnesian composition of carbonatites (cal-citic and dolomitic); high content of fluorine, chlorine, mercury and rare earth elements in the enclosing rocks; discovery of apatites enriched in light and heavy lanthanoids; presence of magnetites rich in magnesium, and poor in titanium and manganese. The vertical zonality in the apatite deposits may be used for determin-ing the present level of erosion of the anomalies, and for assessing the prospects of finding ore bodies at certain depths.

As has been shown by experimental and methodical work, the geophysical methods (magnetic prospecting and high-fidelity gravimetry) are capable of trac-ing out the edges of known apatite bodies, and of detecting new deep-seated ore bodies. The apatite and sphene–apatite ore bodies are indicated by positive grav-ity anomalies on the background of a rather undisturbed magnetic field, whereas alkaline formations and carbonatites, as compared to ultrabasic rocks, are char-acterized by reduced density and magnetic susceptibility. Apatite ores associated with carbonatites are sometimes distinguished by increased radioactivity.

Methods of work

General surveys at scales of 1 : 200 000 and 1 : 50 000 should result in the lo-cation and outlining of intrusions of nepheline syenites and carbonatites, which are then evaluated for apatite abundance. Combined large-scale airborne magnetic and γ-ray spectrometric survey is the main method of prospecting for an ap-atite deposit. High-quality airborne γ-ray spectrometric surveys require standard objects (models) and refinement of the navigational positioning methods. The ap-plication of radiometric methods demands their accurate correlation with magnetic prospecting. The analysis and specification of the results of airborne geophysical investigations should be followed by the on-site verification of the most prospec-tive target areas by the lithogeochemical surveying of secondary dispersion haloes and dispersion trains. This is accomplished on the anomalous sites localized by dispersion trains, or in prospective target areas determined from the integrated geological and geophysical data. The search by dispersion trains may involve different methods, namely the heavy concentrate geochemical, lithochemical, and hydrochemical methods. It is advisable to use the heavy concentrate geochemical method in prospecting for an apatite mineralization supposedly associated with carbonatites, i.e. in those cases when the diagnosis of minerals is frequently impeded. Spectral analysis of heavy concentrates aids in obtaining additional valuable information when ordinary methods of low sensitivity fail to detect the anomalies of Ta, Nb, and rare earth elements. It is also advisable to apply these methods in heavy-concentrate sampling of river valleys. To this end, a sample is taken from the heavy concentrate and subjected to spectral and special analyses. Every second sample is analysed in a mineralogical laboratory to detect apatite and accessory minerals. All heavy concentrate samples should be subjected to detailed analysis only within anomalous zones. By reducing the number of mineralogical analyses it is possible to obtain additional information on the genetic type of prospective deposits, to reduce the time required for such analyses, and to achieve economic efficiency. Being highly sensitive, heavy-concentrate sampling allows detection of single grains of minerals contained in the initial sample.

In the search for apatite deposits by dispersion trains, the sampling grid depends on the distances between the rivers to be sampled and on the spacing of the sampling points along each river concordantly with the detailedness of prospecting. A sampling density of 4–5 points per 1 km with a spacing of 100 m corresponds to a scale of 1 : 50 000 [46]. The best results in prospecting for apatite mineralization associated with carbonatites are obtained by testing coarse fractions, sized 0.5–1 mm. Geochemical exploration involves sampling of bottom sediments in river beds and in their proximity. It is recommended to separately collect mosses and turf from river slopes, and to fractionate the bottom sediments by sieving.

The most reliable results are obtained by combining all kinds of sampling, because they concentrate various elements.

Ground waters and surface waters are the major objects of hydrochemical prospecting. It is necessary to sample small streams fed mainly by subsurface waters and water sources confined to exposed alkaline rocks and carbonatites. Sampling of subsurface water sources is mandatory in areas with thick loose formations, morainic deposits, and stone taluses. In the case of hydrochemical exploration, the sampling grid depends on the length of small streams and rivers, and on the presence of subsurface water sources. One sample is ordinarily collected every 5 km^2 when prospecting at a scale of 1 : 200 000, or one sample per 0.5 cm^2 at 1 : 50 000. Changes in the concentration of elements in the water should be considered (it is advisable not to collect samples during floods and rainy periods). Blind apatite deposits can be detected by sampling fissure waters tapped by boreholes. Structural wells, used to study intrusive rocks, can tap deeply circulating waters which also need testing. Anomalous concentrations of pathfinder elements in such waters indicate the presence of a deep-seated, concealed apatite mineralization that was not intersected by holes. Interpretation of analytical data obtained from heavy concentrate testing results in the delineation of local sites to be examined in greater detail. The iso-concentrations of the indicator elements in the waters and loose sediments, as well as the degree of concentration of apatite, zircon, pyrochlore, baddeleyite, zirkelite, perovskite, barite, pyrite, columbite, rutile, anatase, brookite, and monozite in river valleys are shown in mineralogical–geochemical maps.

The lithochemical method of apatite prospecting by secondary dispersion haloes is the chief geochemical method enabling the identification of prospective target areas. This method is mainly employed in specialized detailed surveys at scales of 1:50 000 and 1:10 000.

The applicability of lithochemical surveys depends on the character of the secondary haloes, which are classed into open, buried, or strongly surface-attenuated types. In the case of exposed secondary haloes, the lithochemical survey gives reliable results if samples of 50–100 g are taken from a depth of 0.2–0.5 m. The non-uniformity of the depth of sampling is conditioned by variation in the thickness of loose formations overlying the bedrocks. In the case of buried or semi-buried haloes, samples are taken from the lower representative layer of loose deposits, in specially drilled holes.

Prospecting routes (at the scales of 1 : 50 000 and 1 : 10 000) are oriented in such a way as to intersect the geochemical anomalies in mutually perpendicular directions. A rectangular prospecting grid is established instrumentally. The prospecting lines in mountainous areas should coincide with the contour lines, because dispersion haloes usually extend downslope and to a lesser degree depend on the configuration of the ore bodies. Indigenous exposures of rocks, ores, and their eluvial–deluvial disintegrated blocks piled up on mountain slopes, are subjected in the field to qualitative analysis for phosphorus. Upon identification of anomalies in rocks, lump samples with a mass up to 5 kg are taken, ground, concentrated, and subjected to mineralogical analysis.

Buried haloes are explored by a combination of geochemical and geophysical methods. Magnetic prospecting and airborne γ-ray spectroscopy make it possible to evaluate considerable areas at a comparatively low cost. The boundaries of rock masses concealed by overburden are verified by surface magnetometric survey, which identifies rock masses by the sharply reduced Δz values of the enclosing rocks. This survey also helps to identify zones of development of carbonatite containing high concentrations of magnetite. A combination of geochemical and geophysical surveys permits more reliable identification of alkaline rocks, carbonatites, and apatite ore bodies stripped by workings. However, application of gamma-ray methods becomes less successful with an increase in the depth of the deposits. If the ore bodies occur at a great depth but within the zone of active discharge of subsurface waters, the indicator elements of apatite mineralization are washed out. It is possible to apply hydrochemical prospecting in these cases. In areas dominated by chlorapatite mineralization, where lithochemical survey is less efficient, loose deposits are tested by the thermal fluorescence method, while cathodoluminescence is used to explore the weathering zones of carbonatites whose mineralogical composition is analysed in the field. In mountains and in areas with extensive fluvial and glacial deposits it is advisable to apply the boulder-glacial method in combination with heavy concentrate sampling.

Fluorescence analysis of heavy fractions makes it possible to record the presence and amount of apatite. A deep-seated concealed apatite mineralization is indicated by a sphene mineralization in ijolite urtites.

It is extremely important to use mineralogical methods in the search for an apatite mineralization associated with carbonatites. The occurrence of poorly exposed carbonatite bodies may be indicated by the presence of apatite, pyrochlore, zircon, baddeleyite, columbite, thorite, perovskite, barite, zirkelite, bastnäsite, and schorlomite in loose formations (in the alluvium of river valleys). Mechanical haloes extend over hundreds of metres or, less frequently, over several kilometres. When stripping indigenous carbonatites, it should be borne in mind that their calcite and dolomite varieties, that are characterized by magnetic anomalies generated by apatite–magnetite ores occurring along them, are prospective for apatite. Combined exploration of rocks and loose deposits along radial profiles is carried out in the crusts of weathering. Soils enriched in P, Nb and Ta due to the redeposition of phosphates should also be subjected to geochemical analysis.

The specific features of the crusts of weathering and their dependence on the nature of the substratum (whose kaolin and nontronite–montmorillonite types occur in nepheline syenites and melanocratic carbonatites, respectively) should be taken into consideration when prospecting for apatite ores. Extensive use of lithological and mineralogical criteria is imperative in the case of horizontal and shallow occurrence of apatite-breccia ores in the crusts of weathering. The development of linear crusts of weathering containing limonite, chlorite, gypsum, calcite, and tourmaline, as well as increased concentrations of zirconium, fluorine, and phosphorus, is a favourable prospecting indication. When evaluating the products of chemical weathering of carbonatites, attention should be paid to the characteristic light to dark-grey, and at times, cinnamon-like colour of sandy rocks composing the crust of weathering. The sandy material contains white crystals of apatite in association with goethite, siderite, magnetite, crandallite, pyrite, vermiculite and chlorite.

Ochreous limonite rocks, developing as the crusts of weathering on apatite carbonatites, contain finely dispersed barium, pyrochlore–pandaite, apatite, magnetite and goethite. They lack carbonates, sulphides, and unstable silicates. Since it is difficult to determine minerals of carbonatite complexes, heavy concentrate mineralogical methods can be used for evaluating the prospectiveness of certain regions. While studying loose deposits, it is advisable to remove finely dispersed goethite occurring in minerals by first treating the samples with hot hydrochloric acid then subsequently leaching them, first with hot caustic soda and then with water. Due to the fact that this process leads to the dissolution of apatite and magnetite and the preservation of niobium, zircon, and titanium minerals as well as monazite, it is recommended to subject the dissolved portion of the rocks (slime) to spectral analysis. The remaining minerals can be detected with the aid of high-frequency ultrasonic cleaning. Magnetic separation and heavy fluids are used to diagnose and quantitatively assess the extracted pure fractions. Mineralogical and chemical exploration of loose formations shows what minerals and elements are concentrated in carbonatites. This allows selection of concrete sites and zones that are deemed worthy of detailed prospecting for apatite.

When investigating the composition and typomorphic features of accessory minerals occurring in the disintegrated blocks of carbonatites piled up on mountain slopes, alongside the preparation of their artificial heavy concentrates by crushing and washing, it is possible to analyse the fractions that are undissolved in the acid, or to roast small fragments of rocks at 950 °C and subsequently quenchthem with water. All these procedures fail to alter the greater part of the accessory minerals.

Surface and airborne radiometric surveys can be employed in the search for carbonatites, because they contain strongly radioactive minerals (pyrochlore and monazite). The prospectiveness of alkaline intrusions, ultrabasic alkaline rocks, and carbonatites for an apatite mineralization is evaluated by (a) petrogeochemical criteria, and (b) primary endogenic haloes. In the first case, the geochemical coefficients, showing the content of alkaline elements in the rocks, are determined. Intrusive complexes dominated by sodium are considered to be prospective for apatite mineralization. The value Na_2O/K_2O (Table (3.1)) and indications

$$N = Na_2O/(Al_2O_3 + Fe_2O_3) - (K_2O + CaO)$$

and

$$K = K_2O/(Al_2O_3 + Fe_2O_3) - (Na_2O + CaO)$$

are used to typify nepheline syenites and to quantitatively assess the contents of potassium and sodium. Rocks with $N > K$ are more prospective for apatite.

The value F/Ca, which reflects the degree of activity of fluorine and calcium, is also used for a substantiated sorting out of intrusive rocks prospective for apatite (Table (3.2)).

Table 3.1.
The content of alkalis in the formational groups of nepheline syenites [12]

Formation	Number of samples	Na_2O	N_2O
Ultrabasic alkaline rocks and carbonatites	14	1.64	0.56
Alkali and nepheline syenites, and gabbroids	21	1.71	0.50
Potash gabbroids and feldspathoid syenites	9	0.61	0.17
Miaskite alkali nepheline syenites with carbonatites	9	1.47	0.29
Granosyenites, alkali, and nepheline syenites	18	1.46	0.36
Potash feldspathoid and alkali syenites and monazites	13	0.41	0.21

Table 3.2.
Fluorine–calcium, and aluminum–potassium and sodium ratios in nepheline syenites of different intrusive massifs

Massif	Ore content	F/Ca	$Al/(K + Na)$
Khibiny	Large deposits	2	0.99
Langesund	Small occurrences	1	Not determined
Il'men	Small occurrences	0.1	0.68
Synnyr	Small occurrences	0.1	0.70
Lovozero	Small occurrences	5	1.40
Ilimaussak	Mineralization is not characteristic	10	1.39

The F/Ca ratio in massifs most prospective for apatite ranges from 2 to 5. In so far as the occurrence of agpaite associations is not always precisely predicted by the $(K + Na)/Al > I$ ratio, it is proposed to use the coefficient $(Na) = Na/(Al - K) = Na/\Delta Al$, indicating the degree of abundance or deficiency in the rocks not of the sum of alkaline elements versus aluminum but only of sodium upon bonding of Al with K at the ratio of $1:1$. Rocks are assumed to be agpaitic if the $Na/\Delta Al$ ratio equals 0.55 or more. In all other cases they are considered to be miaskitic. The growth of the agpaite ratio in nepheline syenites indicates an increase in the content of Na with respect to K (from 0.3 in potassium varieties up to 4.0 in agpaite nepheline syenites).

The resultant correlation of P_2O_5 with the other rock-forming oxides serves as a criterion for evaluating the apatite content in rocks. Prospective zones are characterized by positive correlation between P_2O_5 and Al_2O_3, Na_2O, CaO and TiO_2, and negative correlation between P_2O_5 and MgO and K_2O. The apatite content in alkali syenites grows simultaneously with the accumulation of augite and alkali hornblendes, manifested by a positive correlation between P_2O_5 and CaO.

The prospecting evaluation (at scales of $1:10000$ and $1:2000$) and reconnaissance surveys are aimed at studying the primary haloes of indicator elements, determining zonality indications, and predicting the depth of mineralization.

In the course of this work at a scale $1:10000$, samples are taken from the main ditches, prospecting shafts, and boreholes.

Bedrock samples are collected by the dotted furrow method. Six or seven chip samples of rocks are taken from a sampling interval (from 2 to 5 m) at equal stretches, and combined into one sample of 150–200 g.

The collected and numbered samples are disintegrated to 0.1 mm fraction and divided into two parts. One part is then ground to a powder and sent for analysis, while the other is kept as a duplicate. The samples are analysed for F, P, Ba, Sr, K, Na, Li, Y and Yb. The geochemical anomalies are contoured according to minimum anomalous concentration of indicator elements, computed for each difference of rocks typical of the area under investigation.

In the process of drilling, high apatite concentrations may be anticipated in samples with an increase in TiO_2, FeO and MnO, and a decrease in SiO_2. The neutron-activation method is capable of determining phosphorus levels in rocks in the intervals of drilling without core sampling. Its application also makes it possible to rationally select and reduce the sampling intervals, thus reducing the network of boreholes needed at the stage of preliminary prospecting.

Exploration is generally directed towards: (1) finding new areas of deposits and (2) searching for new deposits associated with known fields. In the second case, exploration will, of course, be guided by the practical knowledge acquired in that field before the new exploration begins. Clues may be sought under three headings: petrological, mineral records, and geological and geochemical information.

Occurrences of alkaline rocks obviously deserve consideration since carbonatites are genetically linked with these rocks, but it should be noted that many occurrences have no carbonatites associated with them. The methods and intensity of prospecting are matters for the geologist to decide in the particular circumstances. However, the collection of material for identification or confirmation in the laboratory is the major task, and in addition to rock specimens, samples of residual soils are most important. This is because they are often more representative of the chemistry and mineralogy of the bedrock as a whole than are the resistant and often heterogeneous rocks which form outcrops. Normally they will be enriched relative to the bedrock in some of the diagnostic elements of carbonatites, notably in phosphorus, barium, rare earths and niobium (strontium is leached out).

In the field it is useful to carry a strong band-magnet, conveniently shielded in a thin aluminium cup, for detecting magnetite rich soils, as magnetite and apatite are commonly associated in carbonatites and phosphorites, and both are concentrated in the residual soil overlying them [25, 40].

Having located carbonatitic alkaline or ultrabasic complexes, the chances of discovering apatite may still be uncertain, and finding it may not be easy. Apatite-rich material may be quite nondescript and variable in appearance, especially when fine-grained or weathered, and chemical or microscopic checks should be applied to all doubtful or suspect rocks coupled with soil sampling. In general, the most deeply eroded complexes appear to be most favourable for apatite deposits of the phosphorite or pyroxenite type. In these, carbonatites may be relatively minor consituents of the complex (at Palabora they form only about 2 per cent of the outcrop area), so a complex should not be underrated just because the area of carbonatite is small. The maximum development of carbonatites appears at a somewhat higher level of intrusion in the form of large carbonatite plugs, and these may yield important residual deposits of apatite, provided climatic and erosional factors are favourable. In the less eroded volcanic complexes one may find apatite associated with highly potassic feldspathic breccias and trachytic rock, as exemplified by the Tundulu deposits and several minor occurrences. At the initial stage of prospecting for carbonatites, the usual aeromagnetic and aeroradiometric surveys are particularly useful, as they reveal anomalies that are outlined distinctly in the regions of development of normal sediments. When checking these anomalies on the ground, the areas of possible occurrences of apatite are selected where reconnaissance prospecting is carried out in order to reveal their possible commercial importance.

Remote sensing is a method that employs electromagnetic radiation as a means of detecting and measuring the characteristics of the target without physical contact with the target. It has been used as an exploration tool, especially at the early stages of exploration, because it has the ability to cover wide areas by means of satellite or aircraft.

The value of the remote sensing technique to exploration lies in its ability to identify certain rocks (or rock types) or zones of clay and mica alteration which are often found around ore deposits.

Stages and sequence of geochemical exploration

Prospecting and exploration surveys for apatite and other types of hypogenic mineralization are carried out in three stages:

(1) geological survey and prediction at a scale of 1:200000;

(2) geochemical surveys at a scale of 1:50000–1:25000; and

(3) detailed geochemical exploration.

The first stage encompasses heavy concentrate sampling and geochemical prospecting along alluvial dispersion trains. Every second heavy concentrate sample is analysed in the laboratory. Hydrochemical surveys is carried out within the confines of anomalous sites. Waters are analysed for P, F, Zn, Cu and Mg. Alluvial formations are analysed for P, F, and a wide range of other elements by spectrographic methods. The data of airborne γ-ray surveys are also subjected to analysis. Finally, the areas to be prospected for individual ore-bearing zones are defined.

At the second stage, target areas are identified on the basis of geological maps at a scale of 1:50000–1:25000, made in combination with lithochemical survey. Heavy-concentrate sampling is integrated with sampling of the bottom sediments along drainage channels, and with areal metallometric survey. Further work is specified.

Prediction at this stage is based on the materials of the structural and petrographic mapping of the intrusives, and on the data of geophysical and geochemical exploration of the rocks.

The overall goal of the third stage is localization of the ore source, its stripping and evaluation. Special mineralogical–geochemical maps are compiled at scales of 1:10000, 1:5000, and sometimes 1:2000. These maps show geochemical haloes of P, F, Sr, Cl and Nb, and magnetic and radioactive geophysical anomalies, as well as concentrations of apatite, pyrochlore, zircon, barite and monazite in heavy concentrates. After a detailed search, the prospective anomalies are stripped. When making holes and workings one should take into account the possible displacement of the epicentres of haloes with respect to the ore bodies.

3.2. PHOSPHORITE

Phosphorite ($P_2O_5 > 6$ per cent) is a sedimentary rock composed of phosphatic substance (mainly apatite) and other minerals. Exogenic and metamorphogenetic deposits of phosphorites are known. The former are subdivided into marine and continental ones. Marine phosphorites fall into geosynclinal, and platform deposits, and those occurring in marginal troughs. Geosynclinic phosphorites form thick strata of phosphatized rocks and productive phosphorite beds. The platform-type deposits are commonly characterized by spreading over an extensive area and by horizontal occurrence of productive, 1–1.5 m thick strata. Geosynclinal phosphorites are characterized by a sophisticated mode of occurrence of dislocated beds up to 80 m thick. Continental phosphorites are confined mainly to the crusts of weathering of phosphorite-bearing rocks. They are subdivided into residual and infiltrating deposits. The former are produced by chemical weathering of weakly phosphatized carbonate rocks. Phosphorites fill hollows and karst cavities in calcareous rocks, thus forming deposits of an irregular shape. Under continental conditions, infiltrational deposits originate in the lower horizons of the crust of weathering due to infiltration of phosphates leached out from different phosphatized rocks occurring in the upper horizons by percolating surface waters rich in carbon dioxide. Phosphorites of this type are notable for their high quality and impersistence of deposits.

Both marine and continental exogenic phosphorites are subdivided into chemogenic, organogenic, and secondary or redeposited ones. Moreover, deposits of the metamorphosed phosphorites are known that are represented by rocks which have undergone metamorphic changes and transformation of phosphate into apatite. Phosphorites can be formed chemically, biochemically, or through the substitution of phosphatic solutions for carbonates. The third way is responsible for the formation of the largest deposits of phosphorites. The prospecting indications of phosphorite deposits are:

(1) presence of marine sediments, represented by siliceous-carbonate, lime-dolomitic formations, and sedimentary strata containing organic residues, silicium, algal calcareous rocks, at horizons of coaly-siliceous shales;

(2) zones of articulation of different structures, autonomously phosphatic formations, and lithologically inhomogeneous beds;

(3) presence of phosphorite formations, manganese ore concretions, crusts of weathering of phosphorite-bearing rocks, accumulations of phosphatic shells, bone breccias, and fish skeletons.

Phosphorites consist of phosphatic substance (the main component of the rock) and various minerals associated with such apatite varieties as dolomite, glauconite, chalcedony, clay minerals, quartz, limonite, opal, zeolite, calcite and feldspar. Phosphorites also contain small quantities of ilmenite, leucoxene, zircon, rutile, tourmaline, garnet, disthene, epidote, pyroxene, sphalerite, sphene, amphibole and biotite. Phosphatic substance is represented by fluorapatite, fluorcarbonate–apatite, carbonate–apatite, hydroxyapatite, and chlorapatite. Investigation of the composition of phosphorites, identification of ore bodies, and their evaluation are impeded by the concealed crystalline and fine-grained character of these formations, by interweaving of phosphatic substance with other minerals, and by the complex diagnostic of mineral phases.

Conditions for the application of geochemical methods in prospecting for phosphorites

Direct indicator elements contained in phosphates and in the most widespread mineral phases, such as P, V, Sr, F and U, as well as indirect chemical elements such as As, Pb, Zn, Be, Zr, B, Mo, Ba, Ag and Sb, have been detected in phosphorites. There is a positive correlation between phosphorus and fluorine contained in fluorapatite. The F/P_2O_5 value in phosphorites ranges from 0.05 to 0.17, whereas the vanadium content fluctuates from 30 to 1000 g/tonne [51]. In the majority of deposits it ranges from 50 to 90 g/tonne. Phosphorite-bearing formations are notable for the polarity (antagonism) in vanadium and phosphorus behaviour. Low-grade ores have higher vanadium concentrations. Certain rocks associated with phosphatic ores are rich in vanadium. Here belong coaly-siliceous and coaly-clay shales, combustible shales, and sedimentary formations enriched in organic substance (up to 2 per cent of V_2O_5) that turn into phosphorites along their strike. Positive correlation is also observed between the contents of strontium and phosphorus. Phosphorites are usually enriched in rare earth elements whose concentrations exceed their average content in sedimentary rocks 2–6 times [51]. However, there is commonly no sharp difference in the total amounts of rare earths in phosphorites which range from 0.03 per cent to 0.13 per cent versus 0.08 per cent (average) and 0.5 per cent (maximum).

Manganese, whose content ranges from 30 to 960 g/tonne, is an important indicator element of phosphorites. Sedimentary formations containing increased quantities of manganese and rare earth elements are prospective for phosphorus. Phosphorites are characterized by an intimate correlation between arsenic and silver. In some cases, both elements were accumulated in the rocks that formed simultaneously with phosphorites but spread over greater area. The content of silver in phosphorite-bearing ores varies from 0.1 to 2×10^{-4} per cent. The quantities of Pb, B and Cr, and the contents of Ni, Co, Cu and Zn in phosphorites are, respectively, above and below their clarke concentrations.

Radioactivity of phosphorites is the most important geochemical factor permitting the application of radiometry in the search for these minerals. The uranium content in phosphorites ranges within great limits (in equiv. per cent): from 0.006 to 0.03 in Morocco; from 0.001 to 0.15 in the USA; up to 0.108 in Great Britain; from 0.0007 to 0.05 in Egypt; 0.2 in concretion varieties in Poland; from 0.01 to 0.3 in the residual-infiltrational crusts of weathering in Senegal; from 0.1 to 0.23 in the nodular ores in the FRG; and from 0.0006 to 0.5 in concretions of the coastal areas in Oceania, Africa, and California. The thorium content in phosphorites does not usually exceed $(5-10) \times 10^{-4}$ per cent, which discriminates them from the apatite ore bodies of the hydrothermal class.

Besides the well-known uranium-bearing phosphates, uranium-bearing laminated alumosilicates (mainly hydromicas and clorites) are also encountered in the section of sedimentary strata. The apatite structure firmly retains uranium and the products of its decay; therefore phosphate-bearing rocks display faint emanation and slight deviations from the state of radioactive equilibrium between uranium and radium. Uranium can hardly be leached out of phosphates by aqueous solutions, therefore, there are no uranium minerals in phosphate-bearing rocks. Laminated alumosilicates, on the other hand, are characterized by intensive emanation and sharp deviations from the state of radioactive equilibrium between uranium and vanadium. Uranium passes easily from them into aqueous solutions, as a result of which the minerals of uranyl (under oxidizing conditions) and the so-called 'primary' uranium minerals occur in rocks with uranium-bearing hydromicas and chlorites.

A distinct correlation between the contents of phosphorus and uranium is usually observed in phosphorites. This regularity is characteristic of marine phosphorites of carbonate and siliceous composition (correlation coefficient is 0.94–0.68). The direct dependence of P_2O_5 on V is not observed in the secondary, residual, infiltrational and redeposited ores. Therefore the horizons richest in phosphorus are not always the most radioactive. Lithogeochemical prospecting for phosphorite, which is aimed at detecting secondary dispersion haloes, is based on the fact that the content of phosphorus in eluvial–deluvial formations overlying sedimentary rocks amounts usually to hundredths of a per cent, whereas it sharply increases to several per cent above the phosphorite deposits. The lithogeochemical prospecting along secondary dispersion haloes is most effective in the areas of poor outcropping, where, as has been proved, by experimental and production work, it provides significant information. The examples are given below.

1. In one of the regions, geochemical prospecting on a grid of 100×20 was carried out in the areas with phosphate formations; phosphorite-bearing horizons were contoured by secondary haloes of phosphorus (0.3–10 per cent). Karst phosphorites formed isometrically shaped haloes containing from 0.3 to 1 per cent phosphorus.

2. Phosphorites localized in carbonate rocks overlain with up to 5 m thick eluvial–deluvial formations are indicated distinctly by secondary haloes with $P_2O_5 > 2.5$ per cent (versus the background value of 0.7 per cent) [75]. The phosphorus content in deluvial formations is directly proportional to its concentration in bedrocks.

3. Rocks of the phosphorite-bearing volcanogenic–siliceous–calcareous formation contain up to 2.8 per cent P_2O_5. Samples, collected at a depth of 0.3–0.5 m from 8 to 30 m thick deluvial deposits, contain from 0.5 to 0.7 per cent P_2O_5, whereas the P_2O_5 content in smoothed watershed areas, where the thickness of loams ranges from 30 to 75 m, does not exceed 0.3 per cent [75].

4. In the geochemical landscapes of the tropics, the stratal phosphorite bodies, occurring in sedimentary deposits and overlain by 3–5 m thick loose formations, are indicated by irregularly shaped geochemical anomalies, about 100 km long and 2–30 km wide [123]. It has been established that phosphorus-rich enclosing rocks are detected if the samples are taken from depths of 0.2, 0.4, and 0.6 m.

5. Karsted secondary phosphorites are identified by seconds dispersion haloes of phosphorus with a content of 0.7–1 per cent. Therefore, prospecting along dispersion haloes may be carried out in various regions. Geochemical survey detects not only rich ore bodies but also phosphatic rocks poor in P_2O_5 (up to 3 per cent) that are not discovered by radiometry which, on the whole, is rather effective and may be used in phosphorite prospecting (airborne radiometric survey, surface field observations employing radiometers, and γ-ray logging). The radiometric methods have succeeded in discovering phosphorite deposits in Angola, the Republic of Zaire, Turkey, Tanzania, Sahara, Kazakhstan, Siberia, and in the Baltic Sea region. An increased content of radioactive elements in phosphorites is the physical basis for the application of airborne radiometry. Airborne γ-ray spectrometry from a helicopter flying at an altitude of 500 m has proven effective in detecting either exposed ore bodies or deposits overlain by a shallow overburden (the intensity of γ-ray emission is appreciable even if the overburden is up to 15 m thick). Figure (3.7) demonstrates the use of this method for detecting phosphorite deposits. The γ-activity of rocks permits a tentative determination of the phosphorus content on the basis of linear dependence of phosphorite radioactivity on the amount of P_2O_5. The bulk concentrations of uranium in phosphorites are determined by

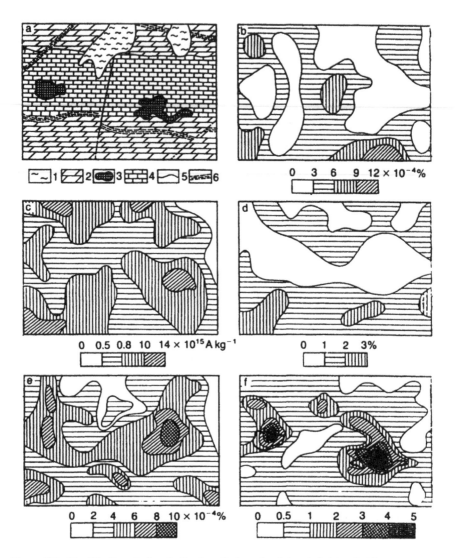

Figure 3.7. Identification of phosphorite deposits by using integrated data of airborne γ-spectrometry [1]. Maps: a — geological; b — of γ-field; c — of uranium (radium) content; d — of thorium content; e — of potassium content; f — of uranium (radium)–thorium ratios. 1 — Tertiary crust of weathering (red-brown phosphatic clays and epigenetic phosphorites); 2 — Mesozoic crust of weathering; 3 — rich epigenetic phosphorites; 4 — calcareous–dolomitic band (poor and light-grey limestones, and calcareous dolomites); 5 — stratigraphic boundaries; 6 — zones of tectonic dislocations.

γ-ray spectrometry, whereas its distribution in minerals is determined by fission tracks, with the help of f-radiography.

Phosphorites normally contain an increased (versus clarke) amount of uranium and a decreased (versus clarke) content of thorium and potassium [67]. The value

U/Th is an evaluation criterion of phosphorites. The phosphorite ore bodies are identified by individual determination of U(Ra), Th and ^{40}K, as well as by the U/Th value. Certain deposits that are not detectable by their total radioactivity are characterized by anomalous U/Th values.

Figure (3.8) presents the results of surface investigations carried out on a phosphorite deposit along a geochemical profile, made to verify an airborne γ-ray spectrometric anomaly. The phosphorite deposits are indicated clearly by the specific distribution of trace elements, and by the U/Th value.

Radiometry helps usually in contouring zones with a phosphorite-bearing crust of weathering. Identification of secondary phosphorites is a prospecting indication for detecting primary sedimentary ores. It is also possible to contour phosphorite ore bodies by radon survey undertaken to measure its contents in soil emanations, if the thickness of the aluvial–deluvial formations does not exceed 3 m. The radon content in the soil gas increases at sites covered with snow or in frozen soils. Radiometric methods are quite useful for approximate evaluation of the quality of the phosphorites. By applying γ-ray spectrometric profiling, it is possible to outline ore deposits and to reduce expensive mining operations.

Sufficiently long P, U, Sr and F dispersion trains are formed in the silt–clay fraction of bottom sediments of both permanent and temporary water courses. Their formation results from the sorption of elements from water solutions, and

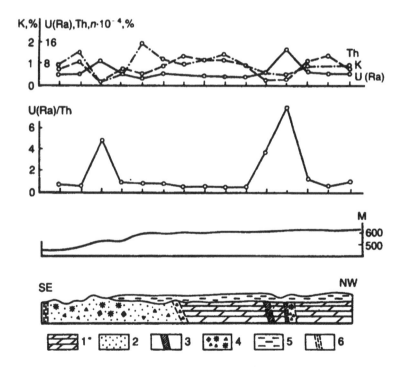

Figure 3.8. Distribution of radioactive elements and uranium (radium–thorium ratios around phosphorites) [67]. 1 — dolomites; 2 — siliceous-phosphate breccias; 3 — siliceous rocks; 4 — sites of increased phosphatization; 5 — eluvial–deluvial formations; 6 — zones of tectonic dislocations.

from the transfer of finely pulverized mineral substance. The hydrochemical prospecting for phosphorites is based on the changes in the chemical composition of natural waters under the effect of ore bodies and associated with them phosphorus, strontium, and fluorine haloes. This method aids in establishing the geochemical specific features of the region being studied, and in selecting smaller target areas for more detailed search. Phosphorus anomalies in waters frequently exceed their background concentrations by 5–10 times. Since phosphorus ion is more stable in water than other elements, its relatively increased concentrations form clearly defined extensive haloes.

The phosphorite-bearing horizons are identified by strontium haloes in water (usually from 8 to 240 mg/l). Strontium concentrations in waters do not depend on the depth of the aquifer and on the content of sulphates. They are controlled by the hydrochemical situation in a sedimentation basin. Broad haloes of the above-mentioned elements are formed in natural waters. Interpretation of geochemical data is impeded by considerable remoteness of elemental anomalies from the sampling point.

An increased uranium content in phosphorites results in raised uranium levels in subsurface waters migrating through the phosphorite-bearing rocks. Uranium applicability in the search for phosphorites depends on the ability of this element and products of its decay (radium and radon) to dissolve and migrate in natural waters. Prospecting along primary haloes is based on the considerable differences between the contents of P, F, TR and V admixtures in phosphorites and in the enclosing rocks. The role of fluorine increases when the phosphorite-bearing beds are affected by hydrothermal solutions. Phosphorites in such cases are transformed into fluorapatite-rich ores. The correlational dependence between F and P in phosphorites makes it possible to evaluate tentatively the content of phosphorus from that of fluorine. The specificity of fluorine distribution in phosphorites and the enclosing rocks in a particular deposit is shown in Fig. (3.9). The fluorine content in phosphoritic ores and enclosing rocks amounts to 2.8 per cent while in the enclosing rocks it is 0.008–0.04 per cent. Fluorine concentration reaches 0.4 per cent only in quartzites and dolomites occurring near phosphorite bodies. The neutron-activation method is a rapid, remote-sensing technique for determining fluorine in the field.

Taking into account the spatial correlation of phosphorite-bearing formations with the horizons of vanadium-rich rocks extending over greater areas than the ores, geochemical prospecting involves also the exploration of vanadium haloes. But it should be borne in mind that vanadium, due to its antagonism with phosphorus in phosphorites, may serve as an indirect indicator of the quality of ores: high-grade ore bodies contain smaller amounts of vanadium. On the contrary, the quantity of rare earth elements is usually dependent on the phosphorite quality and the thickness of ore bodies. Apart from the above-mentioned elements, phosphorites are indicated by Sr, Be, Zr, Y, Yb, Ag and As anomalies. Phosphorus in rocks is determined by qualitative reactions based on the formation of yellow-brown ammonium phosphomolybdate residue resulting from the interaction of phosphate with a solution of ammonium molybdate in nitric acid. Phosphorites

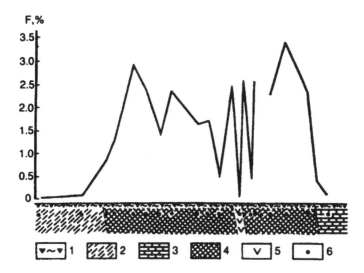

Figure 3.9. Cross-section of a phosphorite deposit with a graph of fluorine content in rocks [60]. 1 — soil-vegetative layer; 2 — quartzites; 3 — carbonate rocks; 4 — phosphorite-bearing band; 5 — porphyrite (dyke); 6 — sampling points.

differ from the enclosing sedimentary rocks by the intensity of their thermoluminescence, which is much higher in the former, primarily due to the presence of apatite. The ores are evaluated for the content of phosphorus from that of fluorine by the neutron activation analysis; due to the close correlation between these elements, the content of aluminum is assessed from that of potassium, and iron is determined by neutron γ-ray logging.

Evaluation of phosphorite deposits by nuclear-physical methods has certain advantages because of the facies changeability of phosphorites and because of difficulties in determining their thickness and quality by chemical analysis. Phosphorite deposits are characterized by mineral and geochemical zonality, stipulated by the differential spatial distribution of minerals and elements. Mineral zonality in some of the deposits is explained by the fact that phosphorite horizons are under- and overlain with manganese concretions. When these concretions occur above phosphorite deposits, they are notable for increased concentrations of P, Hg, U and rare earth elements. The formation of phosphates in sea basins is intensified and slowed down in a relatively gradual way [123]. The phosphatization haloes in the formations overlying ores are traced down to 80 m. The search for concealed horizons of phosphorites is based on the phosphatization of rocks enclosing phosphorite beds, as well as on the vertical geochemical zonality, i.e. a regular distribution of elements relative to the upper and lower horizons of phosphorite deposits. Fluorine anomalies spread in the upper parts of the deposits and on the flanks of the ore zones, and fluorine haloes indicate slightly eroded deposits characterized by non-contrast phosphorus haloes, which makes their identification extremely difficult.

A change in the phase correlation of mercury is also a criterion used in the search for deep-seated beds. The following temperature forms of mercury have been detected in phosphorites: 210–220 °C (first medium-temperature); 280–300 °C (second medium-temperature); 300–400 °C (third medium-temperature); and 518–600 °C (high-temperature). The supra-ore zones were found to contain a great number of mercury forms subjected to sublimation in a broader temperature interval than in the sub-ore zones.

Cu, Ti, Cr and Ni that are accumulated simultaneously with the phosphatic substance, diminish in quantity in the surface zones. Having no definite place in the vertical section of the ore zones, Pb, and Zn are concentrated near the contacts of the phosphorite beds [123]. High concentrations of Ba, Sr, La, U and Ag are characteristic of the upper horizons of phosphorite deposits. Anomalous concentrations of vanadium have been found in the horizons of coaly-siliceous shales that usually underlie the phosphorite horizons. The radioactivity of phosphorites decreases with depth. A single series of zonality is represented by the following upward alternation of elements: P–Sr–Be–Mo–V–Pb–As–Hg. The ratio of elements $(Mo \times P \times Be)/(As \times Hg \times Pb)$ is used as a zonality index whose values change with the distance to the ore bodies, equalling 0.08 and 18.65 at 25 cm and 10 m from an ore bed, respectively, and amounting to 295 within the ore body.

The above-prospecting indications are important for searching shallow ore bodies that are suitable for open-cast mining. For instance, P, Pb, Mo, Mn, Sr and Y anomalies have been detected above concealed beds of phosphorites.

Zonality exploration requires determination of series of accumulation of elements and discrimination of the most contrasting ones for every stratigraphic horizon. According to E. K. Burenkov, spatial distribution of multiplicative values of elemental content and zonality coefficients $(B \times Sr \times Co \times Cr)/(Ti \times Nb \times Cu \times Zn) = K$ change in the following limits: 48–7000 for ore-bearing horizons, and 0.1–13 for ore-free horizons.

There is a distinct correlational dependence between the phosphorus content in rocks and the multiplicative relation $(Co \times B \times Sr \times Cr)/(Zr \times Ti \times Nb \times Cu)$ which makes it possible to use these indices for identification of horizons in cross-sections prospective for phosphorus. Knowing the value of the coefficient K, it is possible to identify the most prospective phosphorite beds, to determine the directions of their outcropping and the increase in thickness, and to assess the quality of ores.

The peculiarities of P_2O_5 and SiO_2 distribution in phosphates are explored to detect high-grade ores in phosphate-bearing rocks. Long-normal distribution of P_2O_5 and SiO_2 contents in the beds containing appreciable quantities of phosphates is explained by the dominant occurrence of these elements in high-quality phosphorite beds. Phosphates may be accumulated in vast areas wherein local concentrations of rich ores alternate with impoverished ores. High and low-grade ores as well as phosphate-bearing rocks are found in the cross-section of stratigraphic horizons. The latter extend over considerably greater areas and exceed the thickness of ore bodies. Geochemical surveys should be aimed at detecting phosphorus-rich rocks, wherein the concentration of this element exceeds by seven and eight times its abundance ratios. Such levels serve as a prospecting indication for the discovery of commercial ores.

Methods of work

The applicability of a certain geochemical method when prospecting for phosphorite deposits depends on the landscape and geochemical conditions, and the degree of denudation of the territory under investigation. Geochemical prospecting along dispersion trains, characterized by greater depth of exploration, is effective in searching for concealed deposits. Vast areas can be surveyed more rapidly by combined bottom sediment and hydrochemical sampling. Reconnaissance and prospecting surveys along dispersion trains are distinguished, depending on the objectives and the detailedness of work. The former are carried out at a scale of 1 : 200 000–1 : 100 000 to obtain general characteristics of the areas being explored, and to identify the targets for a more detailed survey. The latter are conducted at a scale of 1 : 50 000, with the purpose of defining concrete zones and sites favourable for prospecting and reconnaissance surveys. Prospecting along dispersion trains in shallow sediments, being less laborious and time consuming, gives the same information as sampling water courses and suspended solids in water. Since dispersion trains are formed by the material transported to the river bed from a certain catchment area, their prospecting results in defining smaller target areas to be explored by more time-consuming methods. When searching along rivers, consideration should be given to the entire drainage net. However, sampling of large river valleys is less efficient, owing to the formation of slightly contrasting anomalies therein. In regions with weakly developed drainage, sampling of bottom sediments involved in the search at a scale of 1 : 50 000 is combined with metallometric survey. It is advisable to prospect along dispersion trains in mountain-taiga areas characterized by a simultaneous development of mechanical and salt dispersion trains. Salt haloes are mainly formed in flatter territories where appreciably smaller dissection of topography limits the amount of substances to be transported into the streams. The sampling pattern depends on the length of drainage systems and on the scale of work (Table (2.5)). Sampling points in rivers should be spaced every 100–200 m. Dispersion trains with a minimal length of 1 km are identified by several samples. It is preferable to sample bottom sediments in drying-out rivers and ravines that have small catchment areas. It is also recommended that sampling points should be uniformly spaced over the area under study, to sample bottom sediments from various drainage systems. Samples of predominantly fine material should be taken below the water level from the submerged shoals and flood plains. In some cases, instead of inorganic material it is more effective to sample fragments of organic material humified to a different degree. Root systems of mosses are sampled in tundra. Phosphorus and uranium are determined in samples subjected to ashing. Samples should be taken in a part of the river valley more distant from the slope. The reliability of identification of phosphorites is increased by sampling iron–manganese concretions of limonites, encountered in alluvial deposits wherein various admixtures (P, V, As) are concentrated. It is advisable to apply hydrochemical methods of prospecting in mountain regions where channels of drainage systems are devoid of silt–clay sediments. However, the use of this method depends on the nature of the objects being sought and their manifestation in a concrete setting. Hydrochemical surveys should involve sampling of stream, lake, and groundwaters to detect the following indicator elements: P, Sr, Ru, Zn, Pb, U, Cu, F, Cl, SO_4, Y and Br.

The sampling grid in hydrochemical prospecting depends on the scale of work, presence of drainage system, and places of subsurface water outflow. The determination of uranium and phosphorus is an economically express method for locating the target areas prospective for phosphorites. Due to a great length of dispersion trains containing leached-out elements, sampling points in small-scale geochemical surveying may be spaced at considerable distance (0.5–1 km and more). Prospecting at a scale of 1:50 000 and detection of anomalies require sampling intervals of 100–200 m. In the search for phosphorites by uranium, samples can be collected and stored in winter: freezing of water samples in plastic bottles does not affect the preservation of the dissolved radon. If samples are collected in summer, changes in the concentration of elements in water with time should be taken into account. When interpreting the chemical composition of waters, it is necessary to consider the mineralogical–geochemical properties of the ore-bearing rocks. Uranium, strontium and fluorine can go into subsurface waters not only from phosphorites but also from different rocks through which these waters migrate. The computerized analysis of the factor associations of elements makes it possible to have a more unambiguous impression on the hydrochemical conditions of phosphorite formation, and to grade the anomalies. The crusts of weathering where waters are commonly enriched in calcium but impoverished in strontium may effectively be explored by the hydrochemical method. Its integration with the sampling of bottom sediments may result in the identification of local areas to be prospected in more detail by other techniques. The lithogeochemical surveys along secondary dispersion haloes are mainly carried out in promising areas at the scales of 1:50 000 and 1:10 000. The recommended density of sampling is shown in Table (2.5). The profile spacing depends on the geological set-up of the territories, persistence of structures, and thickness of a productive phosphorite-bearing horizon. When prospecting at a scale of 1:50 000, a uniform standard grid can be used in plain areas overlain by thick deposits transported from distant terrains. This work results in the identification of the anomalies of phosphorus and accessory elements, wherein it is necessary to make the sampling grid denser. The promising sites are subjected to drilling and sampling of primary formations. The indicator elements in samples of P, Sr, F, Cl, V, As, Sb and Ba are determined by the spectrographic analyses, and F, As and Sb are determined by spectral methods. Prior to determining the content of radon in soils, samples are submerged in water for a certain period of time.

Mineralogical criteria can be used to solve prospecting problems. Studies are undertaken of phosphate associations (apatite, fluorapatite and carbonate–apatite) with minerals containing manganese, vanadium, iron, rare earths, uranium (glauconite and siderite — indicators of sea basins of normal salinity), as well as calcite, dolomite, quartz, fluorite (especially in carbonate formations). An increased content of organic substance and iron (pyrite) in rocks, giving them a black colour that is well observable in the field, is a favourable prospecting factor. The paragenesis of phosphorites with sedimentary manganese formations extending over vast areas is also very important.

The heavy concentrate geochemical sampling may be used in the search for phosphorites. Heavy concentrate samples assist in determining minerals of the apatite class characterized by a rather high density (3–3.2 g/cm^3) appreciable concentrations (up to 10 per cent), but average stability to weathering agents. The grains of apatites and phosphates that are concentrated in the heavy fraction or grey heavy concentrate, are determined by their reaction with ammonium molybdate in nitric acid. In the second case, mineralogical examination of heavy concentrates is combined with the identification of element characteristic of phosphorites. The method helps to quantitatively assess the contents of phosphorus, uranium, vanadium, and other elements and thereby to obtain integrated information on the concentration of many elements. Washing of a heavy concentrate results in the concentration of elements occurring in minerals thereby increasing the useful signals and contrast of the anomalies. To save money, it is recommended to subject every second sample to mineralogical analysis and to determine elements either in a non-separated heavy concentrate or in different fractions (non-magnetic, electromagnetic). Attention should be given to the morphology of apatite crystals whose isometric individuals occur in the stratified and concretion varieties of phosphorites, whereas their karst varieties contain crystals of various shape and size. If phosphorus anomalies are detected in the samples, but the heavy concentrates contain no apatite, it is necessary to study the following accessory minerals: glauconite, corundum and fluorite.

The species composition of minerals and their paragenesis associations aids a more reliable location of the primary sources. Glauconite is transported over the largest distances from the primary outcrops. An increase in the fluorite, apatite, and calcite contents in heavy concentrates indicates approach to ore bodies. Phosphorite fragments in river beds should be studies. In the majority of cases the degree of their roundness is not great and they are transported over a distance of no more than 2–3 km from the deposits. Samples of sand or aleurolite fines are collected with a subsequent tentative binocular count of characteristic fragments. Sometimes their number is accurately calculated in special samples.

These methods are most effective if phosphorites differ from the fragments of other rocks by their appearance and colour, and can reliably be detected visually. However, in many cases, phosphorites almost do not differ from ordinary sedimentary rocks, and it is difficult to identify them in the field. The composition of manganese concretions is analysed because the discovery of their varieties enriched in phosphorus and mercury makes it possible to form an opinion on the occurrence of phosphate mineralization at depth. Since phosphate minerals are associated with the less dense ones, it is advisable to wash off the heavy fraction from loose formations containing, as a result of rock disintegration, isolated and concentrated quantities of these minerals. It should be emphasized that the absence of phosphorites in river beds does not always indicate the absence of deep-seated ore bodies, which makes the combined application of the heavy concentrate and geochemical methods expedient.

Unequal natural conditions, generally observed over any more or less extensive area shown on a few sheets of a 1 : 50 000 map, necessitate the application of several methods which are efficient in the altered conditions. For instance, areas covered by the products of weathering of local rocks belonging to carbonate or sand–clay formation are outlined. Prospecting along dispersion haloes is most effective if phosphorites are localized in carbonate rocks that are easily leached out and differ from the ore bodies by their composition. Eluvial–deluvial formations in the zones of phosphorite mineralization development are enriched in phosphorus, thereby forming anomalies.

When phosphorite ores are localized in sandstones and sand–clayey rocks, the content of hardly-soluble mineral phases in the formations subjected to hypergenesis increases, whereas soil enrichment in phosphorus decreases. This results in the impoverishment and dispersion of the ore substance in the zone of hypergenesis, and leads to the formation of weakened and slightly contrasting anomalies. To enhance the contrast range of an anomaly and to increase the reliability of interpretation, it is necessary to widen the range of elements to be determined in samples, and to construct multiplicative haloes $(Sr \times Ba \times As)/(B \times Ni \times Co)$. Interpretation of the phosphorus anomalies in the samples associated is impeded not only with phosphorites but also with the enclosing rich-in-phosphorus formations. Element-correctors are used for grading the haloes. Anomalies associated with phosphorites are, among other things, characterized by the spatial coincidence of such elements as P, F, Sr, V, As, Sb, U and Ba.

The geologo-geomorphological specific features of particular regions, their altitude, as well as the thickness and composition of the bedrock overlying loose formations should be taken into account in geochemical prospecting. The low content of phosphorus in the samples taken in the areas with thick (10 m) loose deposits still does not indicate the absence of phosphorites ores. Shallow sampling is capable of detecting a deposit mainly in the upper parts of slopes. The phosphorite-bearing formations, occurring in mountain-taiga areas on the slopes and summits (550–1000 m above sea level) with a loose, up to 5 m thick sedimentary cover, are indicated by haloes with a maximum phosphorus content of 0.7 per cent. If the concentration of this element in the samples ranges from 2.5 to 3.0 per cent, phosphorite ores are unambiguously detected [75].

When phosphorites are overlain by loose sediments whose thickness exceeds 10–12 m, it is advisable to carry out an emanation survey and deep prospecting along secondary dispersion haloes together with hole drilling and collecting samples from the representative horizon or lower part of eluvial–deluvial deposits. Interpretation of the results of geochemical prospecting is greatly impeded when part of the phosphorite body is overlain by more than 70 m of deluvial formations and a stratum of products of weathering poor in phosphate. In a subdued coniform hilly relief, the phosphorite horizons are screened by cover loams. Therefore sampling of the surface layer results in the overlooking of productive bodies. Under such conditions it is necessary to resort to drilling and subsequent sampling of loose deluvial formations and products of weathering poor in phosphate.

Lithogeochemical surveys along secondary haloes allows us to detect a great number of geochemical anomalies whose evaluation is based on the analysis of their geological position and productivity. Airborne γ-ray surveying, undertaken to record γ-radiation at a fixed energy window of 1.7–1.8 MeV, is effective at the stage of reconnaissance and prospecting surveys at scales of 1 : 200 000 and 1 : 50 000. The results obtained may be influenced by the scattering of γ-radiation from thorium decay products in the soil and air. At a high radiation background, use is made of multichannel analysers of the natural radiation spectra with cryogenic polycrystal detectors, as well as airborne γ-ray spectrographic measurements of the individual contents of uranium, thorium and potassium [67], which makes a very precise and reliable determination of the contents of the first two elements and radium in the rocks possible.

The results of airborne γ-ray spectrometry are used to outline the promising areas to be investigated by the following surface techniques: (1) radiometric survey; (2) emanation, radon and gas surveys; (3) survey with the aid of a submerged pulse counter.

The surface radiometric survey is combined with geochemical prospecting. The anomalies registered by radiometric survey may be associated with deep-seated phosphorites enriched in radioactive elements (U + Th + K). The anomalies are graded by comparing the values measured by the counter in the field with those of its standard capsule, having background radioactivity.

In surface surveys, uranium is determined by X-ray fluorescence analysers (in the case of disturbed equilibrium between radium and uranium), and by an alphometer, recording γ-radiation in the field. The radioactivity of phosphate formations (in 10 g samples) may be determined with the aid of scintillation γ-ray spectroscopy (detector: NaI crystals activated by Tl).

Emanation survey is effective in tracing out phosphorites that are localized in carbonate rocks overlain by loose, appreciably thick (> 5 m) formations.

In areas with abundant lakes, phosphorite mineralization enriched in radioactive elements is sought by examining the fission tracks of uranium and its decay products. Radon concentrations in lakes are measured in winter by lowering special cups, containing lead, to the lake bottom. The level of radon concentration is determined several weeks later, when the cups are lifted to the surface. Lake ice prevents radon movement from the deep-seated lake waters.

Nodular and medium-grained phosphorites are prospected abroad by a submerged pulse counter, which records the increased uranium content in ore bodies. Uranium is determined by a special plastic film which is etched by radon, thereby providing information on its content. For this purpose special cups with films are placed in the holes of bouldery sandy clay. This technique is applicable in the search for the ore bodies overlain by 20–160 m thick deluvial formations.

Phosphorite prospecting in boggy terrains is advisable in winter. Sampling cups are placed in January and taken out early in the snowmelt period. Tracking plastic detectors are placed in 0.75 m deep holes drilled on a rectangular grid with a step from 30 to 1000 m [9].

The data of radiometric surveys are used to delineate the prospective areas to be subjected to detailed prospecting. It should be borne in mind that radiometric prospecting for phosphorites is based on the summary γ-radiation, used to determine the total radioactivity that can be produced not only by phosphorites but also by the enclosing rocks, carbonate shales, accessory elements, and potassium minerals. Sometimes slightly radioactive phosphorites are encountered. Therefore, radiometric prospecting should have as its purpose the establishment of the number of phosphorite varieties differing in the content of uranium.

Individual determination of U, Th, Ra and ^{40}K in rocks increases both the efficiency of prospecting and the reliable identification of prospective zones. By using the uranium–thorium ratios that are characteristic of the main varieties of phosphatic rocks, it is possible to evaluate the quality of primary deposits and thereby to cut expenditure on the time-consuming procedures of sample collection, transportation, processing and analysis of a certain proportion of samples.

γ-ray well logging, carried out at the stage of prospecting and reconnaissance surveys, makes it possible to conduct correlation of the cross-sections and to specify the thickness and quality of ores. Logging is performed by the point-to-point method with a spacing of 1 m. Upon tapping phosphorite ores, the distances between the points of measurement are shortened to 0.1 m. This method is effective at poor core sampling. If it is difficult to make γ-ray logging, core samples are examined by radiometers. Logging investigations are accomplished by scintillation spectrometers permitting the determination of U, Th and ^{40}K in rocks at the non-core drilling.

The γ-ray testing of workings is also employed for the lithological subdivision and correlation of the cross-sections, and for the identification of productive horizons coupled with the quantitative evaluation of the P_2O_5 content and calculation of reserves.

Geochemical sampling of bedrocks along profiles oriented transverse to the strike of the known or anticipated structures is advisable at the stage of prospecting surveys in the prospective areas. Sampling should be done on a grid 250×25 m (scale 1:25 000) and 100×10 m (scale 1:10 000). At a change in the rocks, every 200 g sample should include five or six small chip samples of only one rock variety.

Samples should primarily be collected on sites of bedrocks represented by nodular, glauconite, siliceous, and siliceous-carbonate varieties. Prospecting surveys are aimed at tracing out the strike of productive rocks and their sampling for phosphorus.

Geochemical sampling of rocks in workings and drill holes at the stage of prospecting surveys is carried out with the purpose of outlining phosphorite beds and identifying deep-seated bodies. The geochemical methods are used to correct the directions of these surveys. The morphology of anomalies and regular changes in their composition are studied.

Special attention should be given to those indicator elements that form most extensive and stable anomalies, such as fluorine, which is contained in fluorine-bearing phosphates. It should be borne in mind that phosphorites are rather diverse in their appearance, structure and texture, and, therefore, are often hardly distinguishable from the enclosing rocks. This necessitates qualitative chemical analysis of rocks for P_2O_5 directly in the field. The method of express hydrostatic weighing may be used to accomplish in the field quantitative determinations of phosphorus pentoxide content.

In the search for phosphorites, it is advisable to combine geochemical prospecting with the analysis of particle size distribution in rocks. Under standard conditions, this involves sieving and analysis of fractions from 1 to 0.001 mm (eight intervals). Phosphorites and phosphatic rocks are characterized by an unstable, drastically changing mineral composition. Thermography may assist in a tentative determination of the mineralogical composition of phosphorites and their water content.

The thermoluminescent method can be used for express evaluation of carbonate contents in ores. The intensity of peaks at 100, 200 and 310 °C, that are responsible for calcite luminescence in phosphorites, is sufficiently consistent and distinct. In pounded rock samples, phosphates occur in the aggregates in association with light minerals. These aggregates are frequently washed away, which precludes phosphate concentration in mineralogical samples.

Exposure of polished samples to γ-radiation from a ^{60}Co source is an effective method of investigating fine grained phosphorites, whose exploration by petrographic techniques is difficult. The 1–2-cm-thick plates are irradiated in darkness placed on a photoplate and kept for a week. The zones of apatite, calcite and quartz luminescence are clearly seen on a contact print. The photoluminograms obtained are convenient for making mineralogical counts.

Nuclear geophysics capable of the remote determination of the contents of different elements in rocks significantly increases the reliability and urgency of phosphorite prospecting. The nuclear-physical methods help in detecting and differentiating phosphorite bands, in determining the boundaries and depths of ore bodies, and in quantifying not only the P_2O_5, but also Al_2O_3 and SiO_2 contents. The resultant data can be used to determine the parameters of ore bodies, and to calculate the reserves of phosphorite ores.

The combined analysis of the data of radiometry, magnetometry, electrical prospecting, and gravimetry provides most complete and reliable information on the localization of phosphorite mineralization. γ-ray survey does not always distinctly indicate the occurrence of bedded phosphorites along their strike, which, as a rule, are confined to the contact of dolomites with sandstones. Therefore, in selecting a rational direction of prospecting and reconnaissance work, the prospective sites are subjected to an additional geophysical exploration by magnetic and electrical prospecting methods.

Since phosphorites are generally characterized by a horizontal or subhorizontal shape of their ore bodies, the elemental haloes are shallow and protrude along the phosphate horizons. The endogenic anomalies of comparatively deep-seated ore bodies may be overlooked. This should be taken into account when interpreting the results of geochemical surveys carried out on sites that are potentially favourable for locating phosphatic ores. Elements that are in close correlation with the content of P_2O_5 are determined to correct prospecting surveys. Apart from F, these include As, Sb, Ba, Sr and thermal forms of mercury sublimation.

Deep-seated phosphorite deposits may be suspected, based on the discovery of geochemical anomalies spatially coinciding with an increased radioactivity of the bedrock eluvium. In areas where phosphorites have already been detected, the established patterns governing localization of radiometric anomalies in space may assist in tracing out the strike of phosphorite-bearing formations and in discovering the displaced parts of their beds.

Manganese distribution in sedimentary rocks is also studied, bearing in mind that the manganese ore horizons are, to a certain extent, stratigraphically isolated from phosphorite deposits.

The background, anomalous, differentiated and non-differentiated, and supra- and sub-ore elements are determined for zonality investigation. Favourable anomalies are indicated by their intensity and multicomponent composition. Particular multiplicative haloes: $(Sr \times Ba)/(Ni \times V)$; $(Co \times B \times Cr \times Sr)/(Zr \times Ti \times Nb \times Cu)$ are used to evaluate the level of erosional truncation of ore deposits.

Bodies of granular phosphorites are acquiring considerable significance. The keen interest in the prediction of and prospecting for new economic deposits of granular phosphorite can be explained by the fact that this type of phosphate mineral usually forms large deposits of easy-to-treat ores. Due to their special genetic features these phosphates always contain various and higher-than-background concentrations of U, Sr, Ba, rare earth and other admixtures, either in the phosphatic components themselves or in the matrix. Phosphatic (fluorine–carbonate apatitic) material is especially active in sorption with respect to U, Sr and rare earths.

The interpretation of the existing geological and geochemical material obtained for phosphorite-bearing sediments in conjunction with facial and paleogeographic analysis considerably enhances the efficiency of the assessment of the paleogeochemical environment of phosphate deposition in sedimentary basins. It also makes it possible to outline the most favourable area for the formation of phosphorites, and also litho-stratigraphic horizons with a concentrated economic distribution of granular phosphorite bodies. Eluvial phosphate ore in Brazil is localized in mafic rocks in the central part and felsic rocks in the border zone [65]. The rocks are covered by talus and weathered material, on average 30 m thick, and very inhomogeneous in composition. Phosphate mineralization is mainly associated with two levels of the weathering mantle and with fresh rocks. Phosphate is associated with apatite, feldspar, vermiculite, biotite–phlogopite, calcite, pyroxene, amphibole and analcime. It should be noted that in the analysis the complex mineralogical assemblage, as well as some interference between diffraction patterns and variability of composition in some groups of minerals, like feldspars, amphiboles, and pyroxenes, led to the selection of diffraction peaks of low to medium intensity for some minerals instead of major peaks, with some loss of resolution [65].

Planned exploration applies new ideas about the origin and emplacement of pellet phosphorite, especially for areas outside known fields. The recognition of the importance of the movement of phosphate-containing seawater in large basins has been of special value.

The exploration and development of pellet phosphorites in new areas might be separated into four phases as follows [85]:

(1) The search for areas of marine sedimentation.

(2) Within areas of marine sedimentation, the search for beds of chert, or marine diatomite and black shales.

(3) The search for beds of phosphorite in or near the black shales or cherts. Also the search for areas of residual phosphatic soil associated with beds of phosphatic dolomite or limestone.

(4) The determination of the thickness, grade and extent of weathering by trenching and drilling.

The thickness of chert beds, so commonly associated with pellet phosphorites, may be anything from a few inches to over 75 m. Pellet phosphorite commonly contains between 0.001 and 0.002 per cent of U_3O_2. Where the higher percentage is present in exposed beds, airborne scintillation equipment can also be used.

An outcrop of phosphorite will commonly have patches of a thin skin of 'phosphate bloom' where exposed at the surface. Its characteristic light grey-blue colour is readily visible up to distances of several feet. A 10-to-20-power hand lens is indispensable to aid in identifying the small ovoid-shaped phosphate pellets. It has been possible to identify phosphate beds on γ-ray logs on holes. The beds so identified were then found in outcrops some distance away.

Phosphorite deposits are difficult to recognize in the field (the small phosphorite pellet cannot be positively identified). However, a simple field test is possible by placing a drop of acid (sulphuric, nitric, or hydrochloric, not weaker than approximately 20 per cent concentrated acid in water by volume) on the area in question with a medicine dropper and then adding a small crystal or a small amount of powder of ammonium molybdate on the end of a knife blade. The rapid development of a bright yellow colour indicates phosphate. This field test is very sensitive but positive results do not mean a commercial deposit has been found.

Phosphorite is not a very distinctive rock either in a hand specimen or in an outcrop, and can therefore be easily overlooked or incorrectly identified during prospecting operations, particularly in those areas which are relatively unknown geologically or have not previously been considered likely to be phosphate-bearing. In the hand specimen it exhibits a wide variety of physical properties, but within any particular phosphate field its characteristics are generally fairly constant and once identified as phosphatic may usually be recognized without much difficulty.

Geochemical and geophysical prospecting can overcome the problems posed in some regions by the presence of deep weathering, thick soil or glacial drift cover, dense forest and difficult or relatively inaccessible terrain.

Positive identification of samples becomes still more difficult with the lower grades of phosphate rock, particularly when these pass imperceptibly into siltstone or shale with decreasing P_2O_5 content. In such cases the presence of phosphate can generally only be determined by a geochemical field test such as the well-known method using a solution of ammonium molybdate in nitric acid [16, 85]. An easily visible yellow precipitate is formed and, although regarded by some geologists as somewhat oversensitive and therefore unreliable in distinguishing between the higher grades of phosphorite, the method is useful in eliminating samples containing small or trace amounts of phosphate.

A new method, which is being widely used in phosphate exploration work, involves the reaction of a vanadomolybdate reagent with phosphatic material to give a precipitate of phosphovanadomolybdate, the yellow colour of which is then compared with that of reference solutions. This test has been found capable of giving results to the nearest 5 per cent of P_2O_5. Apatite may be tested by shaking a finely ground sample with a nitric acid solution of the reagent; the mineral can thus be readily distinguished from hydrated aluminium phosphate minerals such as wavellite and crandallite, which do not dissolve readily and require boiling with concentrated sulphuric acid and nitric acid before the test can be carried out. Another simple but much less satisfactory field test involves the addition of a few drops of concentrated hydrochloric acid on to a fresh surface of the sample to give a white or grey stain, the density of which is generally roughly proportional to its phosphate content [85]. Best results are apparently obtained on fine-grained compact phosphorites, but even then a stain may not be formed.

The most useful of the indirect methods depends upon the uranium content of phosphate rock, which in the case of many marine phosphorites ranges from 0.005 to 0.02 per cent. Phosphorite, together with certain black shales, is thus significantly radioactive compared with most other sedimentary rocks.

γ-ray logs made in the course of oil-well test drilling provide a most convenient and rapid method of detecting beds of phosphorite within thick sedimentary sequences, particularly in geologically unknown areas.

In the fluorimetric method of uranium analysis a trace of uranium fused in a sodium fluoride bead gives an intense yellow-green fluorescence, the magnitude of which is proportional to the amount of uranium present in apatite. A laser-induced fluorescence method for uranium in solution is reported to have a sensitivity of 0.05 ppb of uranium in natural waters.

Aerial radiometric reconnaissance over large areas, in certain cases by helicopter, can also assist favourable circumstances, as has been done in Tanzania, Angola and Algeria. The investigations are usually carried out at altitudes ranging from 560 to 870 m, depending on local conditions, and appear to be most applicable to a relatively flat terrain where phosphatic rocks outcrop or occur very close to the surface.

Experiment shows that closely-spaced point counting techniques (scintillation) can be used satisfactorily in areas such as that investigated, where the incidence of rocky outcrops militates against the use of a towed instrument due to the likelihood of snagging and loss of equipment. From the bulk of radiation determinations at randomly selected sites, it seems that there is sufficient radiation from rocks and sediments containing more than 2.0 per cent P_2O_5 (normal ore grade is about 30 per cent P_2O_5) to regard this as a threshold in the detection of seafloor phosphorites. It has been established that seafloor scintillation is the best method used so far to prospect rapidly for submarine phosphate deposits [112].

Geochemical exploration, involving the detection of near-surface soil gas radon anomalies has been used in phosphate prospecting. Radon anomalies can be detected at the surface over hidden ore bodies when other surface radiometric techniques are ineffective (using an alpha-track detector and radon emanometer in certain geological areas). The results reveal that the techniques of radon measurement may prove useful for delineating deeply buried ore.

Where there are no outcrops, prospecting becomes more difficult. The best method is first to carry out several geological reconnaissance traverses, starting at the margin of the sedimentary basin.

Apart from γ-ray measuring techniques, there has been very little application of geophysical methods (seismic work, γ-ray measurements) in the exploration of pellet phosphorites to date.

The brown-rock pellet phosphorites of Tennessee are residual deposits formed by the weathering of the underlying phosphatic limestone. This phosphorite is covered by several feet of brown-coloured soil. The soil mantle, coupled with the low uranium content of the phosphorite (less than $1 \cdot 10^{-3}$ per cent U_3O_8), makes airborne scintillation prospecting useless. Road cuttings and the occasional gully will reveal the underlying phosphorite. Even in such places the inexperienced eye probably would not distinguish between the soil and the phosphorite. Prospecting for isolated phosphate deposits is carried out simply by drilling blind at the appropriate horizon, i.e. on the phosphatic limestone.

Prospecting in a little-known area

The less geological knowledge there is available, the more risky is prospecting. Geological documentation should be compiled in any case, in order to eliminate the zones with the least prospects with regard to the presence of phosphate.

In the remaining regions, and when there are many outcrops, the radioactive properties of phosphates can be used as a means of prospecting. Airborne scintillometer surveys are the quickest method, but only give results when the phosphates are clearly exposed. Measurements of radioactivity may be taken on the ground or in rapidly drilled, uncored boreholes. All outcrops should be tested with molybdic reagent and the samples which react positively should be subsequently tested quantitatively by the normal methods.

Phosphorite beds vary laterally in thickness and phosphate content. In some places, as in the phosphoritic formation of the western USA, this lateral variation is small for long distances — several dozens of miles — but in other places and other formations, more extreme variations do occur. In exploration, therefore, extensive lateral exploration should be made even though preliminary exploration may detect only low grade ore. The lateral variation in the thickness and grade of the beds necessitates extensive drilling to ascertain the sometimes tortuous outlines of the minable material.

Even when the exact location is known, pellets or small pieces of phosphorite are difficult to find and recognize. Therefore, it is often better to use harder 'marker' beds, stratigraphically above or below the phosphorite, to map the location of a phosphorite bed than the phosphate bed itself.

Geochemical methods are used to solve various problems in the search for phosphate–iron–titanium ores in the Precambrian autonomous anorthosite associations. Local areas for detailed search are outlined on the basis of diffused fluxes of titanium and phosphorus. Contents of these elements of more than 0.1 per cent in bottom samples are considered to be anomalous; directly on an outcrop of ore bodies they reach 1–5 per cent. Diffused fluxes extend several kilometres.

Primary haloes of Ti, Mn, V, Co, Cu, Zn, P, Ce and La are used to search for ore bodies within the outlined areas. Titanium dispersion haloes are the widest (up to 500 m). Anomalous concentrations of phosphorus are detected only at a distance of 10–50 metres from the ores.

The chemical analysis of borehole or well cutting offers a particularly promising technique in the exploration for phosphate which could usefully be applied in regions where the search for oil or water is being carried out in sediments likely to be phosphate-bearing.

The highest P-levels coincide with those of the highest Fe−Mn−U triad and with the appearance of a facies contaminated by hydrogen sulphide. The existence of complex geochemical anomalies in overlapping series is a direct exploration indication of overlapped phosphorites.

The following geochemical criteria are recommended for the determination of overlapped phosphorite horizons [16]:

1. $Mn \times Ti \times 10^{-4} > 1 \times 10^{-4}$ — to reveal the areas where the depth of the productive horizon is 250–300 m;

2. $(Mn \times Ti)/(Sr \times Cu) \times 10^2 > 0.1 \times 10^2$ — to reveal the areas where the depth of the productive horizon is 100–150 m;

3. $(Mn)/(Ti) > 10$ — to reveal the areas where the depth of the productive horizon is 50–70 m. Geochemical methods can be used successfully and effectively in the search for phosphorite deposits.

The patterns of the distribution of concentrations of chemical elements in sedimentary deposits in humid areas have been taken as the basis of the developed methods, and the general sequence of the geochemical mobility of elements has been established [16]: Mo−Ba−Cu−Pb−Cr−Mn−Zn−Co−Y−Ni−Sr.

Strict regularities in the distribution of concentrations of microelements have been revealed in sedimentary deposits enclosing productive phosphate-bearing series and characterized by the development of sedimentary series over large areas. Structural−facial zones favourable for the formation and localization of phosphorites differ sharply, both in the composition and concentration of microelements, from the adjacent zones.

These differences in the composition of microelements can be used in forecasting potentially promising areas of phosphorites of sedimentary origin.

In known phosphate-bearing areas the phosphorites and productive phosphate-bearing series are characterized by increased concentrations of indicator elements, forming haloes that are significantly larger than the ore bodies, both in strike and in thickness. These haloes, which are very important in the exploration of overlapped phosphorite deposits, occur in deposits overlapping productive series and display a zonal structure. Manganese and titanium have the narrowest haloes, up to 50–70 m, and copper and strontium — the widest, up to 250–300 m. In this connection it seems possible both to correct the direction of exploration and to forecast new areas of overlapped phosphorite deposits.

The use of existing holes drilled for oil in exploration for pellet phosphorite has already been mentioned.

In wet places, tests can be made by washing cuttings from the hole. The cuttings are caught in tubs on the surface. This type of drilling commonly results in the loss of slime and, consequently, the under-reporting of the quantity of slime in the final laboratory tests.

The exploration and development of pellet phosphorites involves the determination of the thickness, grade, and extent of weathering by surface work which may include one or more of the following: trenching, shafts, or drilling. Trenching can be done by hand or with mechanical equipment — bulldozers or backholes. Its purpose is to expose the phosphatic beds for measurement and sampling.

The phosphate ore is usually separated every 5 ft into an individual sample. Each sample is processed in the laboratory using bench tests to simulate plant operations, i.e. a weighed sample is pulped with water and deslimed; slime is any material less than 150 ml sh. The sample is split and analysis made for P_2O_5. The other half is saved until a minable deposit has been drilled. Then one composite is made, proportioning the weights according to the feet of core in each sample. This composite sample is divided down to a working sample of about 6000 g. One half is analysed for P_2O_5, SiO_2, Al_2O_3, Fe_2O_3 and CaO. The analyses are carried out using the old-established wet chemical procedures or X-ray fluorescence. A titration process is used for P_2O_5, Fe_2O_3 and CaO, but the SiO_2 and Al_2O_3 analyses are determined gravimetrically. When exploratory oil-drilling has been carried out, γ-ray anomalies should be recorded. Marine beds rich in organic matter may also be of interest, since the arrival of phosphate-bearing upwelling waters generally stimulates the development of life; however, the association of phosphate with dark facies is certainly not a general rule.

Stages and sequence of work

The search for phosphorites at the first stage is based on the regional geological and geochemical prerequisites. The latter's significance is stipulated by high contents of phosphorus, fluorine and radioactive elements in phosphorites. Field surveys at this stage are carried out with the purpose of defining more accurately the geological structure of the region, and studying the lithology and geochemistry of sedimentary formations. Key sections along river valleys are explored. The microchemical methods are employed for determining phosphatic rocks in the field. The geochemical method is used in combination with sampling of bottom sediments. Areas prospective for phosphorites are contoured on prognostication maps.

Prospecting surveys at a scale of 1 : 50 000 are carried out at the second stage. They are based on the materials of geological surveys accomplished at the first stage, at a scale of 1 : 200 000. Airborne γ-ray surveys, heavy concentrate sampling and exploration of dispersion haloes and trains are employed in the search for phosphorites. Of special interest are heavy concentrates containing more than 10 per cent of phosphatic minerals. Phosphorus, fluorine and radioactive elements are determined in the lithogeochemical samples. The prospective

sedimentary rock horizons are thoroughly sampled and analyses for the distribution and correlation of radioactive elements. The U/Tr value is determined
in order to grade the anomalies. Concrete areas favourable for prospecting surveys are defined by analysing the results of work accomplished at the second
stage.

At the third stage, geochemical and geophysical exploration of boreholes (geochemical sampling, remote-sensing determination of elements in rocks) is carried
out.

3.3. BORON

There are two types of boron deposits: endogenic and exogenic. The former
are associated with skarns of calcareous-skarn and magnesian formations. In the
first case, datolite and danburite are localized in carbonate and carbonate–silicate
rocks in the zones of exo- and endocontact of granitoids. Ludwigite, coccoite
and ascharite ores, formed at the contact of carbonate (dolomite) and alumosilicate rocks in the zones of granite or regional granitization impact, are associated
with magnesian skarns and calciphyres. The said types of boron raw-material
deposits differ in the conditions of formation. The following three stages of
mineralization are characteristic of the objects of the first formation: (a) skarn;
(b) borosilicate; and (c) quartz–carbonate. The borosilicate mineralization (datolite, danburite and axinite) is generally formed metasomatically from chloridic
and carbonic-acid, boron-bearing solutions in the process of the substitution of
minerals of calcareous–ferruginous skarns. The process of skarn formation was
accompanied by the growth of iron content in the mineral-forming solutions and
lowering of temperature. It terminated with a redox potential increase. Borate
mineralization is localized in the external zones of the magnesian-skarn metasomatic column, i.e. in forsterite skarns, marbles and calciphyres [69]. Borate
mineralization resulted from the process of metasomatic substitution that occurred
in magnesian skarns at a high boron activity in the mineral-forming solutions [69].
Halogen sedimentary deposits have been found in the form of borate beds in the
halogenic strata, and among boron-bearing salt formations in the lakes that drain
borate accumulations of volcanogenic formations. Boron is accumulated as precipitate in the course of formation of drainless lakes in arid and sometimes in
high-mountain areas.

The prospecting indications of boron deposits are:

(1) carbonate formations developing along the contacts with granitoids; skarns,
their desintegrated blocks piled up on mountain slopes, and taluses; bleached
marbled carbonate rocks; and calciphyres;

(2) contact haloes in dolomites, as well as xenoliths of dolomites in granites;

(3) marine lagoon deposits with the indications of arid climate; manifestation of
continental volcanism. The presence of potassium facies of sedimentation,
strongly affected by halogenesis, occurring in the pre-geosyncline areas and
in the marginal parts of platforms.

Indicator minerals and elements

Pyroxene, diopside–hedenbergite, andradite, grossular, wollastonite, vesuvianite, and scapolite are formed in the calcareous skarn deposits at the pre-ore stage.

Boron minerals, formed mainly at the stage that follows the skarn formation, are represented by borosilicates: datolite and danburite associated with epidote, clinozoisite, pectolite, siderite, manganocalcite, and ilvaite. Secondary minerals are actinolite, stilpnomelane, chlorite, sericite, albite, prehnite, xonotlite, hisingerite, and chalcedony. Sulphides (galenite, sphalerite, chalcopyrite and pyrite) are frequently encountered.

Quartz, calcite, apophyllite, and fluorite are formed at the third stage. Danburitic ores are generally associated with quartz, calcite, andradite grossular, tourmaline and axinite, whereas association with prehnite, wollastonite, fluorite, scapolite, apophyllite and chalcopyrite is typical of datolitic ores. The danburitic ores occur in the zones of development of skarns, relatively poor in iron, and containing granites and pyroxenes with an iron content of below 50 and 45 per cent, respectively [69].

The prospecting indications for boron are: wollastonite (associated mostly with datolite and andradite), hedenbergite (enriched in manganese and widespread in skarns and nearby formations), and axinite. Under conditions of moderate activity of manganese, the axinite composition and quantity depend on boron activity.

Phlogopite, clinopyroxene, forsterite, clinohumite, pargasite, amphibole, brucite, scapolite, tourmaline, axinite, periclase, magnetite, tremolite, pyrite and chalcopyrite have been detected in magnesian skarns together with magnesium and iron–magnesium borates. Ca, Mg and Mn borates and carbonate borates (including kurchatovite, rauvite, sakhaite, charcerite, borcarite and cahnite) are formed in the process of substitution of magnesian skarns and calcyphyres by calcareous skarns. High boron and iron potentials in mineral-forming solutions are indicated by the combinations of ludwigite or suanite with clinopyroxene, of periclase, brucite, or magnesian carbonate with magnesian ludwigite, of calcite with suanite, and of kurchatovite with kotoite.

The formation of B, F, K, Na, Cl, Cu, Ni, Co, V, Pb, Zn, Ba, Sr, Ag, As, Sb and Sn haloes around mineral deposits is conditioned by the occurrence of skarn, borosilicate, and quartz-carbonate stages (Figs (3.10) and (3.11)). The typical indicator elements of the post-skarn ore stage of development are: B, Cl, F, Hg, P and As. In some cases, the ore bodies are surrounded by highly contrasting, mobile and rather elongated W, Mo, Sn, Cu, Pb and Zn anomalies. It should be emphasized that large datolite and danburite deposits form insignificant Pb, Zn and Cu haloes. The characteristic indicators of boron deposits in magnesian skarns are Cu, Sn, Zn, Mo, W, Be and Ge. These in calcareous skarns are indicated by V, Pb and Li, whereas the alumoborosilicate mineralization is indicated by Sn, Pb, As, Mo, Ba and Y.

Boron-containing minerals in sedimentary-volcanogenic deposits are represented chiefly by their sodium, calcium, sodium–calcium and chlorine-bearing varieties. Hydromicas, montmorillonite, kaolinite, chlorite, halite, anhydrite, quartz, carnallite and carbonate are the accessories of borates in halogen formations. The presence of montmorillonite at a limited occurrence of hydromicas in halogen strata is

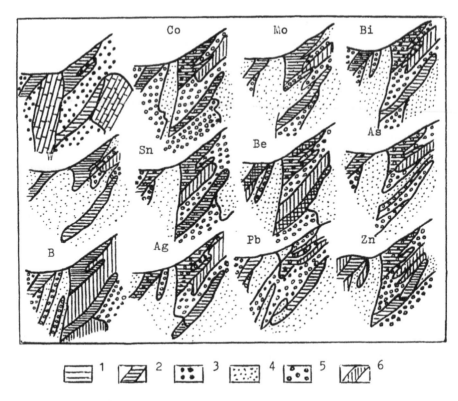

Figure 3.10. Content of elements in boron bodies and primary haloes [128]. 1 — limestones. 2–3 — ore body with content B ($2 < 3\%$, $3 > 3\%$); 4–6 — content elements in haloes (ppm) (4 — B < 10, Be < 1, As < 100, Ag < 0.05, Pb < 13, Mo < 5, Zn < 10, Co < 0.5, Bi < 1, W < 3, Sn < 1; 5 — B 10–1000, Be 1–4, As 100–1000, Ag 0.05–0.2, Pb 13–100, Zn 10–100, Co 0.5–10, Mo 0.5–1, Bi 1–100, W 3–10, Sn 1–10; 6 — B 1000–30 000, Be > 4, As > 1000, Ag > 0.1, Pb > 100, Zn > 100, Co > 10, Mo > 1, Bi > 100, W > 10, Sn > 10).

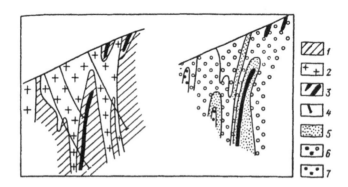

Figure 3.11. Primary haloes around boron bodies [128]. 1 — quartzites, 2 — diorite, 3 — boron veins, 4 — drills, 5–7 — content of elements, ppm (5 — W > 100, 6 — B from 5 to 10, 7 — B from 10 to 100).

a favourable indication for borate prospecting. This is explained by the ability of hydromicas to absorb boron from solution, thus preventing borite formation [87].

The volcanogenic-sedimentary deposits in lacustrine formations are character-ized by high concentrations of As, Sr, Cu and Zn (4 to 30, 6 to 18, 16 to 28, and 54 to 118 g/tonne, respectively) [87]. Boron accessories in clay formations are Sr and I. The data supplied demonstrate the diversity of indicator elements in boron deposits and their possible utilization in geochemical prospecting.

Fundamentals of geochemical prospecting

The geochemical methods of prospecting for boron deposits may be used at dif-ferent stages of prospecting and in reconnaissance surveys. The diverse and rather complicated secondary dispersion haloes of boron have been detected in differ-ent mountain-taiga, steppe, and coniferous broad-leaved forest landscapes. The dimensions of haloes appreciably exceed those of the ore bodies. The contents of boron in secondary dispersion haloes above the calcareous skarns with datolite and ludwigite ore bodies amount to 0.2–0.09 per cent, compared with a back-ground boron content of 0.04–0.06 per cent. The displacement of haloes from the ore zones reaches 100–200 m. Haloes with a relatively low boron content distinctly indicate entire zones of skarned rocks, whereas highly contrasting and local anomalies exceeding the background values by 1.5–3 times, are confined to the exposures of ore bodies. An increase in boron concentration was identified in the high-grade ore zones.

The high resolution capacity of the neutron method, used to detect and delineate boron haloes in soils, was proved by comparing the results of chemical analyses with those of remote-sensing profiling.

Boron is extremely mobile in water solutions. The dissolution of boron and other elements from calcareous and magnesian-skarn ore bodies results in the formation of their extended dispersion haloes. Aqueous haloes may indicate the presence of productive mineralization at a considerable distance from the sampling points in a water course. The subsurface water composition may indicate the occurrence of ore bodies at a depth from 10 to 100 m.

Boron migrates in the form of complex anions BO_3 and B_4O_7 and hardly dissociating orthoboric acid H_3BO_3. The intensity of boron migration in water depends on its acidity and alkalinity, physico-chemical properties, and presence of organic substances. High-intensity boron migration is observed in acid and alkaline waters [99]. Its decrease in a calcareous medium is caused by the formation of hardly-soluble compounds. Organic substances that are responsible for the formation of organo-mineral complexes of boron, mobile in water, significantly intensify its migration even in calcareous conditions. The content of boron in the surface waters of river valleys draining datolitic bodies in a broad-leaved forest landscape amounts to 3 mg/l, compared to a background value of 0.01 mg/l; the halo length is small, being restricted by that of the water course [102].

Mechanical dispersion flows play a significant role in the formation of aqueous haloes. Being transported over considerable distances in river valleys and pre-served in the mechanical products of weathering, hypogenic borates and borosil-icates may form aqueous haloes up to 10 km long. The boron content of waters

near ore bodies occurring in a southern-taiga landscape fluctuates from 0.02 to
5 mg/l (background 0.01 mg/l), whereas in river valley and swamp waters it
amounts to 0.02–0.04 mg/l [99].

Subsurface waters, responsible for the dissolution and hydrolysis of hypergene borates play an important role in the formation of sedimentary-volcanogenic
borate deposits. The boron content of subsurface waters in borite-containing
zones amounts to 10.0–520.0 mg/l compared with background concentrations
of 0.1–30 mg/l [102]. Dispersion trains of boron and its accessories are formed
due to sorption of elements and their compounds from the water solutions containing finely disintegrated minerals. The lithogeochemical sampling of bottom
sediments along dispersion trains is based on the ability of iron and aluminium
hydroxides, clay minerals (clay fraction), and organic substances to sorb boron.

Being a biophilic element, boron is taken up by plants. Its content in the ashes
of plants amounts to 67×10^{-3} per cent, thus exceeding the boron clarke in acid
igneous rocks by 45 times and in limestones and dolomites by 225 times [99]. An
increase in the content of boron in plants is observed with the transition from acid
non-calcareous to slightly alkaline and alkaline calcareous soils.

In principle, geochemical exploration for boron deposits may be carried out in
various regions. Highly contrasting biochemical anomalies of boron are formed
in areas containing diatolitic and boratic mineralization, whereas zones affected
by tourmalinization are practically devoid of boron haloes. In the latter case, the
contrasting anomalies of boron are formed in deluvial formations (Table (3.3)).

In some cases, biogeochemical haloes are more sensitive indicators of boron
mineralization because they reflect comparatively weak endogenic haloes. The
most distinct, contrasting biochemical haloes were detected in datolitic occurrences
of boron in steppe regions and in broad-leaved forest landscapes. In such cases,
boron in plants compensates for the biological damage caused by an excess of
calcium [99].

Biochemical haloes associated with magnesian and magnesium–calcium borates
are less stable. Being well absorbed by plants, magnesium and iron compensate
for boron deficiency. A decrease in boron content is found in plants growing
in landscapes enriched in various trace elements. Within areas containing borate
ores, the boron content in the old parts of plants is generally larger than in the
younger ones.

Primary haloes of B, F, Sn, W, Mo and other elements, exceeding the ore
bodies, have been detected around skarn zones containing borates and borosilicates.
Haloes of different dimensions are formed depending on the mineral composition

Table 3.3.
Correlation between the average contents of boron in soils and plant ashes in background units [99]

Region	Soil			Plant ashes		
	Datolite[a]	Borate	Tourmaline	Datolite	Borate	Tourmaline
Far East	20.0	2.0	8.0	8.0	13.0	1.0
Transbaikal	—	1.4	7.0	—	4.2	1.3

[a] Mineralization.

and genetic type of mineralization. Wider endogenic haloes occur near thick ore zones. The supra-ore haloes extend vertically over 100–120 m. The shape of endogenic haloes is controlled by the peculiar features of skarn-ore bodies. Linear anomalies, developing conformably with the ore bodies, extend beyond the limits of the latter. The maximum content of boron is observed in the outcrops of datolitic and ludwigitic skarns (whose minerals are characterized by increased boron content). These are outlined by anomalies with a much smaller boron content, owing to microdissemination of these minerals.

Wollastonite, hedenbergite and andradite, forming conformable metasomatic zones, as well as calcite and quartz, prevail in the Sikhote-Alin skarn deposit. Useful minerals — datolite and danburite — are distributed irregularly. Skarn (homogenization temperature is 680 to 400 °C), borosilicate (T — 420 to 265 °C), and quartz-calcite (T — 365 to 180 °C) stages are distinguished.

Neutron-alpha-radiographical analysis shows the absence of syngenesis within the intrusion and the enclosing sedimentary rocks of the deposit. Borosilicate mineralization is related paragenetically to the complex of the minor intrusion (alkaline–potassic series) which originated at great depth. An increased amount and strong correlation of Mn–Zn–Ge seem to be characteristic of skarnoid rocks.

The boron-bearing skarnoid rocks differ from those devoid of ore in their B–Li and Bi–Ag correlations and in the interrelation between these two couples; unlike the barren rock, the skarnoid rocks possess higher F, Cl and Hg. The metasomatic rocks have an obvious spatial association with the skarnoid rocks and also show a B–Li correlation with and F-association.

Even though a geochemical field might be poor in contrast, combined data processing by computer still makes it possible, by examining the aureoles of the above-mentioned elements, to trace the known ore manifestations with sufficient accuracy and to outline new areas as promising for boron.

The greatest amounts of rare and trace elements are concentrated in the most remote zones of the metasomatic column; the content of these elements in the skarns decreases with increasing distance from the intrusive.

Some pegmatites and contact metamorphic rocks contain assemblages of various boron-bearing minerals such as tourmaline, ludwigite–paigeite or datolite. These represent a 'concentration' of boron that relates more or less directly to the crystallization of intrusive granitic magma. Analyses show that granites average about 10 ppm of boron with a few exceptions ranging up to 300 ppm. However, boron does not readily enter into the crystal structure of the common rock-forming minerals; hence, when magma crystallizes the boron tends to leave with released water. The conditions of high temperature and fluid under high pressure at a contact also open up the possibility that it may be extracted from the country rocks. In either case, if the boron is not fixed in the contact zone, there is the possibility of its being transported to the surface for release at thermal springs, steam vents, and volcanic fumaroles.

Boron-bearing minerals in one of the borate deposits are concentrated as nests (sized 0.1–0.8 m) that are confined to skarn and calciphyres in the lenses of serpentinized dolomites. The latter contain no economic mineralization but are

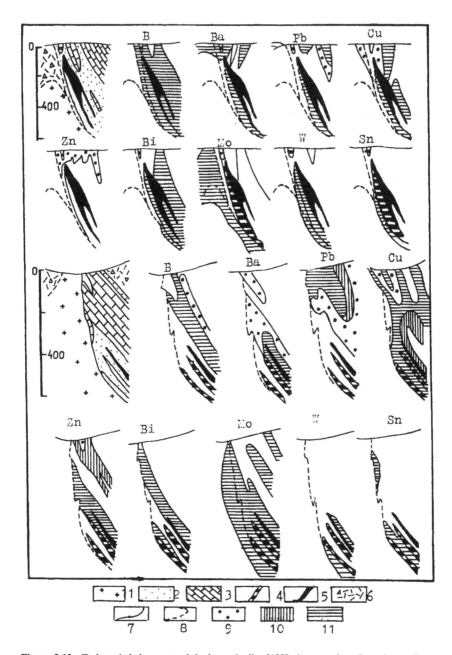

Figure 3.12. Endogenic haloes around the boron bodies [128]. 1 — granites; 2 — skarns; 3 — marbles; 4 — ore body; 5 — sahaite ores; 6 — vulcanic rocks; 7–8 — contact rocks (7 — stability, 8 — non-stability); 9–10 — content elements in haloes, ppm; (9 — B 6–24, Pb 0.07–1.7, Ba 120–500, Cu 4.2–80, Zn 90–140; 10 — Pb 1.7–4.8, Cu 80–240, Zn 140–250, Mo 4.8–7.2, W 3–6; 11 — B > 88, Pb > 4.8, Ba > 1250, Cu > 240, Zn > 250, Bi > 12, Mo > 7.2, W > 6, Sn > 16.8).

characterized by increased concentrations of indicator elements and are quite conspicuous against the background of the enclosing unaltered dolomites and granites, which are poor in boron (0.003 per cent).

Combinations of Sn, Mo, Bi, As, Ba and Pb are found around kurchatovite–sakhaite zones, whereas sulphide–magnetitic ores are surrounded by the combinations of W, Sn, Mo, Cu, B, Ag, Ba and Pb (Fig. (3.12)).

The thermal forms of mercury sublimation are the indicators of boron mineralization. We have established that datolitic bodies contain medium-temperature (270–360 °C) and high-temperature (600 °C) forms of mercury sublimation. Danburite mineralization generally hosts one medium-temperature form (290 °C).

Ore bodies with boron are characterized not only by increased concentrations of boron in the altered rocks of the ore zone, but also by the contact of positive and negative anomalies of this element, which indicate the epicentres of commercial ores (Fig. (3.13)). The remote-sensing nuclear-physical methods enable continuous determination of boron quantities in rocks. The sensitivity and relative error of the analyses are 0.03 per cent and ± 20 per cent, respectively. Favourable conditions for the formation of boron deposits are created at high potential in post-magmatic solutions. The presence of boron removal and introduction zones is characteristic of granitoids whose exocontact hosts ore bodies. Boron accumulation and variance of boron concentration in the marginal facies of boron-bearing granites are 2 to 10 times higher than in the central parts; the average concentration and variance values for different facies of barren massifs are equal [69].

The bulk concentrations of boron in boron-bearing and in barren skarns are rather close. The ratio of boron concentrations in rocks and minerals indicates its potential concentration in skarns.

The boron concentration coefficient in skarns (ratio of average boron concentration in all calcareous boron-bearing and barren skarns) and in boron-containing minerals fluctuates between 6 and 18 (average 10).

An increased (in epidote and vesuvianite) or decreased (in granite and spinel) concentration of boron is not indicative of boron mineralization in skarns that mobilize boron when it is increased in mineral-forming solutions. Pyroxene and wollastonite are the most informative indicators of mineralization. Magnesian-skarn borate deposits formed in abyssal facies environment (suanite and ludwigite bodies) are prospective for boron. A comparatively shallow occurrence of calcareous skarns favoured the development of danburite, datolite, wollastonite, prehnite and apophyllite. Boron deposits in calcareous and magnesian skarns are characterized by vertical zonality, manifesting itself in the spatial alteration of qualitative and quantitative parameters of haloes. Barium, Pb, Zn, As and Hg, are supra-ore indicators, and Co, W, Mo, Sn, Ni, V and Cu are sub-ore indicators, of those deposits that are represented by ludwigite kurchatovite-sakhaitic ore bodies (Fig. (3.14)).

The endogenic vertical zonality is as follows: Ni, V, Cu, W, Mo, Sn, Co $\rightarrow$ Bi, B, Cu, Zn $\rightarrow$ Pb, As and Ba. Index of zoning $K = (Pb \times B \times As)/(Co \times W \times Mo)$ haloes productivities of < 0.1, 1–100 and > 1000 are characteristic of sub-,

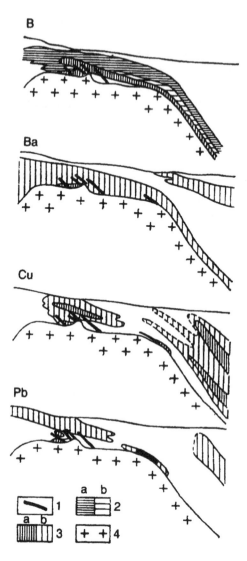

Figure 3.13. Endogenic haloes around the ore bodies of a skarn–datoli deposit [99]. 1 — ore bodies; 2 — negative haloes with contents, per cent (a — 0.0–0.004; b — 0.001–0.011); 3 — primary positive haloes with the following contents of indicator elements (a — Pb 0.01–0.08; Cu 0.006–0.008; Ba 0.1–0.3; B 0.008–0.04; b — Pb 0.001–0.01; Cu 0.002–0.006; Ba 0.003–0.1); 4 — granite-porphyries.

middle, and supra-ore haloes, respectively. The supra-ore zones are notable for distinct boron haloes and for a greater number of mercury temperature forms than the sub-ore zones.

Deposits of sedimentary borates overlain by the intercalations (nodules) of calcium borates produce peculiar haloes around productive ore bodies. The thickness

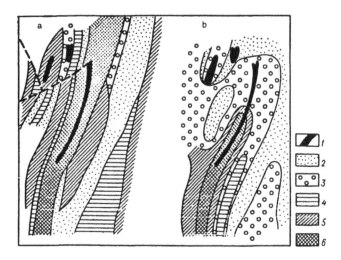

Figure 3.14. Index of zoning K = (Pb × Ba × As × Zn)/(Co × W × Mo × Sn) in rocks (a) and soils (b). 1 — boron bodies, 2–6 content K (2 — < 1, 3 — 1–10, 4 — 10–100, 5 — 100–1000, 6 — > 1000) [128].

of productive horizons increases with depth. Unique deposits (in terms of boron reserves and content) of sodium borates are formed (Fig. (3.15)).

For instance, the following three zones have been identified in the sedimentary deposits in Turkey [87]:

(a) an upper (supra-ore) zone, composed of carbonate rocks with intercalations of blue-green clays and inclusions of inderbarite, kurnakovite and inderite;

(b) a middle (ore) zone composed of brown montmorillonitic clays, ash and kernite; and

(c) a lower (sub-ore) zone composed of clays and carbonate rocks.

The identification of ore accumulations by conventional methods is impeded due by the high susceptibility of borates to hydrolysis and their substitution by carbonates in the process of weathering. Geochemical sampling of the drill-hole core should be performed in order to detect boron, antimony, mercury and its forms, and arsenic, which form the supra-ore haloes. Such sampling increases the efficiency of prospecting.

The sodium borates, or products derived from borax and borax-bearing brines, are the preferred borates for industry because they are soluble in water and combine readily with most other chemicals. Calcium borates (or sodium-free compounds) are required for certain end uses, but in general, non-sodium borates are used only where borax is not available. Only three are currently utilized in any quantity for their borate content. These are: borax (tincal) — $Na_2B_4O_7 \times 10H_2O$; colemanite — $Ca_2B_6O_{11} \times 5H_2O$; szaibelyite — $MgBO_2(OH)$.

Borax is by far the most important mineral for the borate industry. Field studies have added many new minerals to the list of borates identified and, more

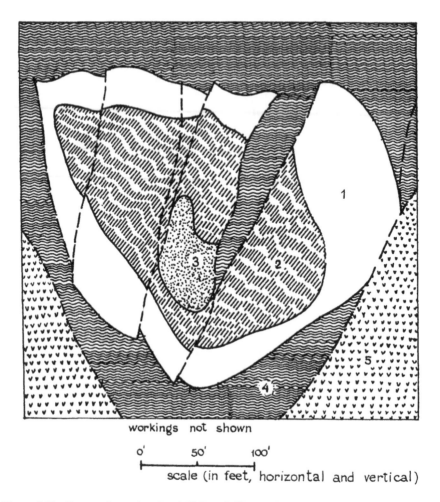

workings not shown

scale (in feet, horizontal and vertical)

Figure 3.15. Cross-section, mine, Death Valley, California [85]. 1 — colemanite, 2 — ulexite, 3 — probertite, 4 — clay, 5 — basalt.

importantly, have defined more accurately how various boron minerals occur in their natural geological environments.

Exogenous boron minerals represented by fluoroborates, alumoborosilicates, borosilicates, boroarsenates, borophosphates and borates have been found in sedimentary and volcanogenic-sedimentary deposits.

Boron concentration in the deposits at the time of their formation attains a value 1000 times higher than the clarke. This is a specific feature of boron ore deposition which predetermines the diversity of sources and the variety of ways in which boron enters sedimentary basins.

Boron mineral deposits are classified simply according to the main mineral of value — borax, ulexite, colemanite, etc. Borate deposits that are mined or potentially minable contain a small part of the total boron in the world. Most of the boron is dispersed in trace amounts in geological situations from which it cannot be recovered economically.

The concentration of boron into minable deposits involved geological processes and conditions that are being noted, as their classification is reviewed.

Boron minerals in marine evaporite basins

Magnesium borate, boracite, and a number of less common borates are known to occur in evaporite beds of Permian age formed by the extensive evaporation of seawater. There are also some borates in the associated gypsum-anhydrite beds.

Boron does not occur in nature as the element. It is found combined with oxygen in about 150 minerals and with fluorine in two. The more numerous minerals are hydrated borates that form under the conditions prevailing at or near the earth's surface. The sequence in which borates form has been established by field observation and laboratory experiment. Among the sodium borates, borax is the primary mineral formed in a lake or elsewhere at ordinary temperatures. Borax dehydrates readily in dry air to tincalconite, and in moist air the conversion reverses readily. Kernite, stable at somewhat higher temperatures than borax, was formed from the metamorphism of borax at Boron (Argentina).

Ulexite is the common primary sodium–calcium borate mineral formed in borate marshes or playas under normal circumstances. The lower hydrate, probertite, has been found only in deposits which were probably subjected to elevated temperatures and pressures, as at Boron and in some of the Death Valley deposits. For the calcium borates, inyoite–meyer–hofferite–colemanite, the lowest hydrate is the common mineral, although it may not form under normal conditions. The highest hydrate, inyoite, has been found as a primary mineral in a few small deposits in South America. Sodium is usually available and, consequently, ulexite is likely to form. The origin of colemanite is somewhat complex, as it is generally found as a 'secondary' mineral with or replacing ulexite. This suggests that the sodium is removed and replaced by calcium-rich waters under conditions in which inyoite is not the stable mineral. The largest known boron deposits originated as evaporites or chemical precipitates interbedded with clays, mudstones, tuffs, limestones, and similar lacustrine sediments. There is evidence that most of the deposits were close to thermal (volcanic) springs at the time of their formation and these are generally regarded as the source of the boron. Numerous hot springs and associated basic volcanic forms throughout the world today contain boron in concentrations of 50 ppm or greater, and several South American springs are actively depositing borates. In addition to the concentrated source of the borates and a 'basin' in which they can collect, an arid to semi-arid climate also seems to be an essential requirement during the deposition and concentration of economic amounts of soluble borates. These concentrations of soluble borates can, in the long run, only be preserved by burial.

Hydrated borates may accumulate in several ways within a non-marine basin. Borates may be deposited in layers in a spring apron around a borate spring, with ulexite, borax, or inyoite as the primary borate mineral [87].

Searles lake provides the example, possibly unique, of an important boron concentration formed in a continental basin by the evaporation of deep lake waters.

The borates there consist of borax crystals interbedded with other crystalline saline minerals, mostly sodium chloride, sodium–calcium carbonates, and sulphates.

Thermal springs in the distant headwaters have been identified as adequate sources of the boron. The stratification of the denser borate-containing salts occurred during several periods in the lake's 40 000-year history, while fresher waters flowed. The borax from these salt layers is now being recovered by pumping the entrapped brines.

The borate lake deposits originated in lakes that were not desiccated during borate deposition; these are chemical precipitates with borates that are not accompanied by an array of other evaporite minerals. Some deposits of ulexite–colemanite (with ulexite the primary borate) may also form in borate lakes rather than on playas.

We have identified four temperature forms of mercury: 270, 310, 380 and 450 °C in borates of sedimentary genesis (ulexite and inyoite) that occur in the near-surface (supra-ore) zones. The most significant borate concentrations, associated with the final stages of halogenesis, are localized in potassium–magnesian salt horizons. The paragenetic associations of elements in rocks productive for boron differ from the admixtures in the salt horizons devoid of boron mineralization. In salt deposits lacking boron mineralization, the bulk of boron and other elements constitute either isomorphic or sorbed admixtures in hydromicas. Boron minerals proper are found in salt-dome structures containing productive mineralization. The parageneses of boron and other elements correspond to the following series:

$$B \rightarrow Mn \rightarrow Zn \rightarrow Ga, \ Ni \rightarrow Ti, \ Li \rightarrow Sr \rightarrow Cu \rightarrow Pb \rightarrow Ga \rightarrow V.$$

In the barren zones, they are as follows:

$$Ti, Zr, Ni, Cr \rightarrow Ga, Mn \rightarrow Cu, Sr, Li, B.$$

Two temperature forms of mercury, medium (230 °C) and high (410 °C), were detected in preobrazhenskite. The potential for boron areas may be forecasted by exploring the nature of boron association with other elements.

Methods of work

The applicability of an appropriate geochemical method in territories potentially prospective for boron depends on the degree of their previous geological study and availability of a pertinent sampling object. Geochemical exploration along dispersion trains and secondary dispersion haloes, may be carried out in areas where bedrocks are overlain by alluvial and eluvial–deluvial formations.

The hydrogeochemical method is applied at the first stage of prospecting to elucidate the general prospects for boron mineralization. It is highly advisable to use this method in mountain regions where stream beds are devoid of the silt–clay fraction that is needed for lithogeochemical sampling, and to employ it also in the search for haloid-sedimentary deposits.

Sampling of rivers and temporary water courses combined with the investigation of bottom sediments makes it possible to rapidly prospect vast areas. During the first stages of prospecting surveys, relatively long water courses are sampled at 250–500 m intervals. If anomalies are subjected to a more detailed survey, the sampling intervals are reduced to 100 m in accordance with the smaller length of the streams. It is advisable to sample 3 to 4-km water courses with a maximum discharge of 0.2 m^3/s, outflows of subsurface water, and ground water, as well as ground and soil waters from workings. The volume of the water sample should be 0.7 l, i.e. sufficient for determining not only boron and fluorine but also the accessories of boron mineralization directly in the field. Consideration should be given to the change of the elemental concentration in waters with time. Water sampling during floods and heavy rains is useless. Colorimetric and photocolorimetric methods utilizing such reagents as carmine and crystalline violet are used in laboratories for the determination of boron.

The quantity of boron in natural waters depends on different factors. The use of hydrogeochemical methods is impeded by the fact that some haloes are associated with dispersed mineralization, and their boron content is frequently controlled by landscape-geochemical conditions of prospecting and by the changes in the concentrations of elements in water resulting from climatic conditions and water-course discharge. Hydrochemical haloes accompany borate and datolitic ore bodies and do not practically develop around tourmaline mineralization. Being easily soluble even in weak acids, calcium borates are intensively leached out in a zone affected by hypergenesis. Therefore, waters are analysed not only for B but also for F, Cl, Cu, As and Zn.

The uniformity of boron and chlorine behaviour is proven by their quantitative ratios and spatial distribution. In every region, parallel sampling of bottom mud and near-bottom waters assists in determining the affiliation of an element to the group of leached out or absorbed ones. If the content of an element in the mud water exceeds that in the near-bottom water, the element belongs to the leached out group, and vice versa.

The hydrochemical method is also used in the search for hypergenic deposits in sulphate–haloid rocks (boron-bearing salt-dome structures with primary and secondary borates). The following values — $(B/Cl) \times 10^4$; $(B/Mg) \times 10^4$; $(B_2O_3/Cl) \times 10^4$; $(B_2O_3/Mg) \times 10^4$ — are used for reliable and confident grading of anomalies and their discrimination from the false anomalies produced by continental salinization of ground water, and by concentration of boron through evaporation. The association of aqueous haloes with the destruction of borates in salt-dome structures is indicated by the $(B/Cl) \times 10^4$ values exceeding 10–15.

Where both the hydrogeochemical and lithogeochemical methods are applicable, the latter should be preferred. Lithogeochemical surveys along dispersion trains are most effective in areas with slightly dissected topography, and in arid climates with drying-out water courses.

Finds of borates and borosilicates in alluvial formations that can be made at a distance of several tens of kilometers from ore deposits are related to direct indications of mineralization.

Heavy concentrate sampling is not performed due to the difficulties in the determination of borates and their relatively small density. Therefore, it is only advisable to use the heavy concentrate geochemical method in the search for boron deposits in closed areas. The weight of the alluvial sample should be 10 kg; fractions of less than 1 mm are most representative. The method of radiography is used to determine boron in the magnetic fraction wherein magnetite and ludwigite are commonly concentrated.

Luminescence can aid in accomplishing mineralogical phase analysis for many boron minerals directly in the field. These determinations are made visually by the characteristic colour of luminescence, or by standard spectra of photo- and X-ray luminescence.

Every heavy concentrate sample should be subjected to spectral analysis, whereas only 50 per cent of the samples should undergo mineralogical analysis. This cuts the cost of work and increases the contrast of anomalies.

A sampling density of 4–5, and less frequently 2–3 points per km^2, corresponds to a scale of 1:50000. A sample of 100–150 should contain a clay fraction or fines of gravel–pebble formations (0.5 mm fraction). Sampling of eluvial–deluvial sediments aids in defining local zones.

The lithogeochemical method of surveys along secondary dispersion haloes is applicable in the areas covered by: (a) eluvial–deluvial deposits resulting from the disintegration of indigenous rocks; and (b) deposits transported from distant localities.

The quantity of boron transported to a given landscape depends mainly on the lithological features of geological formations. Sites with boron-rich rocks should be regarded as special kinds of landscape [98].

Shallow sampling is permitted in the areas covered by the products of weathering of indigenous rocks. The humus horizon A_0 or A_1 is a representative horizon in the mountain-steppe, steppe, or southern taiga landscapes [99]. Samples are taken from the lower part of the humus horizon, which is poor in organic substances. The processes of biochemical transformation are less intensive here. The BC horizon is sampled in the zones of broad-leaved forests and in the northern taiga landscape [99]. In permafrost-taiga landscapes, illuvial horizon B and transitional horizon BC are the representative ones for performing sampling at the survey scales of 1:50000 and 1:10000, respectively.

The role of dispersion train surveys, hydrogeochemical exploration, and drillhole testing is appreciably increased within closed and semi-closed regions. Samples are taken from a certain horizon (commonly crust of bedrock weathering) and from the holes drilled, especially along single lines in the middle and lower parts of the slopes. Sorbed and mineral boron is extracted with the help of acid and water extracts. The weight of the samples is 250–300 g. The 1 mm fraction enriched in boron is sieved and analysed [99].

Samples are also taken of the following boron concentrators: limonites, goethites and iron–manganese concretions occurring in the zone of ludwigite ore oxidation. Boron-rich goethites and hydrogoethites exposed at the surface indicate a possible boron mineralization at depth.

Particular attention in geochemical exploration should be given to the landscape conditions, geomorphology of the area, direction of migration of elements in the secondary haloes, and to the geological structure of the sites being investigated. The haloes of boron deposits are generally characterized by a flux-like shape, elongated along the slope. The lines of maximum alteration of anomalous concentrations (halo–background) run parallel to the contour lines of the topography. For this reason it is advisable to orient the lines of survey profiles along the contour lines, with an average sampling density of 40 points per km^2 [98]. The survey routes are plotted at 450–500 m intervals along the slopes, starting from the watersheds. The sampling interval is 50 m.

The standard sampling technique of prospecting surveys at a scale of 1:50 000 may be employed on plains with an obscure geological structure. The detailed prospecting should be carried out on a grid of 100×25 and 100×50 m. The profiles are oriented up the relief or at a 90 °C angle to the direction of the prospecting lines plotted at a scale of 1:50 000 [99].

Such orientation of the prospecting survey profiles makes the detection of boron mineralization easy and less expensive. It also ensures rational cutting of workings.

The samples are used to determine the content of boron and accessory elements, whose enumeration depends on the suggested genetic type of mineralization. The DFS-13 device is used to carry out spectral analysis with the sample spilling on coal electrodes (sensitivity $0.5 \times 10^{-4} - 1 \times 10^{-3}$ per cent, relative error of reproducibility 10–20 per cent).

Based on secondary dispersion haloes, the type of boron mineralization is determined by the content of boron in a 2 per cent hydrochloric acid extract. $K(\mathrm{H}^+)$ values of > 50 per cent and < 50 per cent in such extracts indicate, respectively, borate and borosilicate (or alumoborosilicate) types of mineralization [99].

Great intensity of boron migration in soils is a reliable criterion for the interpretation and grading of anomalies. Identification of geochemical anomalies is based on the processing and interpretation of the results of various analyses (visual, statistic calculations, trend, factor, multidimensional fields). Highly precise analyses and determinations of boron forms are made to contour weak anomalies. Analytical results are processed by the 'sliding' window (moving average) method. The concentrations appearing as different from the geochemical background with a 25 per cent probability or less should be regarded as anomalies, or as concentrations exceeding the average background by one standard deviation or more [99].

Factor analysis is used to process the data on several elements. Grouping of variables is made upon elucidating the interrelations between these elements. Moreover, the method of fields of natural geochemical associations is employed. The anomalies are graded on the basis of their elemental composition, spatial coincidence of the haloes of different elements, and geological conditions of their localization.

By processing the geochemical information, boron anomalies are differentiated into the following groups [99]:

(a) anomalies in soils and groundwaters that accompany the outcrops of skarns;

(b) complex lithogeochemical anomalies with a spatial coincidence of the haloes of boron and various accessory elements that are not accompanied by aqueous anomalies;

(c) boron anomalies in soils unaccompanied by other accessory elements and dispersion trains but caused by disseminated mineralization; and

(d) anomalies in waters unaccompanied by haloes in soils.

Having plotted the anomalies on a geological map, it is necessary to analyse the individual dependence of their spatial localization on the rock variety and on tectonic dislocations. The epicentres of prospective anomalies are marked inside their contours, and their mineralogical–geochemical and geological nature (fault, ore body, skarn, dyke) is indicated by a sign.

The first (ore) group of anomalies is checked by mining and drilling operations. Additional detailed exploration (sampling on a denser grid, determination of boron forms) is carried out in the zone of prospective anomalies. The third group of anomalies is disarded. The fourth group of anomalies, associated with deep-seated mineralization, is subjected to more detailed exploration. Supplementary determination of boron and mercury forms in the key horizon of rocks is made.

The geological (predicted) reserves of boron may be evaluated from the results of geochemical surveys carried out along secondary dispersion haloes. The areal productivity of boron within the secondary dispersion halo contour, and the change in linear productivity per metre of depth are calculated [46].

The application of biogeochemical method should be confined to areas where bedrocks are overlain by up to 20 m thick alluvial, fluvioglacial, and aeolian deposits, and to territories with buried defluction haloes.

Since the mineral form of boron occurrence determines the degree of its availability to plants, it is advisable to carry out biogeochemical surveys in buried haloes in order to interpret and grade lithogeochemical anomalies. Samples are taken along the perimeter of plants at a constant height for each species. The above-ground parts of grasses are also sampled. The collected samples are subjected in the field to ashing in iron crucibles (metal moulds used to bake bread) heated on bonfires. Electric furnaces with porcelain crucibles are also used. After obtaining black ashes, the samples are sent to the laboratory for further ashing in muffle furnaces at 400–500 °C.

In order to discriminate between boron-bearing and barren granites, it is necessary to consider the specific features of their formation, the variation in boron content, the presence of positive and negative anomalies, and the distribution pattern of total and isomorphic boron. Comparison of boron content in granite should be confined to magmatic complexes of similar composition and origin.

While generalizing the results of geochemical surveys, efforts should be made to elucidate the forms of boron occurrence in skarns as well as the specifics of distribution of the associated indicator elements. Coefficients of boron content in skarns, aposkarn ores, and their minerals range from 4 to 18 (7–10 on average), which indicates boron mineralization in skarns.

At times, high boron concentrations are conditioned not by the isomorphic penetration of this element into minerals but by the finely dissiminated boron mineralization producing endogenous haloes around the ore bodies. Outside the haloes, the boron content in skarns is very small. It is even smaller than in the enclosing rocks.

Moreover, an increased content of boron in calcareous skarns may be observed in insignificant deposits with low-grade mineralization. In such cases, it is advisable to study the pattern of boron distribution in pyroxene and wollastonite, which are the indicator minerals of mineralization. The trustworthiness of conclusions on the prospectiveness of skarn zones is enhanced if they contain not only high concentrations of boron but also of fluorine. The post-magmatic transformation of magnesian skarns results in the formation of hydroxyl- and fluorine-containing minerals, and in high (up to 1 per cent) concentrations of fluorine in hornblende and phlogopite.

The multiplicative indices of haloes as well as the change in the relation of products of linear productivities $(Ba \times As \times Pb)/(B \times Co \times Ni)$ are calculated to give a quantitative characteristic of vertical zonality in the search for blind bodies.

Successful exploration for borates has consisted in core drilling in favourable areas selected after detailed geological surveys in the vicinity of known occurrences of borates in non-marine sedimentary rocks. Some prospecting guides are recognized. Because the bedded borates, borax, and ulexite–colemanite accumulate in lakes, sections of lake beds are studied closely, particularly if they include saline minerals. Tuffaceous and other volcanic materials associated with lake beds are also considered encouraging, although not a direct guide to borates.

Some borate deposits seem to be closely associated with basic volcanic forms, others with tufa cones and similar spring deposits, and so indications of centres of volcanism may be favourable. In arid regions, near-surface accumulations of borates around springs or on playas are possible indicators of older borates in the immediate vicinity or below the surface. The combination of a salt flat with ancient high shorelines suggests the possibility that extensive evaporation may have formed saline beds containing valuable brines. In humid areas, typical associated minerals such as gypsum or secondary calcite may provide guides to areas worth testing for borates that do not show at the weathered surface.

There are two field tests which are applicable to the common borates. In the flame test, the mineral is dissolved in sulphuric acid, mixed with alcohol and ignited. Dissolved in hydrochloric acid, borate will turn turmeric paper reddish- or pinkish-brown. Borax or other borates which will dissolve rapidly in water can sometimes be detected by testing, but this test is not considered reliable. Tests of physical properties are given in mineralogical works, but it should be noted that the borates in older bedded deposits are much harder and denser than the playa varieties usually cited. Some Turkish colemanite and associated secondary calcite are said to have a distinctive yellow-white colour under ultraviolet light, but this is not reported for other colemanite areas.

Geochemical surveys for boron have been reported in works [99, 128]. A number of elements that often accompany boron might be regarded as potentially useful indicators or pathfinder elements in a regional geochemical programme, but boron itself would seem to be the most reliable indicator. Boron is highly mobile, so water sampling would appear to be useful in a search for the soluble borates. As little as 0.5 ppm of boron in water can be detected. Agricultural studies show that some plants are highly sensitive to boron, but surveys of vegetation have only been pursued as a prospecting technique in Russia.

Geophysical surveys using gravity, magnetic, and resistivity methods have been used in borate exploration, mostly to determine the shape of buried structures or the profile of the basin bedrock beneath sedimentary in filling. No method that indicates borates or borate bodies directly has been reported. Shallow seismic methods have been tried with some success as a means of tracing limestone–colemanite zones, or lenses that have a much higher velocity than the enclosing shales. Similar tests on borax showed no indication of the ore. Various well-logging techniques will indicate borate zones in a sequence of sands of shales and show, indirectly, the borate content. However, it is generally easier and more reliable to examine a core or cuttings for borate minerals and to analyse for B_2O_3. Ludwigite–paigeite and the borosilicates datolite, axinite and tourmaline have even lower contents of boron. Their potential as sources of boron has been studied extensively in Russia. When prospecting for boron of sedimentary origin, there is a need to examine the bore hole core (primarily of carbonate clay and lacustrine clay formations) to determine boron, antimony and arsenic.

The geochemical criteria of halogen occurrences are Mg and Ca that precipitate together with boron. Insofar as the content of boron oxide depends on that of $MgSO_4 + MgCl_2$, the prospectives for boron horizons are characterized by consistent correlations of Mg and Ca($MgSO_4 + MgCl_2$) and $CaSO_4$ salts. The correlation coefficient between them and B_2O_3 is 0.95.

Naturally, the conclusions on areas prospective for boron should be based on the integrated exploration of geological formations, taking into account the mineralogical and geochemical peculiarities of facies, paleogeographic reconstructions, hydrochemical indications, and zonality. It is advisable to use the methods of neutron logging and neutron quantitative powder analysis in view of the difficulties of determining many boron minerals. Boron in bedrocks is detected by transportable devices. Magnetic prospecting is used in the search for borate deposits with ludwigite–magnetite mineralization that are overlain by eluvial–deluvial formations.

Preliminary chemical tests are normally confined to a determination of the grade of the ore and the impurities in it. For the grade, both water-soluble and acid-soluble B_2O_3 are determined. The impurities are determined by chemical analyses and mineral identification; they include chlorides, sulphides and sulphates, iron, and other metallic elements. The specifications and tolerances can vary depending upon the borate mineral involved and its prospective use.

Physical tests bearing on the separation of the borates from the gangue, which is generally a clay or limestone, include the standard milling tests to determine the size at which the mineral is most efficiently liberated and the crushing strength of the ore.

Stages and sequence of work

At the first stage, the revisionary geochemical surveys (hydrogeochemical sampling and prospecting by dispersion haloes and trains) are carried out with the purpose of defining areas prospective for searching at a scale of 1 : 50 000. Integrated geochemical surveys are at the second stage confined to concrete areas. Hydrochemical sampling is applied prior to other geochemical methods. Lithogeochemical survey along secondary dispersion haloes and sampling of deluvial piles of blocks on mountain slopes, rock-streams and rock exposed by workings are the major geochemical procedures employed at this stage. Anomalies recommended for verification are defined on the basis of the accomplished geochemical surveys. Detailed prospecting at a scale of 1 : 10 000 is carried out at the third stage to localize, strip and evaluate the primary source.

3.4. BARITE

The minerals barite ($BaSO_4$ — barium sulphate) and witherite ($BaCO_3$–barium carbonate) are the chief commercial sources of the element barium and its compounds whose many uses are almost hidden among the technical complexities of modern industrial processes and products.

Barite, the major ore mineral, is of vital importance to the petroleum industry as a major ingredient of the heavy fluid circulated during the drilling of oil and gas wells.

Barite production has mainly been consumed in the manufacture of barium chemicals and glass and as a pigment, filler and extender. Witherite is much less common and abundant than barite, although it is more desirable in many ways as a raw material for the production of barium chemicals.

Witherite ($BaCO_3$ — barium carbonate) contains 70 per cent Ba (or 77.7 per cent BaO and 22.3 per cent CO_2), and has a calculated specific gravity of 4.2. Its fracture is uneven and its lustre is vitreous to resinous. Witherite is a minor accessory mineral in some barite deposits. It is highly desirable as a source for the production of barium chemicals because of its solubility in acid. Sanbornite ($BaSi_2O_5$ — barium silicate) contains 50 per cent barium and is of potential interest as a source of barium for chemicals. Unlike many silicate minerals, sanbornite is soluble in acid, which permits the recovery of barium by further chemical processing. The mineral, formerly considered rare, has been found in abundance with other barium silicate minerals in contact zones with metamorphic rocks, mostly quartzite, in roof pendants of the Sierra Nevada. The resource potential of sanbornite is virtually unstudied.

Barite occurs in many geological environments in igneous, metamorphic and sedimentary rocks.

Barite is the most common and abundant ore mineral of barium. Pure barite contains 58.8 per cent Ba (or 65.7 per cent BaO and 34.3 per cent SO_3) and has a calculated specific gravity of 4.5, although inclusions of natural barite may reduce this value considerably. The barite in most commercial deposits occurs as irregular masses, concretions, nodules and rosette-like aggregates.

Barite is most commonly associated with quartz, chert, jasperoid, calcite, dolomite, siderite, rhodochrosite, celestine, fluorite, various sulphide minerals such as pyrite, chalcopyrite, galena, sphalerite, and their oxidation products. Ferruginous clays make up a large part of the residual deposits of barite. Barite is a common gangue mineral in many types of ore deposits, especially veins that are mined principally for other types of commodities, including lead, zinc, gold, silver, fluorite and rare earth minerals. Barite is a common constituent of certain siderite-rich iron deposits in Europe.

Barite in vein and residual deposits may contain as much as several per cent of strontium which replaces barium isomorphously in the crystal structure because of the close similarity of the ionic radius of the two metals in their bivalent state.

Barite occurs throughout much of the world and is available from three major geological types of deposits — vein and cavity filling, residual and bedded — in sufficient quantity to meet current demands at competitive prices.

The vein and cavity-filling deposits are those in which barite and associated minerals occur along faults, gashes, joints, bedding planes, breccia zones and solution channels and in various sink structures. The deposits in solution channels are most common in limestone. Many deposits are associated with igneous rocks. Most of the barite is associated with other minerals including those of the metallic ores, as already mentioned. Carbonatitic, hydrothermal, sedimentary, and volcanogenic-sedimentary barite deposits are known. The most widespread are the hydrothermal (medium and low-temperature) deposits, which yield one third of the barite mined throughout the world. The deposits are subdivided into barite proper and complex deposits, depending on the quantitative ratio of barite and metal elements. The former are those where barite is the only mineral. The latter contain ore bodies from which barite is extracted as an associated component. There are two possible schemes of hydrothermal barite formation:

- combined transportation of barium and SO_4^{2-} by highly saline solutions into the mineralized zones with a subsequent crystallization of barite;
- SO_2 influx into the enclosing rocks enriched in barium.

The vertical extent of barite mineralization is limited by the change of the oxygen regime with depth, and retrospective solubility of barite. Barite occurs in equilibrium with both the relatively high-temperature sulphide ores ($> 350\,^{\circ}C$) and with parageneses that originate at temperature below $250\,^{\circ}C$. Sedimentary and volcanogenic-sedimentary deposits resulted from the reaction of sea-water sulphates with barium ions. The prospecting indications of barite deposits are:

(1) Post-magmatic hydrothermal occurrences of carbonate rocks (calcite–quartz veins) associated with the Hercynian and Alpine tectono-magmatic cycles.

(2) Dislocations, with a break in continuity accompanied by tectonic breccias with the indications of hydrothermal alterations; zones of jointing of faults extending in different directions.

(3) Volcanism in the superimposed geosynclinal troughs; interbedding of volcanogenic, carbonate (enriched in iron) and sand-clay rocks persistent at their strike.

(4) Polymetallic and copper mineralization indicating sulphide-baritic and barite-haematitic deposits.

The grade of the barite ore differs according to deposit and within deposits. The deposits commonly have sharp contacts with the wall rocks; large-scale replacement of the host rocks beyond the ore-controlling structures is rare. Within individual districts, the deposits are commonly scattered and irregular and range in thickness from a few inches to a few feet and, in length from tens to hundreds of feet.

On weathering, vein and cavity-filling deposits may form valuable bodies of residual ore.

The barite and other minerals of this type of deposits are typical of an epithermal suite precipitated from a low-temperature hydrothermal solution.

Residual barite deposits occur in unconsolidated material and are formed by weathering from pre-existing deposits. The barite in these deposits ranges in size from microscopic particles to irregular boulders weighing hundreds of pounds.

Mineralogy and indicator elements of barite deposits

Witherite, calcite dolomite, siderite, fluorite, quartz, gypsum, celestine, iron, copper, lead and zinc sulphides (chalcopyrite, galenite and sphalerite), haematite, sulvanite, cinnabar, anthraxolite and limonite are the major accessories in hydrothermal barite deposits proper. A certain trend in the change of the chemical composition of mineralization with time is observed in the overwhelming majority of deposits. The recrystallization of limestones occurs at the pre-ore (quartz–calcitic) stage; the bulk of barite undergoes crystallization at the ore (baritic) stage, which is replaced by the calcite–sulphidic, fluoritic, and late baritic stages. The circum vein alterations result in dolomitization and silicification of carbonate formations, and in kaolinization and sericitization of sandstones and shales. Being chemically resistant, barite is accumulated in the zone affected by hypergenesis. The complex barite–chalcopyritic, barite–polymetallic, barite–fluoritic, barite–haematitic and barite–manganese deposits contain various paragenetic associations of minerals, including witherite, calcite, fluorite, quartz, pyrite, chalcopyrite, galenite, sphalerite, haematite, celestine, pyrolusite, psilomelane, braunite, goethite, cinnabar, arsenopyrite, jamsonite, tetrahedrite, cerussite, anglesite, malachite and azurite. Barite accumulations in the form of 'sypuchka'* sometimes appear in the zone of intensive oxidation in sulphide-baritic deposits.

Within carbonatitic complexes barite occurs in association with quartz, ankerite, dolomite, calcite, chlorite, fluorite, and sulphides. Owing to its stability, barite is found in heavy fractions of heavy concentrates.

The manifestation of adularization, alunitization, argillitization (secondary quartzites), dolomitization, and sideratization (carbonate rocks) is characteristic of the volcanogenic formation of hydrothermal barite-polymetallic deposits. Alunitic secondary quartzites are regarded as an important indication of barite-polymetallic mineralization.

---

*Russian term denoting a loose body of fine-grained sand containing barite, quartz, or pyrite.

Ca, Mg, Na, K, HCO_3, F and Cl, as well as Hg, Cu, Zn, Pb, CO_2, N_2, H_2 and hydrocarbons were detected in gaseous–liquid inclusions in barite. Proceeding from the real composition of gaseous–liquid inclusions and mineral composition of baritic zones, it is possible to distinguish the following typomorphic indicator elements of barite deposits: Ba, Cl, F, K, Na, Hg, Sr, P, Ti and S. Barites contain the following admixtures (in per cent): titanium (0.01), phosphorus (0.13), lead (0.12), copper (0.02), zinc (0.02), strontium (0.3–8.0), manganese (0.13), yttrium (0.008), molybdenum (0.0002), silver (3×10^{-4}), chromium (0.03), gallium (0.001) and mercury (1×10^{-4}). Bismuth, tin and indium have also been detected in barite, which is characterized by a distinct correlation between the content of silver and that of lead and boron.

The process of barium and strontium mobilization from the enclosing rocks has been found to occur in barite deposits. This is evidenced by rather broad limits of variation in the contents of these elements and in their ratios in the altered rocks, and also by negative anomalies of chemical elements. The average Sr/Ba values in the altered carbonate rocks of one of the deposits fluctuate from 0.002 to 13.4 owing to the redistribution and migration of these elements. Increased concentrations of chlorine (0.04 per cent), fluorine (0.02 per cent), and mercury (1×10^{-3} per cent) were detected in the enclosing rocks around barite veins. Of special interest is the presence of mercury in barite and in the enclosing rocks.

Chert, jasperoid, and drusy quartz are common in many deposits, small amounts of pyrite, galena and sphalerite occur in or on some of the barite, and locally the lead and zinc minerals are recovered as a by-product. The shape of many residual deposits reflects that of the original ones. Deposits derived from solution channels and vein systems tend to be elongate and those derived from sink structures tend to be circular. The depth of residual deposits differs according to local conditions.

The bedrock in many areas of residual barite deposits contains veins filled with barite, fluorite, calcite, and quartz and locally some sulphide minerals including pyrite, chalcopyrite, galena, and sphalerite.

Bedded deposits include those in which barite occurs as a principal mineral or cementing agent in stratiform bodies in layered sequences of rocks. The major deposits of commercial value of this type are bedded concentrations of fetid, generally dark-grey to black, fine-grained barite. The beds of barite are commonly intercalated with dark chert and siliceous shale and siltstone. Some barite-rich zones of these rocks are more than 35 m thick. The beds of barite are commonly laminated. Conglomeratic beds consisting of nodules and fragments of barite, chert, phosphate (apatite), and fragments of rock in a fine-grained matrix of barite have been found in Nevada. Many beds of ore in these deposits consist of 50 to 95 per cent barite. The chief impurity is fine-grained quartz, the amount of which is inverse to that of barite. Small amounts of clay minerals and pyrite are common. Carbonate minerals are rare. Analyses of many samples indicate that the composition of these barite deposits is extremely low in calcium and magnesium — the total of these two elements is generally less than 1 per cent. The most abundant minor element is strontium; it occurs in amounts of up to about 7000 ppm (0.7 per cent). Both manganese and vanadium content are generally less than 50 ppm. The content of cobalt, copper, chromium, nickel, yttrium and zirconium is less than 20 ppm each. The suite of trace elements is limited, and the amount of each element is usually small as compared to the suite and amount of the trace elements in vein deposits.

The beds of black barite contain several per cent of organic matter and characteristically give off the odour of hydrogen sulphide (H_2S) when struck with a hammer. The black bedded barite deposits are probably of sedimentary origin and were formed at virtually the same time as the enclosing rocks by organic and inorganic chemical processes of concentration and deposition. The source of the barium in these deposits has not been established, although some suggestions have been made.

Barite deposits in China are often associated with polymetallic deposits in the region: sedimentary barite deposits are generally associated with Mn-, P-, Cu-, Ni-, Mo-deposits; vein deposits are associated with Pb-, Zn-, fluorite- and siderite-deposits [127].

Benstonite carbonatites are a new type of carbonate strontium–barium ores. The main rock-forming mineral is represented by a strontium variety of benstonite, including 11 per cent SrO and 32 per cent BaO in an easily extractable carbonate form. Together with fenite and calcitic carbonatite they form a single carbonatites complex, genetically connected with the syenitic massif (east Siberia).

Anomalies of Cu, Pb, Zn, Sb, Ge, As, Sn and Ag (i.e. of the elements that are characteristic of metallic minerals of this class) are formed around vein bodies in those zones where barite mineralization accompanies sulphide–copper and polymetallic mineralizations. Fluorine and strontium were detected in calcite–barite and fluorite–barite occurrences.

The presence of barium (in the form of barite) and fluorine (in the form of fluorite and ratofkite) is a distinctive feature of volcanogenic-sedimentary deposits, by which they differ from sedimentary deposits proper. Vanadium and germanium are the typomorphic admixture elements in sedimentary-volcanogenic deposits while antimony, mercury, and silver are the admixture elements in hydrothermal lead–zinc, nickel and cobalt are admixture elements in hydrothermal chalcopyritic complex barite deposits.

Fundamentals of geochemical methods of prospecting for barite deposits

Geochemical methods have been applied in the search for barite deposits on a small scale. However, the materials obtained make it possible to define a broad range of geological problems to be solved by these methods, and to justify their employment in the search for barite. Secondary haloes of barium (0.6–1.0 per cent) associated with carbonate formations, as well as anomalies of phosphorus (1–3 per cent), strontium (0.2 per cent), zinc (0.01 per cent), copper (0.01 per cent), and molybdenum (0.02 per cent) were detected by lithogeochemical surveying in the Kuznetzk Alatau. This work resulted in the discovery of a barite deposit.

In the search for barite deposits mercury was found to be an effective indicator of mineralization. The content of mercury in the samples of deluvial deposits overlying barite veins reaches 200×10^{-7} per cent (compared to background concentrations of 9×10^{-7} per cent). The concentration gradually increases in the direction of productive barite horizons.

Volcanogenic-sedimentary and hydrothermal deposits are generally accompanied by hydrothermally altered enclosing rocks associated with intensive haloes of cobalt and molybdenum (in quartz–sericitic metasomatites of volcanogenic-sedimentary deposits), and zinc, lead, and arsenic (in chloritic metasomatites).

The occurrence of distinct Ba, Cl, Hg, B, Sr and F haloes around monomineral barite veins was established by analysing the available material on the use of geochemical methods in prospecting for barite zones. Endogenous haloes of Ba, Pb, Zn, Cu, Si, Bi, As, Mo and Hg are formed around ore bodies of barite–polymetallic deposits. In both cases, generally linear anomalies of these elements propagate parallel to their contacts with veins.

Mercury forms the widest and most elongated haloes around baritic bodies. Its positive geochemical anomalies were detected even at a distance of 20–30 m from the productive zones. The content of mercury in the zones that contact ore bodies amounts to 2×10^{-5} per cent, compared to background values of 7.5×10^{-6} per cent. The quantity of mercury at the very contact and inside the barite vein is sharply reduced, and reaches the level of a probable negative anomaly (3×10^{-6} per cent).

The content of Ba and Sr increases in the direction from the pre-contact part to the centre of the deposit. An opposite trend is characteristic of titanium, nickel, vanadium and chromium. A combination of positive and negative anomalies of the said elements was detected at the boundary between the transitional and barite (productive) zones. It should be pointed out that the barite zone is characterized by haloes caused by the introduction of Ba, Sr and Hg, and by removal of Ti, Ni, V, Cr and other elements. Vertical zonality, reflecting changes in the mineral composition and the geochemical properties of the vein bodies in space, is characteristic of barite deposits. An upward increase in the contents of calcite, barite, cinnabar, haematite, arsenopyrite, and alunite is generally observed in the ore zones, whereas an opposite trend is characteristic of ankeripyrite, sphalerite, galenite, anhydrite and magnetite. For instance, in a deposit in Kazakhstan the average concentration of barite at depths of 0–20, 40, 60 and 100 m diminished from 84 to 82.3, 75 and 74 per cent, respectively.

Low-temperature conditions are most favourable for the crystallization of barite. Anhydrite originates from Ca-, Ba- and S-rich solutions. The associations of barite with anhydrite may serve as a temperature criterion because the latter is usually preserved at low temperatures. In complex barite deposits, the vertical mineral zonality is displayed in an upward transition of sulphide–baritic zones into fluorite–sulphide–baritic and calcite–baritic ones. Different genetic classes of complex barite deposits are generally characterized by the following trends of vertical zonality:

(a) localization of barite, calcite and fluorite veins and vein zones in the upper parts of productive bodies; and

(b) disposition of the thread veinlets of the stated minerals above the concealed barite-bearing ore bodies.

The quantity of admixture elements in the vertical series of baritic bodies also changes on a regular basis. The following two groups of elements are distinguished:

(a) with a positive geochemical gradient increasing in the upward direction: I, Br, Hg, Ba, Cl, Ag and As; and

(b) with a negative gradient whose content decreases in the same direction: Co, Mo, Zn, Cu and Pb.

Vertical zonality is also characteristic of the structure of primary haloes of barite deposits. The following change of different associations of elements (from the rear to the outer zones) is observed in pyritic barite–polymetallic deposits:

$$Mo \rightarrow (Mo + Co) \rightarrow (Cu + Mo + Co) \rightarrow (Zn + Cu + Co + Mo)$$
$$\rightarrow (Ba + Pb + Zn + Cu + Co + Mo)$$
$$\rightarrow (Ba + Pb + Zn + Cu).$$

The concentration of barium in altered rocks tends to increase up the dip of ore bodies, which is an agreement with a more intensive manifestation here of barite–carbonate stages of mineralization and their dominant development in the upper zones. Barium migrates from the circum-ore enclosing rocks. It should be noted that the content of barium at the sub-ore level generally equals the local geochemical background value. There is a general tendency towards barium and strontium content growth in minerals from early to late stages of ore formation. However, at the final stages, barium forms an independent phase (barite), but strontium is mainly dispersed. A non-uniform distribution of strontium is characteristic not only of barite veins but also of separate individuals from these veins. The greatest concentrations of strontium are found in the central part of this mineral.

Methods of work

It is advisable to apply geochemical methods in the search for barite deposits at the following three stages:

(1) identification of ore-bearing regions;

(2) detection of barite deposits; and

(3) exploration of deposits.

The sequence of work is as follows: geochemical surveys along dispersion trains, heavy concentrate sampling, soil metallometry, taking of bulk samples, examination of 'artificial' heavy concentrates from shallow wells, study of primary haloes, geophysical investigations, drilling and mining operations. Heavy-concentrate, lithochemical, and geochemical methods can be employed in the search along dispersion trains.

After a sufficient previous geological exploration of the main mining regions, exploratory surveys are carried out at scales of $1:50\,000$ and $1:25\,000$. Table (2.1) contains information on the detailedness of search along dispersion trains. Owing to barite stability to hypergenesis, the heavy concentrate method can be used in combination with lithochemistry.

Chemical methods can be used to detect barite in mineralogical samples. A heavy concentrate sample is heated in a solution saturated with Na_2CO_3 and, upon pouring the carbonate solution out, it is again heated, in a 4–5 per cent solution of alkali bichromate (e.g., $K_2Cr_2O_7$). The $BaCrO_4$ film formed on the surface of barite grains has a characteristic yellow colouring. This is a selective reaction because minerals giving yellow insoluble chromates in such conditions are commonly not encountered. This method, permitting examination of a large number of samples, is especially important for analysing heavy concentrates.

When identifying deep deposits within the boundaries of groundwater action, it is advisable to conduct a hydrogeochemical survey. The individual specific features of the ore zones and the local conditions specify the potentiality of combining various methods.

Particular attention in the search for barite deposits proper by hydrochemical methods is given to detecting F, Cl, S and Mo haloes. Polymetallic, copper and other deposits containing sulphides are identified by determining zinc, lead, copper (analysed separately or as a sum of metals), and such leachable elements as sulphate ion and arsenic. The above-stated elements are also determined in the samples of bottom sediments. The results of surveys along dispersion trains and generalization of all available data are used to define concrete areas to be subjected to lithogeochemical surveying along secondary dispersion haloes.

Secondary dispersion haloes in barite deposits are identified by methods similar to those employed in the search for fluorite deposits. They depend on the degree of outcropping, thickness, and nature of loose formations.

Alluvial–deluvial deposits are sampled along the profiles oriented across the strike to detect geochemical anomalies. The scale of sampling (1 : 50 000 and 1 : 25 000) generally depends on the type of deposit. Samples with a mass of 200–300 g are taken from a representative (in real conditions) horizon of loose sediments. In the areas covered by products of weathering of indigenous rocks, samples are generally taken from a depth of 10–20 cm, sometimes 30–50 cm, below the humus layer. In buried and semi-buried haloes samples are taken from workings and boreholes. In such cases sampling is confined to the horizon wherein the distribution of indicators is most intimately associated with the primary sources of hypergenic haloes: ore bodies and bedrocks.

Interpretation of secondary haloes is based on the data obtained in the process of exploration of a broad range of indicator elements. It is advisable to construct multiplicative haloes, which differ from monoelemental ones by greater contrast. Where the territories under investigation contain the original outcrops of rocks and areas covered by eluvial–deluvial formations, both of them are sampled. This results in the identification of slightly contrasting anomalies, and in greater representativeness of geochemical sampling.

In the search for barite deposits, lithogeochemical surveys along dispersion haloes may be combined with geophysical exploration (electric and magnetic prospecting). The possible utilization of fluorine as an indicator of barite deposits allows us to investigate its haloes by field activation methods (according to γ-radiation of isotopes).

Prospecting for a blind mineralization based on the data of metallometry makes it possible to outline anomalies represented by the supra-ore haloes of barite mineralization. To this end, use is made of the vertical zonality of haloes, multiplicative indices, and calculated productivity ratios of anomalies. Stripping and geochemical sampling of bedrocks are recommended within anomalies prospective for a blind mineralization. At the stage of prospecting surveys within the favourable site, bedrocks are sampled along the profiles oriented across the strike of the already known or conjectural structural on a grid of 500×50 m (scale $1:50\,000$). All the samples collected in the process of geochemical exploration are subjected to spectral analysis for the indicator elements characteristic of the barite proper or of complex barite mineralization.

Average contents of indicator elements in geochemical anomalies constitute a sufficiently reliable criterion for determining the type of barite mineralization. The criteria for grading anomalies into productive and un-promising are their complex nature, elemental concentrations, contrast range, and non-uniformity of distribution. The latter implies a combination of zones of the introduction and removal of elements, and variation in their concentrations.

Electrical profiling on a grid of 100×10 m is conducted at the stage of detailed exploration of deposits at a scale of $1:10\,000$, with the purpose of tracing out barite mineralization, contouring the vein bodies, and determining characteristic features of their occurrence. The barium content of rocks tapped by boreholes can be determined by the γ-ray method, allowing reduction of the sample volume and time of analyses.

Particular attention in the search for a blind mineralization and in the process of interpretation of geochemical surveys is given to the identification of anomalies represented by supra-ore haloes. This is achieved by:

(a) determination of the ratios of pairs of indicator elements and antagonists;

(b) determination of the zonality coefficient and change of multiplicative indices of haloes for the following groups of elements: Ba, Sr, As, Sb, Hg, Cl, Cu, Ni, Co, V and Mo;

(c) calculation of the products of linear productivities, e.g. (Ba $\times$ Sr $\times$ As $\times$ Sb)/ (Cu $\times$ Ni $\times$ Co $\times$ Mo);

(d) finding the coefficients of rank correlation between all the contents of elements, forming endogenic haloes, by the 'sliding' window (moving averages) method; and

(e) determination of the temperature forms of natural compounds of mercury.

The multiform pattern of mercury occurrence with the predominance of low-temperature forms is characteristic of the supra-ore zones.

The stripping of primary barite deposits is followed by their sampling. The spacing between the sampled faces should not exceed 20 m. Samples are analysed for BaO, SiO_2, and Fe_2O_3 in barite proper, quartz–barite, iron–barite and clay–barite rocks. The calcite–barite and fluorite–barite formations are additionally analysed for CO_2, CaO and CaF_2, and the sulphide–barite ones for Pb, Zn, Ag, Cu and Mn.

Evaluating the commercial potential of a specific barite prospect is a complex process that requires a careful study of the relation ship between the pertinent geological, technological, geochemical and economic factors. A knowledge of the zoning, texture, grain size and chemical composition of the ore minerals and the associated gangue is also critical for the geological evaluation.

Geological information puts a limit on technical factors such as the choice of mining and beneficiation methods to be used. Barite prospects are generally evaluated geologically by detailed geological mapping and surveying as well as by digging test pits and trenches for the further examination and sampling of the rock body.

The success of future prospecting will be increased if prospecting is guided by a knowledge of the geological association and the role of barium in various geological environments, especially in a sedimentary environment.

Deposits should be sought in residuum overlying carbonate-rich rocks, especially those of Cambrian and Ordovician age. Other areas of well-weathered rocks that have been hosts for veins or nodules of barite should also be examined for residual deposits. Fetid black beds of barite occur abundantly in sequences of dark siliceous rocks of mid-Paleozoic age. The rocks are easily confused with impure limestones, except for the noticeably greater weight of the barite-rich beds.

Geophysical techniques have not been used widely to explore for barite deposits. Gravity surveys for residual barite deposits have been undertaken in Missouri where they indicated that large concentrations of barite could be outlined and the tonnage computed with an expected error of 35 per cent or less.

The gravity anomalies reflected not only the presence of ore, but also the existence of pinnacles of bedrock, flint bars, and unknown concentrations of mass and also probable errors.

In East Germany, the use of geophysical techniques to locate hidden deposits of barite and fluorspar was described as less than ideal [7]. Useful geochemical techniques are available for exploration. A relatively simple and inexpensive turbidimetric test for barium was applied in a region of bedded barite deposits in western Arkansas. It was concluded from the study that barium anomalies could be found and that targets for further study could be outlined. Geochemical studies can be carried out using emission spectrographic and X-ray fluorescence techniques, but the initial cost of the equipment is high.

Barium is removed during the ore formation process from near-vein areas of country rocks on the Daut and Elbrus deposits. When the barite stages proper are displayed on the boundaries of the upper horizons, barium depletion is compensated by the formation of primary barium aureoles. The regularity established in the distribution of barium may be used in practical exploration work to determine the level of shear of the country rock in relation to ore bodies.

Soils in the area of barite deposits in Kenya (Coast Province) were analysed for Pb, Zn, Ba, Cu and Hg. Chemical Pb, Zn, Ba and Hg in soils show a good correlation with the mineralized barite vein. Copper defined the vein to a lesser extent, and mercury where it occurred displayed very broad dispersion haloes. Physically dispersed barite shows very good correlation with the location of mineralized faults, although the extent of the physical dispersion is less restricted than that of chemically dispersed Ba. This would imply that in barite-associated mineralizations it may be quicker and more cost-effective to analyse for physically dispersed barite, with associated galena, cinnabar and chalcopyrite.

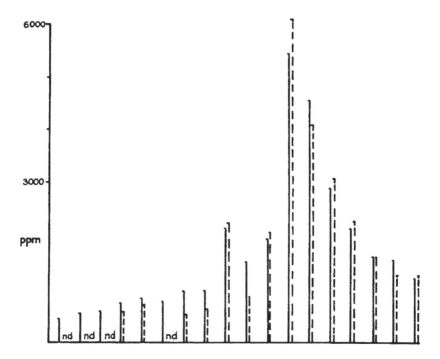

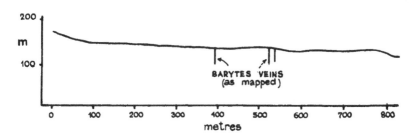

Figure 3.16. Analytical results for Ba in a soil sampling traverse over mineralized ground. Hatched lines represent the mineral analyses determinations and full lines the values obtained by conventional XRF [7].

A method is described for determining the Ba and Sr content of soils in the field laboratory using a portable, isotope source X-ray fluorescence analyser [119]. The technique of radioisotope X-ray fluorescence analysis is based on the excitation of characteristic X-rays using either radioactivity or X-rays generated in a radioisotope source.

The application of these techniques to exploration for vein-type barite and stratiform celestite deposits in the west of England are compared in Fig. 3.16 and Fig. 3.17. Satisfactory accuracy is achieved in comparison with other methods,

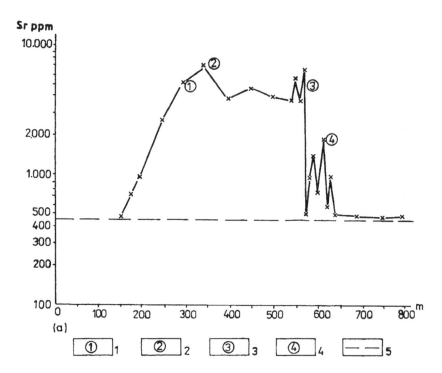

(a)

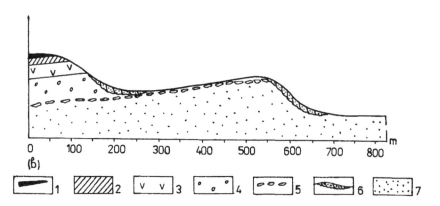

(b)

Figure 3.17. (a) A generalised plot of a traverse across the Severnside Evaporite Bed with geological interpretation. 1 — Increasing overburden; 2 — Backslope; 3 — Outcrop of severnside; 4 — Talus slope with detac bed blocks of celestine; 5 — Threshold level. (b) Generalised geological cross section showing the relationship of Severnside Evaporite Bed to topography. 1 — Lias; 2 — Rhaetic; 3 — Tea Green Marl; 4 — Kauper Marl; 5 — Severnside Evaporite Bed outcrop on backslope; 6 — Talus; 7 — Gently stratiform nodular celestine [7].

along with considerable economies of time and effort. Detection limits (2-x standard deviation) are 200–300 ppm for Ba and about 500 ppm for Sr.

3.5. FLUORITE AND CRYOLITE

Fluorite is one of the most important non-metallic minerals. It forms under a great range (from high-temperature endogenic to hypergene) of physical and chemical conditions, which specify the diversity of genetic and mineral types of fluorspar mineralization. Deposits of rare earth rock-crystal-bearing fluorite formations (pegmatitic group) are represented by shallow granite pegmatites containing cavities with crystals of optic fluorite. Productive bodies in pegmatitic fields are encountered comparatively seldom; their reserves are, as a rule, small but in certain cases may be significant.

Deposits of rare earth apatite–fluorite formations (carbonatitic group) belong to a new genetic type of commercial fluorspar sources whose significance has become quite obvious in recent years.

Only those deposits of apocarbonate rare earth-fluorite formations (greisenic group) that have formed as a result of superimposition of the greisenic process over carbonate rocks caused by multiphase intrusions of subalkaline and alaskitic granites have independent commercial importance. The most important micaceous-fluoritic, topaz-fluoritic, and micaceous-topaz-fluoritic metasomatites form bodies of irregular or tubular shape that are localized at the exocontacts near mother granites. Deposits of fluorite formation (hydrothermal group) are the most extensively utilized commercial sources of fluorspar.

Quartz–calcite–fluorite, quartz–fluorite, quartz–barite–polymetallic fluorite, and sulphide–fluorite mineralizations have much greater commercial value than other representatives of various mineral types of the fluorite formation.

Cryolite is a sodium aluminum fluoride, Na_3AlF_6. It is a rare mineral which has been found in commercial quantities in only one place in the world, Greenland. Small amounts of it are shipped from stock in that country, and its place in industry has been taken over almost completely by synthetic ryolite. Because of the relative unimportance of cryolite, discussion of it has been relegated to a short section at the end of this chapter, which is principally concerned with fluorite.

The Greenland deposit (Ivigtut) was a pegmatite in porphyritic granite in which the cryolite was accompanied by fluorspar, siderite, pyrite, arsenopyrite, galena, topaz, molybdenite, and a number of rare aluminum fluoride minerals. Approximately 90 minerals have been found in the deposit, several of which have never been found elsewhere.

Originally cryolite was used mainly in the production of soda, alum, and aluminum sulphate by a method devised by a chemist.

Indicator minerals and elements

There are direct and indirect indicators of mineralization. The former comprise minerals and elements reflecting the composition of commercial minerals, whose spatial distribution and quantitative parameters are directly dependent on the fluorite bodies. The dependence has not been established for indirect indicator elements.

Direct indicators of the rare earth fluorite formation are fragments and crystals of optic fluorite, whereas its indirect indicators include crystals of quartz, phenacite, beryl, xenotime, topaz, bertrandite, helvite, feldspar, micas, and such chemical elements as beryllium, yttrium, cerium, and rare earth elements contained in these minerals and in fluorspar.

The indicators of the rare earth apatite–fluorite formation are more diverse and are represented mainly by the products of the late stages of mineralization of carbonatites (apatite, haematite and sulphides, zircon titanium, niobium and tantalum minerals) and fluorcarbonates of rare earths.

Deposits of rare metal fluorite formation are indicated by minerals of skarn-fluoritic micaceous-fluoritic, topaz-fluoritic and quartz-rare metal sulphidic parageneses, which may be slightly altered in concrete deposits.

At times, the major direct indicator mineral is supplemented by wolframite and molybdenite. Indirect indicator minerals are rather diverse. The most characteristic of them are such hypogene minerals as topaz, cassiterite, scheelite, chrysoberyl, beryl, phenacite, euclase, bertrandite, helvite, bismuthine, lithium muscovite, lepidomelane, zinnwaldite, siderophyllite, margarite, sericite, pyrite, sphalerite, galena, chalcopyrite, cosalite, ephesite, diaspore, strontianite, celestine, sellaite and garnet. Such hypergene minerals as gypsum, limonite, gearksutite, creedite, wulfenite, pyrolusite, wad, kaolinite, dickite and jarosite. Fluorine as well as beryllium, rubidium, lithium, molybdenum, tungsten, bismuth, tin, lead, zinc, copper, potassium and sodium are the indicator elements of mineralization.

The affiliation of mineralization to quartz–fluoritic, quartz–calcite–fluorite, quartz–barite–fluoritic, or calcite–barite–fluoritic types, generally characterized by different but always insignificant contents of ore minerals, is determined by various ratios of the most widespread minerals: quartz, calcite and barite in the fluorite formation (at a constant appreciable amount of fluorspar). The above-mentioned minerals are represented by galena, sphalerite, pyrite, markasite, arsenopyrite, haematite, and sometimes by adularia, montmorillonite, hydromicas, zeolites, and such hypergene minerals as gearksutite, kaolinite, halloysite, nacrite, cerussite and pyrolusite. Grains of faded ores, molybdenite, cassiterite and argentite (?) are also encountered in quartz–calcite–fluorite deposits, whereas quartz–barite–fluorite deposits sometimes contain manganocalcite and such accessory and rare minerals as manganite, romanechite, and opal. Carbonate–barite–fluorite deposits are notable for an extremely rare occurrence of molybdenite, azurite, wulfenite, smithonite and cerussite.

Albite, serucite, chlorite, calcite, quartz, fluorite and sometimes pyrite occur in the zones of hydrothermal metamorphism of rocks enclosing deposits of these groups.

The sulphide–fluorite (with quartz and calcite) and quartz–barite polymetallic fluorite mineralizations contain, apart from gangue minerals, which are of major significance, appreciably increasing quantities of sulphides and barite, respectively.

The sulphide–fluorite (with quartz and calcite) and quartz–barite polymetallic fluorite mineralizations generally represent the later-formed members of the group of affiliated ore formations, which, according to the time of their origination, directly replace the polymetallic proper and other similar formations wherein fluorite is present as an admixture.

The sulphide–fluoritic ores contain various quantities of gangue minerals, but quartz predominates in them over calcite, whereas barite is encountered only sporadically. Sometimes they contain from 5 to 10 per cent feldspar. Arsenopyrite, hournanite, proustite and native silver are present in addition to lead, zinc, copper and iron. Besides the above-enumerated minerals, antimonite, argentite, magnetite, greenockite, faded ore, bornite, cuprite, ancerite, gold, cinnabar, molybdenite, pyrargyrite, stephanite, chalcocite, ilmenite and others are encountered in quartz–barite polymetallic fluorite mineralization. These minerals can also be detected in sulphide–fluorite (with quartz and calcite) deposits.

The fluorite-bearing deposits whose ore component is represented mainly by galena or galena and sphalerite, are the most widespread of the types of mineralization being considered.

Less frequently, their base consist of copper (tetrahedrite) and chalcopyrite minerals associated with galena, arsenopyrite, jamesonite, dickite, dolomite, calcite, malachite, azurite, chrysocolla, pyrolusite, covelline, bornite, native copper and silver, aurichalcite, scorodite and halloysite.

The pre-ore stage of the hydrothermally altered rocks in sulphide–fluorite (with quartz and calcite) and quartz–barite polymetallic fluorite deposits is generally characterized by the development of epidote, chlorite, sericite, and, less frequently, of orthoclase. The process of mineralization is accompanied by albitization, silicification, fluoritization, carbonatization, and pyritization of the substratum, and sometimes by hydromicatization (illitization) of rocks (at the post-ore stage).

All the types of fluorite mineralization are directly indicated by fluorite which, in the case of sulphide–fluorite (with quartz and calcite) and quartz–barite polymetallic fluorite mineralization, is accompanied by galena and, less frequently, by sphalerite. If sulphides occur without fluorite they are considered not to be indicators of fluorspar mineralization because they can indicate proper polymetallic occurrences. Other minerals should be regarded as indirect indicators of mineralization of the appropriate types. Of major significance are the most widespread minerals, such as quartz, calcite, barite and some of the ore minerals (pyrite and marcasite for pyrite–marcasite–fluorite deposits; and copper-bearing minerals for sulphide–fluorite and quartz–barite polymetallic fluorite deposits on a base of copper). Fluorine is the principal indicator element of the above fluorite-bearing formations. Its role is greater in hydrothermolites. However, even an insignificant increase in the content of admixture minerals in ores widens appreciably the range of indicator elements, e.g. in quartz–calcite–fluorite deposits.

It should be noted that fluorite deposits are indicated not only by the elements composing their own minerals but also by those which are present in them as admixtures. Manganese, chromium, rare earths of the cerium and yttrium groups, copper, lead, zinc, beryllium, lithium, germanium, gallium, ytterbium, silver, and others have been found in fluorite. The share of some elements can be considerable: beryllium up to 0.05–0.01 per cent, barium 0.01 per cent, strontium up to 3 per cent, and rare earths 4–5 per cent. The contents of potassium and sodium in fluorspar from deposits of fluorite formation reach 0.35 and 0.087 per cent, respectively.

Many fluorites have been found to contain abundant gaseous and liquid inclusions (seldom with a solid phase), wherein CO_2 is present together with salt solutions (concentration of 10 to 50 weight per cent), volatile compounds and elements (Cl, H, F and others).

Mercury is permanently present in ores and major minerals. It is predominantly concentrated in barite, faded ores, sphalerite and galena. Fluorite contains the most low-temperature chloridic and fluorine forms of mercury. We have unified the groups of indicator elements typical of each type of mineralization of the fluorite formation in Table (3.4).

The more complicated the composition of the ore bodies, the greater is the number of indicator elements. Some are suitable for all cases, but, in the majority of instances, they are quantitatively differentiated, depending on the type of mineralization. This circumstance, as well as appropriate sets of chemical elements, should be considered when searching for fluorspar deposits. In the process of such surveys, the samples should not be analysed only for fluorine. It is also advisable to determine barium, mercury, lead, silver, bismuth, molybdenum, copper, antimony (for mineralization of the fluorite formation), as well as those elements which characterize geochemically other types of calcium–fluoride deposits.

Fluorspar resists chemical weathering and therefore tends to persist in the soil overlying weathered veins. Brightly coloured cleavage fragments washed clean on slopes by rainwater or exposed in anthills or spoil piles from animal burrows have drawn the attention of prospectors to possible fluorspar deposits in many situations. Test pits, trenches, and auger borings are commonly employed to follow the weathered zones along the apices of veins. Where veins are silicified,

Table 3.4.

Indicator elements of mineralization of the fluorite formation

Type of mineralization	Indicators		
	Direct	Indirect	Depending on local conditions
Quartz–fluorite	F, Hg, Li, Be	Yb	Mo, W, Ti
Calcite–barite–fluorite	F, Ba, Hg	Pb, Mo, Ag, Bi, Sr	P
Quartz–calcite–fluorite	F, Ba, Pb, Hg, Mo, Y, Be, Ag	Zn, Sr, La, Li, Yb	Cu, Zr, As
Quartz–barite polymetallic fluorite	F, Pb, Ba, Hg, Cu, Zn, Ag	Bi, Mo, Sn, Sb, Sr, Be	W, Li, Y, Yb, Mo, Ti, P, N
Sulphide–fluorite			
on galenite–sphalerite base	F, Pb, Cu, Zn, Hg, Ag	Ba, Sn	Mo, Bi, Cr, Be
on pyrite–marcasite base	F, Mo, As		
on copper (with bismuth) base	F, Cu, Bi, Ag, Hg	Pb, Zn, As	

they frequently stand up like a 'reef' and are easily noticed. Bedded deposits are less conspicuous, but where there is sufficient topographic relief the outcropping edges of deposits or slumped fragments of ore are sometimes found. At other times the first indication of bedded deposits may be noted incidentally during the study of samples taken from boreholes drilled for water or other mineral deposits. Owing to its softness and extreme cleavability, fluorspar does not last long in the beds of stream and ordinarily cannot be traced by panning.

Once a hint of a vein or bedded deposit has been gained from surface showings, prospect shafts and drifts and core or churn bore holes are used to intersect and follow the ore structures. Photogeology and geochemical methods can be of assistance in locating and tracing the outcrop of a fault along which veins or the extensions of veins may occur at depth, as well as the areal distribution of rock types or stratigraphic units which are known to be most favourable for the occurrence of veins or replacement deposits.

Conditions for the application of geochemical methods of prospecting

The applicability of geochemical methods in the search for fluorite deposits has experimentally been proven in different regions of Russia.

Explorations along secondary haloes have resulted in the discovery of many fluorite-bearing objects in Kazakhstan, Central Asia and in the Transbaykal area.

Rare metal–fluorite apocarbonate-greisen types of mineralization and associated geochemical anomalies are variable in their composition and are reliably detected in loose, up to 5 m thick deposits. The widest haloes (up to 200×100 m) are those of fluorine, whose concentration reaches 30 per cent. The zone of its accumulation is almost always characterized by an increased content of beryllium. Apart from fluorine and beryllium, the rare metal-fluoritic metasomatites are sometimes superimposed by comparatively small nickel haloes, which are absent in loose deposits in areas with skarns and granites.

The quartz–calcite–fluorite type mineralization, represented by large blanket-like deposits, is characterized by the widespread anomalies of fluorine in loose deposits. Some of them extend over 3 km². The content of fluorine within the haloes generally fluctuates, from 0.2 to 0.5 per cent and more. The existence of such large anomalies is explained by the occurrence of closely spaced elongated and thick commercial fluorite-bearing deposits, and by the general contamination of rocks (in the ore fields) with fluorspar. Some of the haloes are confined to the zones wherein fluorite-bearing bodies were detected at a depth from 40 to 200 m. The fluorspar mineralization is also accompanied by local secondary haloes with anomalous contents (per cent) of: lead (0.005), zinc (0.015), copper (0.0004), molybdenum (0.004), boron (0.001), vanadium (0.002), tin (0.0005), and beryllium (0.05).

Secondary haloes of fluorine, lead, zinc, copper and molybdenum (ranging in width from 100 to 180 m), as well as tin and beryllium (from 60 to 80 m wide) have been detected in a vein quartz–barite polymetallic fluorite deposit. Maximum concentrations of fluorine, lead, zinc, copper, molybdenum, tin, and beryllium in geochemical anomalies reach 10, 0.3, 0.15, 0.03, 0.004, 0.007 and 0.002 per cent, respectively. When the thickness of proluvial–deluvial deposits exceeds 3 m, secondary haloes are not detected in areas containing ore bodies. The overwhelming majority of quartz–fluorite deposits and deposits close in composition, occurring in the most diverse rocks, are accompanied by secondary fluorine haloes, which are commonly several hundred metres long and several tens of metres wide (sometimes they are greater in size). These haloes can partly outcrop in creek valleys, thus forming 'medallions', which play a significant role in prospecting.

The ore bodies are identified by contrasting fluorine anomalies occurring in loose deposits directly on the sites with quartz–fluorite veins and elongated in the direction that coincides with the orientation of tectonic dislocations with a break in continuity.

Experimental surveys carried out on one of the quartz–fluorite deposits (with an admixture of sulphides) resulted in the contouring of a complex lead, copper, zinc, silver, molybdenum, arsenic and bismuth anomaly that was strictly confined to the site of quartz–fluorite mineralization (with the exception of Cu and Bi haloes), even though the zones of lead, zinc, molybdenum, arsenic and silver accumulation sometimes differed appreciably in size contrast, and location.

Having proved the high efficiency of lithogeochemical surveys in a number of areas, the explorations carried out in recent years, have highlighted some negative features of this method. Along with the haloes that are superimposed in space with ore bodies and ore zones, secondary geochemical anomalies are known whose source is obscure and the secondary haloes, as compared with the primary ones, are less contrasting, more diffuse and wide. Moreover, they are almost always displaced downslope from the primary sources. For example, on one of the deposits, wherein a large group of haloes (with a fluorine content ranging from 0.1 to 1.0 per cent) several hundred metres long and several tens of metres wide has been identified, the geochemical anomalies are not distinctly confined to fluorspar bodies. Only a small part of them is tentatively associated with the stripped veins and mineralized zones of crushing. Successive detailed exploration of geochemical anomalies in some cases failed to detect the epicentres of mineralization, and reproduced only fragments of the original weak haloes. At times, it was difficult to distinguish genuine geochemical anomalies above the ore bodies from false ones.

Layer-by-layer sampling of slope sediments has shown that the secondary fluorine haloes, developed by 0.3 to 6.0-m-third quartz–fluorite veins containing from 20 to 60 per cent CaF_2, belonged in the majority of cases to the buried type and produced at the surface only weak anomalies, limited in size and displaced by 70–150 m. These anomalies were most contrasting at a depth of 0.5–1.0 m. Mixing of the material moving downslope sometimes brings the haloes to the surface. In addition, the formation of exposed geochemical anomalies is favoured by salt enhancement of the haloes of mobile elements. This is characteristic of flooded solifluction deposits.

Exposed train-shaped fluorine haloes can be detected on strongly eroded steppe slopes containing tectonically crushed ore bodies. The hypergene anomalies proper (mechanical, salt and mixed) are of varying prospecting significance. Mechanical haloes formed as a result of disintegration of fragmental material (veined and veinlet-disseminated fluorite formations), are of utmost importance. These haloes are mainly confined to the positive features of the topography. Thus, sufficient steepness of the slopes in mountain regions ensures gravitational migration of the rock debris from the occurrences of fluorspar bodies.

Mechanical haloes occur only in some localities in low mountains and steppe areas with hilly topography. They are not formed at all on flat areas where the thickness of the loose deposits reaches many metres.

The application of mineralogical methods of prospecting over vast areas is restricted by the rapid disintegration of fluorite to small fractions 0.5–1.0 mm. However, their effectiveness increases when prospective sites are subjected to large-scale exploration. The presence of numerous fluorite grains in heavy concentrates indicates a nearby occurrence of their primary source.

In monomineral fluorspar veins fluorite is disintegrated more intensively than in quartz and quartz–baritic ones. This phenomenon is responsible for the differences in the intensity and contrast of secondary haloes and dispersion trains of fluorine and its accessory elements.

The widely distributed salt haloes reflecting, in particular, the degree of absorption of fluorine compounds by loose rocks, are confined as a rule to river and creek valleys, wherein they may reach a high contrast (up to 20 per cent of fluorine). False anomalies are not unusual among them.

Registration of the haloes formed by upward-migrating salts in defluction slopes may be very important for refinement of geochemical methods of exploration. In large-scale surveys the neutron-activation method can be used to detect fluorine anomalies in loose sedimentary cover up to 3 m thick. This method makes it possible to measure the fluorine content in the cross-sections of boreholes and workings, as well as in bulk and powder samples. It also aids in an unambiguous identification of the ore-controlling structures among numerous dislocations with a break in continuity and in their urgent testing by mining and drilling operations.

Hydrogeochemical, biogeochemical, and, especially, geobotanical methods are still applied in the search for fluorite on an experimental basis. The feasibility of the use of hydrochemistry in the search for fluorite mineralization is conditioned by the fact that the content of fluorine in the waters flowing around deposits and ore occurrences becomes several times higher. For example, groundwater sampled from a fissured zone in granites hosting quartz–fluorite veins contains up to 7.5 mg/l of fluorine, which significantly exceeds the content of fluorine in similar waters in other types of deposit. At the same time, serious consideration should be given to the hydrodynamic conditions of the area being prospected, where intensive circulation of surface water may be responsible for exceptionally low fluorine concentrations.

The efficiency of the hydrogeochemical method is also greatly dependent on the geological structure of the region, which in many respects determines the ground water chemical composition, genesis and regime. Thus, preliminary collection of information on the lithology and structure of the region, as well as on the composition and dynamics of the waters, is imperative to make a correct decision on the applicability of this method in the search for fluorite mineralization.

Bedded deposits commonly occur at the intersection of minor structures such as faults with slight displacement or joints or shear zones with certain favoured strata in any given stratigraphic series. Consequently, the search for these deposits usually involves finding a mineralized structure and following it with vertical bore holes drilled to the appropriate depth to intersect the 'favourable' beds.

Geochemical methods have been employed with varying degrees of success in the search for flourspar deposits. Fluorine anomalies have been sought in ground-waters and surface streams using the specific ion electrode method. Soil samples and stream sediments have also been examined for anomalous concentrations of various non-fluorine elements which might prove to be guides to the location of fluorspar deposits.

Some success has been reported in the use of lead and zinc as guides.

Todorov *et al.* (1972) in Bulgaria reported the advisability of applying hydro-geochemical methods in the search for fluorites only on sites composed of alkaline rocks, where waters are rich in sodium and poor in calcium. These authors empha-sized the low efficiency of hydrogeochemical explorations in the areas composed predominately of sedimentary and acid magmatic rocks. In Czech Republic, on the contrary, hydrogeochemistry with quantitative determination of fluorine, lead, zinc, barium, copper and other components is used in prospecting for fluorite–barite ores, and has already resulted in the discovery of commercial deposits in the Krushny Hory (Iron Mountains).

The results of surveys carried out in the Madoc area (Ontario, Canada) with veins of similar composition have shown that the position of mineralization bodies is well outlined by fluorine dispersion in ground water. Hydrochemical fluorine haloes, reaching sometimes several hundred metres, have also been detected in the Transbaykal region. For example, the anomaly in sub-permafrost waters, which is associated with one of the deposits, reaches 1.5 km, and its intensity amounts to 2.2 per cent [89].

Surface water, containing from 0.2 to 0.23 mg/l fluorine, displays a small res-olution capacity in the search for fluorspar, mineralization as compared to fracture water. Ya. D. Fedorenko believes that background concentrations of fluorine in the surface waters of the Transbaykal region do not exceed 1 mg/l. According to other investigators, ground waters in the fluorite-bearing areas of the same province contain from 2 to 7.4 mg/l fluorine, versus 0.1 to 0.9 mg/l in barren areas. An increased fluorine content (0.8 mg/l) was observed in the waters of small rivers, originating in the same fluorite-bearing areas, along their entire length. Fluorine levels reach 1.1 mg/l where streams that have crossed fluorite deposits flow into the rivers.

Of some interest are the as yet rare data on the possible delineation of fluorine anomalies from the analyses of ice bodies formed at points of outflow of subsurface waters that are confined to tectonic zones. Fluorine concentrations of 1.2–1.6 mg/l were determined in ice sampled on some of the sites of fluorite deposits, as against background levels of 0.4 mg/l. In arid regions it is probably advisable to sample ice bodies in winter.

Many plants have physiological barrier to high concentrations of fluorine in soils. Nevertheless, the potential of utilization of biogeochemical surveys along closed haloes has been proved by investigation. On the spoil banks of quartz–fluo-rite ores, special attention is attracted by *Crepis* sp., whose morphological changes are less conspicuous on other sites. Small-flowered columbine (*Aquilegia brevistyla*) is well-controlled by the zones of hydrothermal alteration of rocks [89]. The maximum area of columbine growth extends 20 m upslope and 60 m downslope. In the area under investigation, seven species of plants growing above ore bodies and argillized zones accumulated 2–3 times more fluorine (0.06 per cent) than similar plants growing on barren sites.

Where slope deposits are up to 8 m thick, biogeochemical sampling is generally more effective than lithogeochemical surveys. The fluorine concentrations in plant ashes average $(10–50) \times 10^{-4}$ per cent, with a maximum of 600×10^{-4} per cent. The maximum contrast range is 15. Fluorine from fluorite is more intensively absorbed by plants than fluorine from amphiboles and micas.

According to A. L. Kovalevsky, the barks of birch and pine have relatively high threshold concentrations of fluorine in ashes: 10 and 6 per cent, respectively, compared to background values of 0.06–0.08 and 0.02–0.03 per cent. The fluorine content in birch and pine bark does not depend on the height of sampling (within the interval from 0.5 to 15 m) on an orientation with respect to the cardinal points. However, it should be emphasized that above ore bodies fluorine concentration in the external layer of pine bark was three times as high as in its inner layer.

The depth of biogeochemical survey depends on the thickness of the overburden, the length of the root systems of plants, and the ground water table. According to V. V. Polikarpochkin, the root systems of plants and shrubs in the Transbaykal region can extend to a depth of 5 m and in rare cases down to 10 m. On the average, the root systems of trees penetrate down to a depth of 3–5 m, shrubs reach 2–3 m, and grasses 1.5 m. A 2–4 m thick overburden is pierced by a dense network of roots that penetrate into the bedrocks and ores and form local supra-ore or slightly displaced biogeochemical haloes. If the thickness of overburden ranges from 5 to 15 m the root systems cannot reach the bedrocks and ores. In such cases, plants uptake predominantly salt solutions and form less local and contrasting biogeochemical anomalies. It is supposed that the maximum depth of the allochtonous cover (including that on slopes), restricting biogeochemical surveys, ranges from 20 to 50 m in steppe and desert regions, from 0 to 30 m in forest zones with humid climate, and from 3 to 10 m in permafrost regions. This technique can be used for the identification of the part of buried deposits.

Primary haloes develop around fluorite mineralizations of various genetic types. Negative fluorine haloes, enveloping productive chambered pegmatites in granites, have been detected in the occurrences of rare earth, rock-crystal-bearing fluorite formation. In plan, anomalies have a ring structure and generally accompany pegmatitic bodies. The outer part of the fluorine halo is actually a negative anomaly, located between the inner maximum and the surrounding background. Fluorine accumulation in the individual zones of pegmatite (up to 0.03 per cent in graphic granite, and up to 0.09 per cent in feldspar zone) is indicative of the inhomogeneity of the geochemical field of this element above the productive body. Halo intensity depends on the dimensions and shape of the pegmatite. If it is 15×20 m in size, a fluorine anomaly can be detected at a distance of up to 100 m from the productive body, which can be discovered at a depth from 10 to 80 m by negative fluorine haloes.

Apocarbonate−greisen rare metal−fluorite mineralization is accompanied by fluorine and beryllium haloes, and by slightly pronounced anomalies of chromium, titanium, manganese, nickel, lithium and other elements. Fluorine haloes are much bigger in size than those of other elements in the bedrocks. Fluorine anomalies are also very conspicuous at the contacts of limestones with granite−porphyry and quartz−wolframite veins affected by rock greisenization. The contrast range of the considerable, absolute and relative (with respect to background) concentrations has been observed.

Good correlation between fluorine and beryllium is observed in zones influenced by rare metal−fluorite mineralization. At the same time, contrary to rare metal−fluorite mineralization superimposed on granite−porphyries and skarns, apocarbonate metasomatites are characterized by an increased amount of nickel and complete absence of bismuth, lead, zinc, copper, arsenic and silver haloes in their vicinity.

The following geochemical zonality was detected by G. Yu. Kolomensky on one of the apocarbonate−greisen deposits of rare metal−topaz−fluorite composition:

$$\text{(upward) } TR \rightarrow Mo + Be_I + Sn \rightarrow Mo + Sn \rightarrow Be_{II} \rightarrow Cu + Pb + Zn;$$

and

$$\text{(from west to east) } Cu + Pb + Zn \rightarrow Be_{II} \rightarrow W + Bi \rightarrow Mo + Sn.$$

Rare metal−fluorite (with accessory beryllium) occurrences of ore in the greisenized limestones are characterized by the following row of zonality:

$$Cu \rightarrow Pb \rightarrow Li \rightarrow F \rightarrow B \rightarrow Be \rightarrow Mo \rightarrow Zn.$$

Table (3.5) contains zonality indices characteristic of different levels of primary haloes associated with ore occurrences.

The above-stated stable associations of elements, rows of horizontal and vertical zonality, and indicator ratios and their values should be taken into account when widening the known fields with rare metal−fluorite mineralization, and in searching for and evaluation of deposits in new areas.

A typical quartz−calcite−fluorite mineralization, formed with a considerable participation of the process of fluorine metasomatism, may be exemplified by the deposit wherein positive anomalies of fluorine, barium, molybdenum, lead, zinc, silver, arsenic, yttrium, ytterbium, and strontium, as well as negative anomalies formed by the elements of alkaline (potassium and sodium), and ferric (manganese, titanium and chromium) groups are detectable (data of the author, and

Table 3.5.

Zonality indices [59]

Indicator ratios	Level of primary haloes		
	Supra-ore	Ore	Sub-ore
$Li \times Pb/Zn^2$	$n \times 100$	n	$n \times 0.001$
$Fe \times Pb/Zn^2$	$n \times 100$	$n \times 100$	$n \times 10$

L. S. Puzanov and M. A. Zubov). The negative primary anomalies result from the removal of the stated elements from the zones of hydrothermal alteration, where they were present in the host rocks in amounts exceeding their concentration in mineral-forming fluorine-bearing solutions.

Decreased potassium and sodium contents are distinct in a zone of intensive acidic leaching of the substratum, wherein rocks are affected by the most enhanced processes of solicification and fluoritization. The leaching zone is characterized by a distinct negative halo formed by manganese, titanium, and chromium, which are liberated by the destruction of minerals contained in terrigenous rocks. The negative haloes of elements are generally characterized by small dimensions (somewhat larger for haloes of alkaline elements); close relationship with the ore body and the major solution-supplying channel; occurrence in the supra-ore rocks; and practical absence in the underlying carbonate formations. It should be noted that the halo of the alkaline group of metals is more contrasting compared to the anomaly formed by elements of the ferric group, at potassium and sodium concentrations of 0.01–0.1 per cent, and also thus the haloes of manganese, titanium, and chromium, amounting to 1×10^{-5} per cent. Fluorine haloes, associated with rather elongated quartz–fluorite type veins, are commensurable with the zones of hydrothermal alteration of rocks.

Primary haloes of fluorine, beryllium, molybdenum, arsenic, antimony, and tungsten have been detected on the sites of one of the fluorite-bearing zones in Central Asia. Being outlined by the minimum anomalous values, they do not generally extend beyond the occurrence of circum-ore altered rocks. Unlike fluorine, beryllium, molybdenum, antimony and arsenic haloes probably result from the dispersion of corneous quartz containing increased quantities of these elements. The fluoride vein extends over 1 km and its wall rocks are notable for increased contents of fluorine, molybdenum, arsenic, and antimony [32]. The last three elements are mainly concentrated in pyrite. Greatly increased concentrations of fluorine decrease drastically with distance from the vein body. A fluorine halo extends generally over 4 m, but increases up to 12 m at the places of vein branching. Being a rather mobile component of the ore-forming solutions, fluorine is capable of forming haloes that can be detected at a distance of hundreds of metres (in the vertical direction) from productive deposits. The very existence of primary haloes of such dimensions and shape favours their utilization in the search and evaluation of fluorite deposits of various types.

The ore zones of fluorite formation display a regular distribution of subsequently-originated mineral paragenesis, which is responsible for the vertical zoning of mineralization. Preference should be given to those fluorite-bearing veins, occurring at the same hypsometric level, whose zonality originated from substance deposition in the process of repeated opening of the host structures.

The calcite–barite–fluorite deposits of Central Asia are characterized by the change of quartz–calcite to calcite–fluorite composition along the strike of veins, and by the accumulation of barite in the upper ore horizons, coupled with the diminution of this element (until it almost disappears) in the lower horizons. Regular changes in the composition of quartz–fluorite veins from one flank to the other have been detected on deposits in the Transbaykal region. In the ore

Table 3.6.

Vertical zonality of primary haloes [22]

Zone	Thickness, m	Characteristic elements*
Upper	Not determined	As, Mo
Supra-ore	50–60	F, As, Be, **W**, **Sb**, Mo
Ore	150–200	F, As, W, Be, Sb, Mo
Sub-ore	40–50	F, As, **Sb**, **W**, **Mo**
Lower	Not determined	As, Sb, **W**, Mo

*Bold type denotes mutually interchangeable elements.

bodies of similar composition, occurring in the same region, A. A. Kotov *et al.* distinguish three vertical zones: the upper one containing fluorite, barite, kaolinite, dickite, quartz and iron hydroxydes; the lower one containing quartz, pyrite, hydromicas and carbonates; and the middle one with an intermediate composition. Not infrequently, the elongation of veins along their dip and strike, as well as the downward extension of mineralization, are generally proportional to the length of ore columns and the fluorine content therein (data of surface surveys).

In other types of fluorite deposits, vertical zonality, which can be direct, inverse, or combined, is by no means less contrasting. This diversity of vertical zonality impedes the evaluation of the downward development of mineralization and makes it imperative to resort to mining and drilling for its specification.

The data on vertical zonality of haloes associated with veined quartz–fluorite mineralization are presented in Table (3.6).

As has been reported by S. A. Domoryad, the quartz–fluorite mineralization (in the same region), associated with quartz–barite–polymetallic fluorite and other metalliferous complexes, is characterized by an indistinct vertical zonality of elements ($F \leftarrow Be \leftarrow Li \leftarrow W$) whose selective capacity of accumulation in the upper horizons and on the flanks of ore-bearing flows increases in the direction from right to left. The index of zonality $(F \times Bi)/(Li \times W)$ for the erosional truncation of ore bodies and zones is expressed by the following values: $(n \times 10^2) \div (n \times 10^3)$, $(n \times 10^0) \div (n \times 10)$, and $(n \times 10^{-1}) \div (n \times 10^0)$ for the upper, middle, and lower horizons, respectively.

The following zonality is typical of quartz–barite–polymetallic fluorite deposits:

$$Sr \leftarrow Ba \leftarrow Ag \leftarrow F \leftarrow Pb \leftarrow Be \leftarrow Bi \leftarrow Li \leftarrow W.$$

Strontium and barium are predominantly accumulated in the supra-ore and above-ore horizons; silver, fluorine, lead, bismuth and beryllium accumulate in the middle and lower horizons; and lithium and tungsten accumulate in below-ore and sub-ore horizons. These is an upward increase in the zonality index of the halo space (ratio of linear productivities $(Ba^2 \times Sr)/(F \times Li \times Be)$) from $n \times 10^{-4}$ up to $n \times 10^{-1}$. The concentration of tungsten is increasing in the opposite direction.

M. A. Zubov gives the following rows of chemical indicator-elements of primary haloes in one of the quartz–calcite–fluorite deposits:

Total: F ← Ag ← Ba ← Bi ← B ← Li ← La ← P ← Zn ← Be
            ← Sr ← W ← Pb ← Ni ← Ga ← As ← Co ← Mo ← Sn
            ← Sc ← Cu ← Yb ← Y ← Ge

Supra-ore: F ← Ag ← As ← Sr ← Ba ← Co ← Pb ← Mo ← Co ← Zn
            ← Ni ← Sn ← Be

Ore (upper part):   F ← Sn ← Zn ← Pb ← Mo ← Co ← Cu ← Ba ← Sr ← Ag
            ← Be ←Ni ← As

Ore (central part): F ← Co ← Mo ← Pb ← Cu ← Ni ← Sn ← Ba ← Zn ← Sr
            ← Ag ← Be ← As

Sub-ore: F ← Sn ← Be ← Co ← Mo ← Pb ← Zn ← Sr ← Ni
            ← Ba ← Cu ← Ag ←As

These rows, adjusted by the predominance of one element over another as to the value of the relative content or productivity, reflect the degree of accumulation of the major halo-forming components in the anomalous sections (zonality), and allow us to determine erosional truncation of geochemical anomalies.

Methods of work

Lithogeochemical surveys along secondary haloes of fluorine and its accessories may extensively be carried out in semi-closed areas of fluorite-bearing provinces, wherein bedrocks are overlain by eluvial–deluvial deposits that preserve secondary elemental haloes resulting from the destruction of ore bodies, enveloped by primary geochemical anomalies. Due to their considerable size, some of the hypergene lithogeochemical haloes are sufficiently well-detected by fluorinemetric survey at a scale of $1:50\,000-1:10\,000$. If loose deposits are 2–4 m thick, it is advisable to take samples on a grid of 100×200 to 100×10 m.

Optimum results are obtained, as a rule, in areas where the thickness of deposits (except those brought from distant territories) do not exceed 3 m, ensuring the most complete preservation of the direct relationship between the chemical elements of the ore-bearing and overlying rocks in the conditions of contemporary hypergenesis. This relationship diminishes with an increase in the thickness of deposits, and practically disappears when it reaches 6–10 m and more. For example, the following varieties of secondary haloes, whose specific features depend on the landscape conditions, may be formed in loose local formations (up to 5 m thick) of the Transbaykal region:

(1) residual eluvial–deluvial, open;

(2) residual leached out;

(3) residual impoverished;

(4) residual complicated; and

(5) residual (possibly) false.

Halo formation on slopes can be appreciably affected by the processes of defluction, solifluction, and creep, which frequently result in the appearance of concealed or insignificantly outcropping secondary anomalies produced by an ore source located at the middle and lower levels of the slopes. Such haloes can also be formed on sites with relatively thick loose deposits.

Experimental surveys help elucidate the optimum depth of sampling. This depends on the genesis and position of the enriched horizon in loose deposits. Samples should be taken in dry regions with weakly developed vegetation cover right at the surface or from a depth of 10–20 cm. In areas with strongly leached soils sampling should be from a depth of 30–40 cm and more, while in humid areas samples should be from a depth of 30–40 cm; and in the arid zone and all mountain regions from a depth of 10–20 cm.

The development of diffusional and diffusional–defluction haloes in the positive features of topography is characteristic of the steppe zone, where they are quite effectively detected by shallow sampling (down to 20 cm). Buried and semi-buried diffusional defluction haloes, detectable by similar pattern of sampling mainly on the watersheds and upper parts of slopes, are typical of the forest zone. The same is true of the transitional forest-steppe zone, where deposits are detected by lithogeochemical surveys mainly on the watersheds and southern slopes with steppe vegetation, whereas on northern forested slopes they are identified only in their upper parts. Shallow sampling is applicable in the tundra, characterized by the development of closed and semi-closed haloes occurring at a small depth. In mountain tundra zones, sub-bold mountain, and middle-taiga areas, samples of frozen non-podzolized soils should be taken from the upper horizons, from directly under the roots of plants. The same sampling technique is applicable in moderately dry and dry steppes. However, in southern taiga and forest-steppe landscapes it is advisable to sample the illuvial horizon, occurring commonly at a depth of 30–50 cm.

The application of lithogeochemical surveys is impeded by:

(1) The occurrence of podzolized soils in mountain tundra, mountain taiga and, partly, in forest-steppe zones.

(2) The presence (in the upper part of the soil profile) of 10 to 40-cm-thick humus horizons characteristic of the southern-taiga and, to a lesser extent, of forest-steppe and middle-taiga landscapes. Sampling of podzolized and humus horizons yields distorted data, therefore, material required for the identification of hypergene haloes should be sampled about 10 cm below these horizons.

(3) The occurrence (on the slopes of positive forms of topography) of loose formations with a binary structure of profile resulting from the superimposition of local deposits by the material brought from distant areas that also decreases the efficiency of the lithogeochemical methods; and

(4) The presence of waterlogged landscapes formed on thick loose deposits in river valleys where surveys along secondary haloes are inapplicable. In the middle and southern taiga areas, notable for the occurrence of a rather

thick (40–50 cm) moss and peat layers, the greatest contrast of haloes is provided by samples taken from humus horizon A of soils.

Deep lithogeochemical sampling is the major technique used to discover fluorite-bearing bodies in semi-closed or closed areas with hypergene haloes weakened at the surface of buried. Such sites should be subjected to layer-by-layer sampling of slope deposits in the surveys and test workings, and to blast-hole and drill-hole sampling. Manual samplers and UPB-25-type units can be used to perform relatively shallow (down to 3 m) sampling at the first stage.

Sampling of prospective areas favourable for fluorite mineralization prospecting along secondary haloes, is generally performed along the profiles oriented across the strike of the known ore-bearing structures, and on a grid that depends on the scale of work. Control sampling (3–5 per cent of total samples) is performed along the same profiles. Commonly, samples of small (up to 0.2 mm), 200–300 g fractions taken from the representative horizon are ground and subjected to chemical analysis. In the arid zone Sochevanov [110] recommend sampling of 1–3 mm fractions areas overlain by loess-like deposits, because this fraction contains more ore elements than the small (0.25–0.1 mm) and fine (< 0.75 mm) fractions.

There is certain specificity in the evaluation of secondary haloes identified in considerable amounts by lithogeochemical surveys. It is especially difficult to assess closed and buried secondary haloes occurring in areas with a complicated structure of thick loose cover containing sedimentary formations brought from distant areas. Exploration of such haloes involves drilling. The easier cases are those when the environmental conditions permit lithogeochemical surveying of fluorite-bearing sites along loose formations and bedrocks.

The surveys result in the compilation of fluorometric maps showing primary and secondary haloes. Mutual correlation of these haloes, based on their contrast, dimensions, and geological position, makes it possible to more objectively identify the prospective haloes and to establish the priority of their evaluation. It should also be borne in mind that secondary geochemical anomalies reflect quite distinctly, but in a considerably smoothed shape and without displacement, the structure of endogenous haloes in a terrain with a horizontal surface but a steeply dipping ore body. In other cases, secondary anomalies are displaced with respect to their primary sources at a distance depending on the migrational capacity of the elements, the complexity and inclination of the topography, the elongation of slopes, the thickness of overlying rocks, the dimensions and number of disintegrating ore formations, etc.

The displacement value S of a steeply dipping ore body may be determined from the formula proposed by A. P. Solovov [46]:

$$S_\alpha = Ah^2 \sin \alpha,$$

where A is the parameter depending on local conditions, h is the thickness of eluvial–deluvial deposits, and α is the angle of slope surface inclination. The secondary halo associated with a gently dipping ore body is displaced up the dip at a value of $S_\beta = Bh\cot \alpha$, where B is a coefficient that depends on rock alteration by weathering, and β is the angle of ore body dip. The total displacement of a halo from a vein projection on the surface is determined by the geometrical sum $S_\alpha + S_\beta$. Displacement of secondary haloes can be determined from the dependence (for slopes of 8° and more) found by M. S. Solodyankin

$$l = m \times \tan\left(90 - \frac{\alpha}{2}\right),$$

where l is the value of displacement, m is the thickness of overburden, and α is the angle of slope inclination.

Grading of anomalies of fluorine and accessory elements should be based on the elucidation of the nature of haloes. The following indications, characteristic of the areas in the Transbaykal region, should be taken into account:

(1) Secondary haloes above granodiorite–gneiss type rocks are appreciably elongated, have persistent distribution of fluorine concentrations (0.1–0.3 per cent) and an increased phosphorus content.

(2) Fluorine concentrations above granites do not exceed 0.01–0.1 per cent in wide areas, limited by contours of intrusions.

(3) Fluorine concentrations of 0.01–0.3 per cent are characteristic of the fields of pegmatites, greisens, and other high-temperature formations, but fluorine is not uniformly distributed in the haloes and is accompanied by other chemical elements usually contained in the above-stated endogenic products.

(4) Haloes above ore deposits are notable for an insignificant contrast range of fluorine (0.01–1.0 per cent) and increased concentrations of other ore elements.

(5) False geochemical anomalies of fluorine, associated with its secondary accumulation, are formed in the low-lying parts of the topography (valleys and depressions).

(6) Haloes, prospective for fluorspar, are of small extension (hundreds of metres–1 km) but high fluorine concentrations (0.1–10 per cent).

Fluorite anomalies are more reliably, compared to fluorine haloes, detected in eluvial–deluvial deposits. Being a mechanically unstable mineral, fluorine does not always form considerable fragmental accumulations. Nevertheless, the thermoluminescent method, permitting the determination of fluorite content in the range from 0.02 to 10.0 per cent, may assist in detecting crushed fluorite and in outlining the area of its occurrence.

In the opinion of M. S. Solodyankin, evaluation of secondary lithogeochemical anomalies may be based on the totality of the following indications: (a) the quantitative ratio of quartz and fluorite grains in coarse fractions (upto 1 mm) sampled within the halo; (b) the contrast range of anomalies; (c) the width of the halo containing fluorite and quartz grains with an overburden thickness of around 2 m; (d) results of thermal luminescence analysis; and (e) data of chemical analysis for

Table 3.7.

Evaluation indications of secondary haloes

Priority of evaluation	Fluorite–quartz ratio	Contrast range of fluorine anomalies	Halo width, m	Thermal luminescence	Fluorine content, per cent
I	1	$C_b - 3\sigma$	More than 20	Intensive	2 and more
II	1	$C_b - 3\sigma$	20	Intensive	1–2
III	1	$C_b - 2\sigma$	Less than 20	No	1
IV	0		Up to 10		Up to 0.5

Samples that do not emit light are studied under microscope because some of the fluorites do not fluoresce; C_b is the background content of fluorine.

fluorine. Based on these indications, secondary haloes are subdivided into priority groups for their evaluation by mining operations (Table (3.7)).

If the origin of secondary geochemical anomalies is not clear it is advisable to subject the site once again to lithogeochemical survey at the stage of more detailed prospecting in order to elucidate the reproducibility of haloes. To obtain more reliable results, secondary anomalies are defined from the flourine–phosphorus ratio.

Grading of haloes is significantly assisted by the quantitative evaluation of the predicted reserves of fluorite raw materials. Calculations are made by utilizing the data on the secondary fluorine haloes delineated in the process of detailed fluorometric surveys. The quantity of the useful component in the halo, expressed in m^2 per cent, is compared with its quantity in the erosionally truncated ore body outcropping under eluvial–deluvial deposits. The ratio of these values, i.e. the coefficient of proportionality K (major parameter for calculations) is fairly constant in various geologo-geomorphological conditions. The halo productivity (in m^2 per cent) is determined from the known formula:

$$P = \Delta x \times l \times \left(\sum_{x=1}^{N} C_x - NC_0 \right),$$

where Δx is the sampling step along the profile; l is profile spacing, m; $\sum_{x=1}^{N} C_x$ is the fluorite total content in the halo, per cent; N is the number of observation points within the halo; and C_0 is the on-site normal geochemical background. The units of m^2 per cent are converted into tonnes according to the ratio

$$q = (P \times d) \times 10^{-2},$$

where q denotes the reserves in a layer 1 m thick; d is the volume mass; and 10^{-2} is the scaling factor of the concentration, expressed in per cent per tonne. The predicted reserves in an ore body may be evaluated from the formula:

$$Q = \frac{1}{K} qH,$$

where H is the depth of a halo extension equal to half or one quarter of its length, calculated, respectively, by the triangle or rectangle methods. K — local coefficient

$(1 < K > 1)$. It should be noted that all calculations are made proceeding from the assumption that the halo is formed by an ore body having standard parameters of thickness and concentration at the surface.

In evaluating the predicted reserves of fluorite in a primary source by secondary haloes, the following requirements should be observed:

(1) ore reserves should be evaluated only by residual haloes of fluorine that were subjected to detailed studies at a scale of $1:10\,000$ or less when the presence of mineralization at a certain depth had been confirmed;

(2) it is not advisable to calculate the amount of ore in fluorite-bearing bodies from levels in their root or in slightly eroded parts because in the first case the reserves are usually overestimated, whereas in the second case they are appreciably underestimated;

(3) since determination of the reserves is done without taking into account the standard concentrations, the figure obtained should be reduced. The correction factor is applied in much the same way as at the known fluorspar deposits in the given region. If these are absent, the value of the estimated reserves of large, medium, and small fluorite-bearing bodies is reduced by 10–20, 30–35 and 40–50 per cent, respectively.

Of special interest are fluorine haloes coinciding with geophysical anomalies outlined from the combined data of electrical and magnetic exploration.

Based on their significance, we have systematized the major indications of secondary fluorine anomalies that make it possible to evaluate and tentatively grade the haloes associated with quartz–fluorite and quartz–calcite–fluorite mineralization (Table (3.8)). In principle, these factors may be used to evaluate secondary haloes associated with other types of fluorite deposit. Account should only be taken of the appropriate indicator elements, forming complex geochemical anomalies, and of the fact that large metasomatic bodies (particularly, the closely spaced ones) often from secondary fluorine haloes that are appreciable in size.

When secondary haloes are studied by the neutron-activation method, activity determinations should be made in identically shaped exploratory pits. On the surface of bedrocks this analysis is performed without making such pits. Observational grid density depends on the assumed dimensions of fluorine haloes and the required detail of investigation. Attention should be given to small anomalous concentrations of fluorine in rocks, especially in tectonic zones of known fluorite-bearing areas. If the zones of fluorine and other indicator element accumulation are narrow, the contrast and dimensions of geochemical anomalies may be enhanced by constructing multiplicative haloes.

Their application is especially rewarding in the areas overlain with thick loose deposits, or characterized by the development of coarse or lumpy deluvium, taluses, placers, stone streams, shallow, often hummocky bogs, and by waterlogging, all which make the use of lithogeochemical methods difficult. The hydrogeochemical method is inapplicable in the drainless intermontane depressions with salty lakes where it is difficult to identify, among false anomalies, those that are associated with mineralization.

Table 3.8.

Indications for the positive evaluation of secondary fluorine haloes

1.	Small spatial distribution (several hundred square metres 1 km^2)	+
2.	Elongated shape	+
3.	Absence of indications of geochemical barriers contributing to fluorine accumulation and formation of false anomalies	+
4.	Recurrence of haloes at a more detailed lithogeochemical sampling	+ +
5.	Coincidence of secondary haloes with the primary ones	+ +
6.	Absence of increased quantities of chemical elements that are not characteristic of fluorspar bodies (P, Mo, Be, etc.) in the lithogeochemical samples	+ +
7.	Contrasting and high fluorine contents (0.1–10 per cent)	+ + +
8.	Presence of small fragments of fluorine in the lithogeochemical samples taken inside the haloes	+ + +
9.	Coincidence of haloes with dislocations with a break in continuity playing the role of ore-localizing structures	+ + +
10.	Considerable predicted ore reserves in the primary source that were calculated from the halo productivity	+ + +
11.	Coincidence of haloes with geophysical anomalies that were interpreted as dislocations with a break in continuity, or zones of silicification	+ +

The number of + signs reflects the degree of qualitative enhancement of the significance of these indications. Hydrogeochemical methods are advisable in areas with a sufficient number of groundwater sources and surface water courses.

It is expedient to carry out hydrogeochemical prospecting for concealed fluorspar mineralization even though it can also be employed for detecting fluorite-bearing objects exposed at the surface.

It should be emphasized that the efficiency of hydrogeochemical prospecting surveys is largely dependent on the natural conditions in the areas containing fluorite mineralization. Such surveys can be incorporated into explorations at any level of detail. But they are predominantly involved in searches at a scale of (1 : 200 000 – 1 : 50 000) (less frequently in more detailed work), especially in combination with other prospecting methods.

Hydrogeochemical methods make it possible to explore a vast area and to define the most prospective sites. In order to locate and outline geochemical anomalies of fluorine and accessory elements it is necessary to sample surface (rivers, streams, bogs and lakes) and subsurface (springs, wells, pits and drill-holes) waters on the territories of fluorite-bearing provinces, zones, fields and deposits, as well as in the zones of faults that may be associated with mineralization. Melt-waters are sometimes sampled in areas with a weakly developed drainage system.

Biogeochemical surveys should be carried out in areas with extensive development of waterlogged landscapes, stone streams, thick deposits, semi-deserts, and strongly podzolized soils because they may yield positive results, and especially in tracing out the known ore bodies extending under the overburden. The biogeochemical method can successfully be applied if the maximum thickness of the sedimentary cover does not exceed 15 m. It is most effective in areas with a 3–10 m thick overburden, and in landscapes, characterized by excessive moistening or binary structure of deluvium (presence of allochthonous deposits in the upper layer makes lithogeochemical sampling of loose rocks useless). The efficiency of litho- and biogeochemical methods is commensurable in many landscapes. Biogeochemical prospecting is carried out on site with thick alluvial deposits located in large river valleys, because grassy meadow vegetation (due to limited length of root systems) does not permit reliable determination of the elemental content at a great depth.

The work is done in two stages:

(1) Choosing and practical employment of a method (basing on the example of especially prospective areas within the known fluorite-bearing fields) at the continuation of ore-bearing structures, and on the lithogeochemical fluorine anomalies associated with productive bodies; and

(2) Extensive use of biogeochemical methods in all the areas where tracing out of fluorspar mineralization is advisable. These include waterlogged landscapes, areas with strongly podzolized soils, and portions of sites with a binary structure of loose sedimentary cover. The maximum depth of allochthonous deposits (including those of the slope type), restricting biogeochemical prospecting surveys, should be determined experimentally in concrete conditions. In combination with other geochemical techniques, it is advisable to use the biogeochemical method in the areas requiring deep lithogeochemical exploration by sampling loose formations, and where the vegetatite cover is suitable for biogeochemical testing.

Biogeochemical sampling may be sufficiently effective in semi-closed areas with an overburden thickness 3–10 m. It can be randomly carried out in closed areas with an overburden 10–15 m thick, and may also be performed in winter. In a number of cases, application of this method may be restricted by the absence of sampling medium on the sites, and by the availability of plant species with shallow root systems.

The use of biogeochemical methods for detecting and tracing fluorine-bearing or other fluorite-containing mineralization is recommended only in forested areas dominated by birch and pine which concentrate fluorine most completely. A layer of pine bark should be sampled to distinguish maximum contrasting haloes that must be more intensive than those detected through lithogeochemical sampling in the plant-nutrition horizon (usually in soil horizon C, at a depth of about 1.0 m).

The sampling grid parameters and orientation are controlled by the geological situation and scale of work. It should correspond to the grid of lithogeochemical sampling of the loose cover. Taking into account that in many cases biogeochemical haloes above the fluorspar bodies are only several tens of metres wide, the sampling step for their reliable detection should not exceed 10–20 m and should be cut to 5 m in the process of detailed exploration of the anomalies. Ashing may be performed by any technique, including sample burning in the field in special furnaces or on a bonfire because the loss of fluorine is insignificant. The indicator elements are determined by spectral analysis according to standars imitating an average composition of plant ashes. Their anomalous concentrations are checked with a two–four recurrence. Biogeochemical anomalies are identified upon reduction of the results to standard values and time of sampling of different parts of plants.

Primary geochemical haloes are explored in areas with good outcroppings of sampling bedrocks in sections or profiles oriented across the strike of ore-localizing structures on a grid of 500×50 and 250×20 m, when prospecting at scales of $1:50\,000$ and $1:25\,000$. More detailed surveys are run over a closely spaced grid of 50×10 m.

Geochemical samples with an interval of 5 m are taken by the dotted furrow method along the lines of profiles (bedrocks, core, drillhole, walls of workings). Five or six rock chips are taken from each sampling interval and are combined into one sample with a mass of $100-200$ g. In processing the results of exploration, special attention is given to the zones with a combination of positive and negative haloes whose coincidence in the supra-ore strata of the sites with favourable geological features makes the discovery of concealed fluorspar mineralization more probable. At the same time the presence of fluorspar mineralization at a depth of no more than 90 m may be suggested from the increased concentrations of other elements and detected concentrations of beryllium and lithium at the surface, along with distinct negative anomalies of sodium and potassium. The discovery of positive primary haloes of fluorine, barium, molybdenum, silver, arsenic, lead, zinc, strontium and elements of the rare earth group, forming wide and more elongated anomalies, should be considered as an indication of ore body occurrence at a greater depth. Multiplicative haloes are regarded as a criterion for assessing the level of erosional truncation of geochemical anomalies.

3.6. SULPHUR

There are sedimentary and volcanogenic deposits of sulphur. Sedimentary deposits are classified into syngenetic and epigenetic. The former are characterized by large quantities, contain about 90 per cent of all the prospected reserves and yield approximately 95 per cent of sulphur mined throughout the world. The major part of epigenetic sulphur deposits was formed the from *in situ* alteration of gypsum–anhydrite. The first stage of this process results in the formation of H_2S and calcium carbonate, then during the second stage H_2S is oxidized to native sulphur. The presence of hydrocarbons (oil and gas) gives rise to the reduction of sulphates, whereas H_2S oxidation occurs under the impact of liquid water solutions or with the participation of bacteria.

Volcanogenic sulphur deposits are formed at the interaction of volcanic gases, containing hydrogen sulphide, sulphur dioxide and sulphur anhydrite, with atmospheric oxygen, with each other, and with the enclosing rocks which, at sulphur deposition, are subjected to intensive alteration under the effect of sulphuric acid resulting from the dissolution of SO_3, oxidation of H_2S, and high F potential.

Prospecting for sulphur deposits is based on the following conditions:

(1) The presence of carbonate rocks with the dissemination of gypsum, anhydrite, and bitumens; a dark blue or brown colour of the weathered enclosing rocks.

(2) The presence of large elevations at the junction with troughs, parts of the curves of brachyaxial folds and anticlines, as well as monoclines gently

inclined in the direction of troughs and depressions. Intensive fissureness of rocks, bends of layers and their articulation with fractures.

(3) Availability of bitumens, ozokerite, anthraxolite, oil, gases, chloridic and alkaline brines, and hydrosulphuric springs in the prospective zones.

(4) Presence of volcanic nest and nest-linear type chambers together with the finds of monomineral sulphur accumulations.

Indicator minerals

Exogenous sulphur deposits are characterized by diverse mineral composition, represented by: pyrite, marcasite, galena, sphalerite, chalcopyrite, wurtzite, chalcosine, covelline, celestine, barite, calcite, aragonite, siderite, quartz, opal, chalcedony, tridymite, halite, fluorite, alunite, gypsum, anhydrite, anglesite, jarosite, magnetite, ilmenite, haematite, tourmaline, realgar, enargite, smithonite, goethite, carnotite and calamine. Of greatest prospecting value are such minerals paragenetic to sulphur as calcite, aragonite, celestine, gypsum, anhydrite (substituted by sulphur), barite, bitumens and sulphides that occur in substantially larger areas around the deposit. Gypsum is of particular importance owing to its wide occurrence in the zone of sulphatic weathering. Manifestations of mercury mineralization (cinnabar and metacinnabarite), associated with haematite and hydromica, have been detected in certain sulphur deposits.

Modern and ancient deposits of sulphur are associated with various types of volcanic forms, mainly effusive cones and calderas. The sulphur and other deposits accumulated in the depressions may be buried with the post-ore products of volcanic activity; sulphur may be melted down into veins and impregnations and hydrolized into alunite- and sulphide-bearing ores.

Volcanogenic sulphur deposits are indicated by the following minerals, associated with the products of volatilization in volcanic foci: realgar, orpiment, antimonite, ammonium chlorite, haematite, gypsum, anhydrite, alunite, alum, zeolites, pyrite, barite, quartz, chalcedony and opal. Melnikovite, cinnabar, marcasite, pyrite, opal, clay minerals, chalcedony and gypsum develop in sulphur occurrences localized in kaolinized and silicified rock.

In the interval from 20 to 1000 °C, all of the most widespread minerals of sulphur deposits are thermally active and can be quantitatively calculated from curves showing their mass loss on heating. The diverse mineral composition of deposits is due to the high mobility and migrational capacity of sulphur in geochemical processes.

Conditions for the application of geochemical methods in the search for sulphur

Based on the mineral composition of sulphur-bearing sediments, it is possible to single out the following indicator elements of sedimentary deposits: S, Sr, Ba, Mn, Zn, As, Sb, Cu, V, Ho, U, P, Cl, F, K, Na and Li.

Commonly, the intensity of sulphur mineralization directly depends on Sr content in rocks. Limestones and clays occurring beyond the contours of ore bodies contain minimal quantities of this element. Barium is detected in primary (minimum concentrations) and secondary (maximum concentrations) sulphur ores. In all cases, Sr > Ba in sulphur. The values Sr/Ba > 10 and ~ 1.7 are characteristic of the primary and secondary ores, respectively. The content of Ba, associated with calcium carbonate, is higher in clays than in secondary ores. Sulphur deposits in carbonate–sulphate rocks are accompanied by gypsum-bearing formations with anomalous Ba and Sr concentrations, which have not been detected in the zones containing gypsum formed during the period of pre-sulphur recrystallization.

An increased manganese content in sulphur-bearing formations (up to 3 per cent), is commonly stipulated by the dissemination of an independent manganese mineral — hauerite — formed at high H_2S concentrations. This is a promising prospecting indication. Exogenic sulphur deposits are enriched in Se, whose concentration in ores reaches 10 g/tonne, and in mercury (40 g/tonne). Mercury has been detected in bitumens, phosphates, and sulphides of sulphur deposits.

Sulphur deposits are also indicated by F and Cl; these elements are responsible for the aggressiveness of solutions and for the redistribution and migration of elements around sulphur deposits. Deposits are generally notable for the removal of Ca and Mg, and introduction of C bound in carbonates. The process of sulphur formation was characterized by the removal of the iron group elements from and influx of Ba, B, Ga, Pb, P, Ti, and others into the enclosing and host rocks. Many elements contained in the substratum rocks displayed appreciable mobility in the sulphatic medium, which gave rise to the various mineral neoformations (gypsum, anhydrite) around sulphur deposits. Under the impact of acid solutions, clay rocks were enriched in Al, Ca, Mg and K.

The indicator elements of volcanogenic sulphur deposits are Hg, As, Sb, Ba, Sr, Cu, Zn, Pb and Se; the most informative of them are Hg, As and Se.

Application of hydrogeochemical methods in the search for sulphur deposits should result in the detection of their haloes and dispersion trains. The dimensions of haloes, wherein water courses with increased concentrations of direct (sulphate ion and hydrogen sulphide) and accessory indicators are detected, are not equal for various components and depend on their stability in water. Changes in the chemical composition of the water decrease gradually with distance from the productive bodies and associated haloes. The most demonstrative prospecting indications of sulphur are hydrocarbon gases, with high concentrations of CO_2 and H_2S detected in boreholes, and the presence of chloridic and alkaline–earth brines outpouring as springs. Hydrosulphuric waters with a slowed-down water exchange are indicators of sulphur deposits.

The presence of water-impermeable layers (preventing water exchange) at the bottom and in the roof of a productive horizon is unfavourable for sulphur formation.

The spatial coincidence of sulphur-bearing and bituminiferous formations is observed in a number of regions. The latter have appreciably greater spatial distribution than sulphatic deposits, thereby allowing us to specify the contours of target areas prospective for sulphur. The enrichment of bitumens in sulphur (from 6.8 to 12.1 per cent compared with a background content below 6.6 per cent) is a prospecting indicator of native sulphur.

Groundwater circulation in exogenous sulphur deposits results in the oxidation of hydrogen sulphide, which migrates from the productive deposits into the overlying horizons of gypsum–anhydrite and loose formations. Therefore positive anomalous values of the redox potential and alkaline–acid properties of the enclosing rocks are determined above ore bodies. The Eh value in the vicinity of sulphur deposits fluctuates from 20 to 200 mV (compared with background values of 5–10 mV).

Sulphur deposits overlain by deluvial formations are accompanied by Ba, Sn, Cl, Br, Na, Cu and Pb secondary haloes. Reaching 200 m in size, these haloes considerably exceed the dimensions of the sulphur deposits. Anomalies of V, Ni, Co, Cu, U, Ga, Pb and Zn have been detected in the ash residue of bitumens. The sulphur deposits near the Mendeleyev volcano are accompanied by sodium-chloridic, hydrosulphuric and sulphatic waters containing Na, Cl, SO_3, Pb, Zn, Hg, Sn and As. Zones of: (a) practically unaltered liparites; (b) secondary opalization; (c) insignificant sulphur mineralization in quartzites; and (d) sulphur occurrences are distinguished according to their distance from sulphur deposits. Mercury in the ore bodies and surrounding altered rocks is sublimated at 250, 290–300, 340–360 and 520–620 °C. A transition to low-temperature forms of mercury is observed as the distance from the outcrops of sulphur deposits increases. The 290 and 620 °C forms of mercury are found 2.5 m from the sulphur ore body; 290 and 360 °C forms occur 5 m from the ore body; 250, 360 and 520 °C forms at a distance of 25 m, and 250 and 340 °C forms at a distance of 100 m. The total content of mercury in rocks diminishes in the same direction. In supra-ore zones, mercury has greater number of forms and undergoes sublimation within a broader temperature spectrum than in sub-ore zones.

The differences in the density and in the magnetic and electric properties of sulphur ores and their enclosing rocks constitute the prerequisites for the application of geophysical methods in the search for volcanogenic sulphur. Sulphur ores are characterized by the density defect (up to 0.8 g/cm^3), low (zero) magnetic susceptibility, increased electrical resistivity, and negative values of natural electric field.

According to G. T. Sakseyev, the following patterns are observed in the isotopic composition of carbon in the carbonate rocks of sulphur deposits:

(a) decreased quantity of ^{12}C at the coefficient σ^{13} C27‰ and $^{12}C/^{13}C$ value equal to 88.5–89.5 which is characteristic of primary limestones occurring beyond the contours of sulphur deposits;

(b) increased concentrations of light isotope ^{12}C at the coefficient $\sigma^{13}C$ ranging from 32 to 65‰, and the value of $^{12}C/^{13}C > 90.5$, which characterize secondary limestones genetically associated with gypsum–anhydrites and sulphur deposits contained in them;

(c) fluctuation of $\sigma^{13}C$ value from 11 to 57‰ in the intermediate (between the first two types of rocks) supra-gypsum formations (calcitized limestones without sulphur).

The method of isotopic analysis of carbon in carbonate rocks may be used in sulphur prospecting.

The presence of secondary calcite in limestones is a prospecting indication for detecting gypsum–anhydrites and associated with them sulphur-bearing formations. The association of primary limestones with epigenetis calcite is also an indication of sulphur deposits in favourable conditions. Both the isotopic composition of sulphur and $^{32}S/^{34}S$ value depend on the conditions of reduction of sulphates, their concentration, and on the temperature and rate of reaction. The following three varieties of sulphates are distinguished:

(1) primary, corresponding to those in the sea;

(2) residual, enriched in ^{34}S isotope (this variety is regarded as a residue of bacterial biological fractionation); and

(3) secondary 'lightweight' sulphates resulting from the oxidation of sulphur, and having almost similar isotopic composition but more enriched in ^{32}S isotope than those in the sea.

The discovery of biogenic sulphites and sulphur may be anticipated in the presence of the sulphates of the third type. The absence of the second and third varieties of sulphates among beds of gypsum–anhydrite and associated limestones indicated that they are devoid of sulphur deposits now and were devoid of them in the past.

The data on mineralogical–geochemical zonality characteristic of exogenous sulphur deposits may give an impression of the position of sulphur accumulations in space. The following zones alternate from the periphery to the centre of deposits: (a) gypsum, gypsum–anhydrite; (b) sulphur–gypsum; and (c) productive celestine–calcite–sulphur deposit proper. The highest sulphur concentrations are generally detected at the contact between the central and sulphur–gypsum zones. A certain pattern governing the vertical change in the mineral composition of the ore bodies is observed. There is an upward decrease in the content of celestine (in paragenesis with native sulphur) and anhydrite; an opposite tendency is characteristic of barite, whose monomineral accumulations are detected in the top of a sulphur deposit (those of celestine occur at its bottom). The root portions of deposits are commonly indicated by anhydrite.

This geochemical zonality mirrors the mineral one and is closely associated with it. Maximum concentrations of strontium have been detected in anhydrites and gypsum underlying an ore shoot, whereas the barium content diminishes upwards. Thus, the Sr/Ba ratio in the upper parts of sulphur-containing limestones in one of the deposits appreciably exceeds unity. It increases with depth, reaching 10 or more at the bottom of deposits.

The regularities responsible for the changes in the sulphur content in vertical direction have been found. Thus, almost identical quantity of sulphur is contained at the base of ore shoots (sulphur-free gypsum) and in the upper horizons (sulphur-free limestones). Maximum sulphur concentrations are observed in the centre of the ore deposit and at its contact with the enclosing rocks. The epicentres of ore deposits are characterized by a combination of high and drastically reduced sulphur concentrations in the rocks.

Supra-ore 'caps' resulting from sulphur oxidation near the earth's surface, serve as prospecting indications of sulphur occurrence at depth. Sulphates and secondary gypsum in zones of weathering are enriched in the light sulphur isotope. The concentration of this isotope is increased due to the activity of chemosynthesizing thiophilic bacteria. In this case, the prospecting indication for sulphur is the presence of gypsum and limestones with an anomalous ratio of stable sulphur isotopes. Blind and deep-seated sulphur deposits are indicated by the isotopic composition of sulphur and carbon in rocks and ores. The content of carbon with $\sigma^{13}C$ in sulphur-enclosing limestones above, within and under gypsum fluctuates from 3 to 6‰ [73]. By these indices they distinctly differ from the sedimentary and other carbonates, with $\sigma^{13}C$ up to 3‰. Frequently, limestones above a sulphur deposit do not contain sulphur even though they are identical with mineralization-enclosing limestones in terms of isotopic composition of carbon. Geochemical methods can be more extensively applied in the search for sulphur due to the wider occurrence of ore-free sulphur-bearing limestones compared to sulphur-bearing ores.

Methods of work

In the majority of cases exogenic sulphur deposits are overlain by eluvial–deluvial formations and are not exposed at the surface, which impedes their prospecting by conventional areal geological surveys. Such deposits can be detected by geochemical prospecting surveys along dispersion trains and haloes. Dispersion trains are examined by hydrogeochemical methods at the first stage of prospecting, whereas at the subsequent stage dispersion haloes are explored in the areas defined by the geological and geochemical data. Hydrogeochemical methods are most extensively employed in the search for sulphur. The salinity and ionic composition of surface and subsurface waters are analysed in a field or stationary laboratory. Special attention is given to the distribution of SO_4^{2-}, F, Cl, Br and I in water, and to the presence of productive oil strata water, sulphur-bearing crude oil, bitumens and hydrosulphuric springs. Zones of slowed-down water exchange in the sites of washing out and crushing of water-impermeable gypsum–anhydrites are detected.

Methods of hydrogeochemical investigation of sulphur do not significantly differ from those used on other types of deposits. The major objects of exploration are open water courses; subsurface waters deeply circulating, outpouring, or tapped by drill-holes the latter are tested separately. Among surface water courses it is advisable to sample those streams that are fed by groundwater.

Zones with sulphatic hydrosulphuric waters, containing high anomalous quantities of sulphate ion, are designated on geochemical maps by a special sign. Eh is determined in the field at a depth of $0.1-1.0$ m in the process of making workings. Then ΔEh, the gradient equal to the difference in Eh values at depths of 0.1 and 1.0 m, is determined in the fine fraction of samples of mass $50-100$ g.

The soil-hydrogeochemical method, together with sampling of deluvial and bituminiferous formations, envisages preparation of water extracts from rocks, and determination of indicator elements and ΔEh. Microbiological investigations aimed at determining sulphate-reducing and thionic bacteria in waters and rocks are sometimes advisable.

Areas with outcrops of gypsum substituted by sulphur and calcite, and sites where sulphur has filled cavities resulting from the dissolution of carbonate deposits, are identified at the stage of detailed surveys at a scale of 1:50 000. Special attention should be given to outcrops of rock horizons with calcite and sulphur intercalations persistent along the strike.

When analysing a multicomponent mixture of minerals, the concentrations of water, sulphur, pyrite, dolomite and calcite are calculated, by the following methods: the amount released at $100-200\,°C$; oxidation and liberation of SO_2 at $200-400\,°C$; the area of exo-effect of oxidation at $500\,°C$; separate registration of CO_2, liberated due to dissociation of $MgCO_3$ at $750\,°C$ and $CaCO_3$ at $900\,°C$.

Rapid determination of the mineral composition of sulphur-bearing rocks and ores can be made by the thermographic method, which allows a reduction in the number and volume of analyses. Samples with a mass of 400 mg should be analysed in a derivatograph: they should be allowed to uniformly heat up to $1000\,°C$, and then sent the differentially thermal curve, the change of mass, the rate of mass change, and the temperature, should all be measured simultaneously.

This method can be used to determine the concentrations of gypsum, calcite, dolomite, clay minerals, pyrite, sulphur and sulphur–calcite ores. Thermographic analysis is advisable for making rapid, mass determinations of the mineral composition of sulphur-bearing rocks and ores, thereby reducing the number of chemical analyses. In determining the gypsum quantity in rocks in the presence of sulphatic minerals (anhydrite, celestine, and barite), preference is given to thermographic but not to chemical analysis.

The Sr/Ba value is an indication of vertical geochemical zoning of ore bodies. It is < 1.0 in the above-ore zones, equals 1.0 in the middle-ore bodies, and > 10 at the bottom of deposits. It should be noted that sulphur-free limestones of light isotopic composition are detectable in the roof and on the edges of ore occurrences.

At the surface in the zone of oxygenous weathering, oxidation of sulphur results in the formation of alum and bleaching of the enclosing rocks and ores. Secondary gypsum, thenardite, epsomite, copianite, sulphurite and hydrotroilite are also formed there. Finds of these minerals at the present level of erosion may indicate the occurrence of sulphur at depth.

The zoning of hydrothermal phenomena limits the distribution of the main mineral deposits. On the upper parts of volcanic cones there are only solfataras and cold ultrafresh or slightly warm waters. On the slopes there are hydrosolfataras, acid and subneutral sulphate waters formed by the emission of hot deep-seated sulphate–chloride water. The basic types of hydrothermal water are combined in calderas. The environment for both the accumulation of large masses of sulphur, sulphides and clays and their preservation under the cover of lava and tuff occurs in calderas.

The sulphur isotope ratios in various minerals, gases and solutions make it possible to observe the chemical transformation and concentration of sulphur through the migration routes [130].

The data on the composition of waters in volcanic formations of fumarole condensates and soil gases, analysed for Hg, As, B, CO_2, SO_4 and SO_3, may be used in the search for volcanogenic sulphur deposits. The phase status of a deposit may be judged by the ratio of gases and vapour in fumaroles.

Geochemical prospecting for sulphur deposits along primary haloes is aimed at: (a) detecting sulphur accumulations; (b) determining the level of ore bodies and prospects of their occurrence at depth; and (c) evaluating the predicted reserves. Reliable solution of these problems depends largely on the complexity of the geological structure of the deposit, the size and morphology of sulphur accumulations, their relative position, and conditions of occurrence. Prospecting for and evaluation of ore bodies by primary haloes involves sampling of bedrocks in exploring openings, subsequent analysis of samples, and outlining of anomalies. Special attention is given to the following indicator elements, which contrasting haloes: Ba, Sr, F, Hg, As and Se. Interpretation of geochemical data may be both quantitative and qualitative. The latter results in the identification of positive and negative anomalies by drastically differing elemental content. Of particular interest are negative Na and Mg, and positive Zn, Pb and Cu haloes in limestones underlying ore deposits. Geochemical analysis of the borehole core may be carried out in the search for buried sulphur deposits.

Geochemical and geophysical methods are combined in prospecting and reconnaissance surveys. Use is made of magnetic prospecting, the electric field method, and vertical electric sounding. The contours of deposits, predicted by geochemical and geophysical methods, are confirmed by drilling.

The composition of carbon and sulphur or carbonites and native sulphur is analysed for the purpose of prospecting. Limestones and calcites, enriched in light isotope ^{12}C, compose considerable strata and, therefore, are rather easily detected by prospecting surveys. The lightweight sulphate 'caps' are regarded as prospecting indications of a sulphur deposit. A value of $^{13}C = -1.5\%o$ and less for carbonate carbon indicates the presence of sulphur-enclosing limestones.

The isotopic composition of sulphur is studied for the same purposes. Commonly, the ratio of the most widespread isotopes, ^{32}S and ^{34}S, as well as $\sigma^{34}S$ (the difference between the isotopic composition of sulphur from the sample being studied and the sulphur taken as a standard) are measured.

Integrated investigation of the specific features of the geological and mineralogical composition of the deposits being studied is advisable for interpreting the data of isotopic analyses. Application of isotopic methods makes it possible to obtain data on the stepwise character of the mineral formation processes, the duration of the intervals between stages, and the depths of deposit formation.

Geochemical investigations may extensively involve the quantitative thermographic analysis of rocks and ores of exogenic sulphur deposits.

Stages and sequence of work

Work at the first stage, geological surveys at a scale of 1 : 200 000, is primarily concentrated on the establishment of favourable geochemical indications of sulphur mineralization — the presence of hydrosulphuric springs, gas shows, an availability of oil and bitumen enriched in sulphuric compounds. Efforts are made to detect such minerals paragenetic to sulphur as celestine barite, as well as dispersion trains and haloes of copper, lead, zinc, chlorine, bromine, mercury and fluorine. Areas with direct and indirect indications of sulphur manifestation are defined to carry out specialized prospecting surveys at a scale of 1 : 50 000–1 : 25 000. The second stage is devoted to geological survey coupled with hydrogeochemical and lithogeochemical sampling. The geochemical investigations at this stage are undertaken in order to outline concrete zones and sites, prospective for reconnaissance and prospecting surveys. Special attention is given to the zones of sulphur mineralization, characterized by a combination of sharply contrasting (with respect to redox processes) sites and geochemical barriers. Anomalies of hydrogen sulphide in ground water, outcrops of carbonate rocks (marls and limestones), gypsum and anhydrites, druses of spathic secondary gypsum, and haloes of hydrosulphuric weathering are located. Favourable mineralogical indications, including the presence of calcite, celestine, chalcedony, oxidized sulphur-bearing crude oil, as well as calcitization, silicification and alunitization of rocks, are recorded. Especially great significance is attached to the evaluation and grading of secondary haloes in loose formations.

The third stage encompasses prospecting and exploratory surveys at a scale of 1 : 10 000–1 : 2000, core drilling and geochemical sampling of bedrocks.

3.7. SALTS

The following types of salt deposits are known: (a) deposits of fossil rock and potash salts; (b) contemporaneous salt deposits (in lakes, lagoons and bays); and (c) natural brines and salt waters in the earth's entrails or outpouring on to the surface. Fossil salt deposits are of the greatest commercial significance.

By their chemical composition, potash salts are subdivided into:

(a) sulphate-free, containing chlorous salts of potassium and magnesium; and

(b) sulphatic, containing (along with chlorides) potassium, magnesium, calcium and sodium salts.

Soluble potash (sylvite) and potash–magnesian salts (carnallite, kainite and polyhalite), sometimes closely associated with rock salt, are the major raw material of the potash industry.

Geologically, salts is found in solution or in a solid state: a) deposits with solid state salt, b) oceans (playa lakes), c) lakes (bedded deposits), d) groundwater (salt domes, diapirs).

Salt is found in solution in oceans, lakes and springs. The oceans contain 4.5 million cubic miles of rock salt — a volume half as great as the entire North

American continent above sea level. The actual salt content of the oceans varies from 1–5 per cent but averages about 3.5 per cent. Today, in areas along the coasts where conditions are ideal for solar evaporation, salt is produced from ocean waters. The water is evaporated during dry spells leaving a thin salt crust (Mexico). A marine saline is a salt lake which is fed by underground seepage from the ocean (Larnaca, Cyprus).

Lakes are formed when water accumulates in a topographic depression. Waters flowing into the depression carry material in solution. The composition of this material will depend upon the type and solubility of the rocks in the watershed traversed by the meteoric waters. Thus, it is possible to have lakes that very greatly in composition. Mineralized lakes are as follows: sodium–chloride; magnesium salts; sulphate–chloride; carbonate–chloride; and chloride–sulphate–carbonate. Normally, mineralized lake deposits contain little or no potassium. Sodium and calcium are the predominant cations and the most common minerals are the sodium sulphates, mirabilite, thenardite, burkeite, gaylussite, trona and halite.

Ground waters (springs) vary in salinity from practically pure water to those that are almost completely saturated. Connate waters are those trapped in the rock during the time of formation. Rain water is fresh initially but picks up soluble material as it travels through the air, soil, rocks, etc. Connate waters usually reflect the composition of the water in which the rock was formed. Groundwaters range in chemical content from practically nothing to as much as 254 000 ppm. Some brines are exploited for their bromine and magnesium contents.

Solid salt is found in play a deposits, as bedded salt deposits, and diapirs and domes. Playa deposits are fairly common in arid regions and saline playas result from the evaporation of a mineralized lake.

Playa salt is the sandy, salty mud-covered floor of a desert basin. During the rainy season a playa may be covered with rainwater which dissolves part of the soluble material. When the water evaporates, it leaves behind the soluble material, which may be concentrated in one area of the lake.

The composition of the material in a playa lake varies greatly (salt, calcium chloride, sodium carbonate).

Bedded salt deposits are true sedimentary rocks and as such are found in sedimentary sequences with shales, limestones, gypsum, anhydrite, etc. Salt beds can be as much as 450 m thick. Due to the manner in which they are formed, thick salt beds characteristically contain thin bands or lenses of anhydrite and clay, with a salt interval 2–4 m thick. The formation of such thick salt beds has intrigued geologists for many years (terrestrial type, marine type).

Salt is very vulnerable to pressure. Thus, as pressure is exerted on a salt bed, due to the static weight of overlying sediments or tectonic forces, the salt 'flows' plastically, first bulging to form salt domes.

The majority of the known salt-bearing series were formed in sea basins, connected with the ocean but having limited water exchange. Very occasionally, deposits were formed in drainless depressions of the continents. The prospecting indications of salt deposits are as follows:

(1) Presence, in the stratigraphic section, of halogen formations (salts, gypsum and anhydrite) that are paragenetically associated with halopelites, argillites, marls, carbonate rocks aleurolites, and sandstones.

(2) Layered structure of sedimentary formations and presence of thin intercalations of salt-bearing and gypsiferous clays in homogeneous rocks.

(3) Occurrence of halogen formations, brachy-folded and dome structures, and negative tectonic features of platforms (syneclise depressions, troughs, avlakogenes and grabens).

(4) Availability of highly saline brines and saline groundwaters.

Conditions of the application for geochemical methods in prospecting for salts

There is a rather great variety of salt minerals. However, they are dominated by such rock-forming chlorides as halite, sylvine, carnallite, rinneite and chlorcalcite, whereas potassium and magnesium sulphates (polyhalite, langbeinite, kainite, löweite, schoenite, epsomite and kieserite) occur less frequently. Terrigenous formations are mainly represented by clay–aleuritic minerals. Gypsum and anhydrite are present in rock-salt deposits. The mineralogical prerequisites to be considered in the search for salts are as follows:

(a) outcrops of gypsum and anhydritic rocks that are common accessories of fossil salts;

(b) presence of celestine, fluorite, and sellaite, that are salt accessories in salt-bearing clays and carbonate–clay–salt containing rocks; and

(c) presence of considerable amounts of magnesite in sulphatic rocks.

The data on the mineralogical composition of salt formations help in compiling a list of the following indicator elements of salt deposits: chlorine, bromine, fluorine, potassium, sodium, calcium, magnesium, lithium, rubidium, strontium and boron. Of special significance are those indicators which do not form their own mineral phases but isomorphically substitute the key elements in salts. Among them are bromine and rubidium, which substitute, respectively, chlorine in chlorides and potassium in salt minerals. The content of Br in salts settled from solutions is directly proportional to that of KCl and NaCl. The Br concentration in halite, formed at the initial stages of precipitation, amounts approximately to 0.007 per cent, at $(Br \times 10^3)/Cl = 0.1$. At the end of the halitic stage, prior to sylvine formation, these concentrations equal 0.025 per cent and 0.4 and, at the final stage, 0.07 per cent and 0.95, respectively. Thus, the Br content in halite grows in the later-formed mineral phases with an increase in seawater condensation. Being a sensitive indicator of the halitic stage, the value $(Br \times 10^3)/Cl$ makes it possible to determine the position of salt horizons in the barren strata and the level of their erosional truncation. Sylvine crystallization is accompanied by an increase in Rb concentration in the liquid phase, whereas carnallite formation results in the accumulation of this element in the solid phase, which is in agreement with the experimental data (in the first case $I_{Rb} < 1$, and in the second $I_{Rb} > 1$). Upon crystallization of carnallite, rubidium is practically undetectable in the solution. The Rb/K ratio in carnallite of primary sedimentation (at its initial stage of settling) amounts to 24.7×10^{-4}, whereas in sylvines formed in the process of decomposition of carnallite deposits, it equals 10.33×10^{-4} [121]. The distribution of rubidium makes it possible to judge the sequence of the processes and systems of equilibria at the stage of potash salt formation. The Br and Rb contents in salt strata increase from the bottom to the roof. Sylvine dissolution and recrystallization from brines do not alter the trends of Br and Rb distribution. Compared to the enclosing rocks, salt-bearing formations are characterized by a drastic reduction of Ti, V, Cr, Mn, Co, Ni, Cu, Ba and Ge contents.

The typomorphic indicator elements of salt deposits are: Br, Cl, K, Sr, B and Rb. The applicability of hydrochemical surveys (the major method used in the search for salts) is based on the easy solubility of salt minerals and alteration of the chemical composition of waters under the effect of salt strata. Mineralization is indicated by chlorine, bromine and potassium that are not removed from the system by the solutions migrating in the rocks because they fail to form stable minerals in such conditions. Unlike chlorine and bromine, Na, Ca, Mg and SO_4 may form anomalies even in the areas devoid of salts.

Salt deposits contain mother (sedimentation) and descending (leaching) waters. By their chemical composition these are grouped into carbonate, sulphate (sodium sulphate, magnesium sulphate and magnesium chloride), and chloridic waters [121]. Sedimentation waters belong to the magnesium sulphate and chloridic types, commonly enriched in H_2S, Br and I [121]. Leaching waters are related to the sulphate and sodium sulphate types. Sedimentation waters, commonly preserved under salt-bearing deposits, can be used in the search for salts (potash salts, in particular). Descending waters, formed by infiltrating atmospheric precipitation, contain mainly the components of solid deposits and an insignificant admixture of mother brine elements. While percolating through salt-bearing horizons, these waters dissolve salts, thereby giving rise to the formation of dispersion trains of indicator elements. In certain cases, leaching waters are accumulated in intermediate collectors. The presence of potash salts in the salt-bearing formations contributes to the enrichment of water in potassium and magnesium.

In the process of seawater evaporation, Br is concentrated in the liquid phase (until precipitation of bishofite), K is accumulated in the liquid phase only until the crystallization of sylvine and carnallite, whereas in the eutonic solution its quantity is small. The K/Br ratio in sea-basin brines ranges from 4.5 to 7.8, remaining stable at their condensation up to the moment of precipitation of potash salts. At the stage of their formation, this ratio sharply decreases and, at the last stage of bishofite formations equals 0.1 [74].

The composition and concentration of brine is judged by the results of investigating its inclusions in salt minerals, authigenic quartz, and its remains in buried clay intercalations. The temperature of homogenization of solid and gaseous inclusions aids in determining the nature of brine: whether it was bottom brine or brine that existed at the time of salt recrystallization [92].

Patterns, governing the changes in potassium concentration in the solutions of inclusions in halite down the halogen strata cross-section are shown in Fig. (3.18). An increase in the potassium concentration of solutions in inclusions is observed

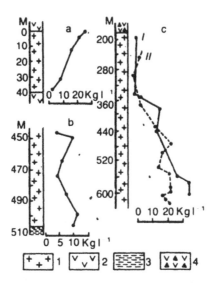

Figure 3.18. Regularities governing the changes in potassium concentration in the solutions of inclusions in halite down halogen strata cross-section [92]. a — Permian, Bryantsev bed; b — Tortonian, Ciscarpathion foredeep; c — Devonian, Romensk salt stock. 1 — rock salt; 2 — anhydrite rocks; 3 — salt-bearing clays; 4 — capping rocks; I, II — boreholes.

with the approach to productive salt horizons which is stipulated by potassium halo formation around them.

On the Siberian platform there are numerous tectonic distortions, in which one can observe thinning and ruptures in the continuity of the salt layers — from the thinnest, traced only through a microscope and due to a specific colour, to the gigantic with haloes of Hg, F, Cl, I and Br. The migration of fluid in these distortions is evidenced by a number of factors [83]: contrasting variation in the geoelectric field, increased radioactivity of spring waters, the presence of hydrothermal mineralization, appearing in the near-surface zone of brines with a temperature of 35–50°C, which is analogous to that of rock on the basement surface, and laterally local geothermal anomalies which do not exist beyond the zone of distributions, but penetrate through the entire cover in a vertical direction. The dispersion haloes 'overhang' a deposit of useful minerals.

For the first time the formerly unknown hydroepigenetic vertical zoning of minerals has been established in salt-bearing regions. The composition of minerals filling cracks changes successively from less soluble in the upper part to more soluble in the lower part of the infiltration stage: hydroxides of iron and manganese–calcite–gypsum–halite. This zoning is a result of variations in the mineral composition of the subsurface waters. New distinctive geological boundaries (the interfaces between zones with a different composition of the secondary minerals filling the cracks) have been revealed. These boundaries are actual hydrogeological boundaries both in significance and origin. They subdivide the cross-section into aquiferous and impermeable horizons and have been derived from the chemical and dynamic activity of the subsurface waters.

Hydroepigenetic mineral zoning is significant for the study of the lithology, stratigraphy, tectonics and hydrogeology of salt-bearing regions and also in the solution of applied problems. The most secure locations for industrial and hydrotechnical constructions can be defined, and criteria for prospecting for commercial chloride–sodium brines and the method for determining the directions of vertical modern movements can be worked out. It is supposed that there are hydroepigenetic hydrocarbon traps in the sedimentary section.

A study of the distribution of neomineralization in the section has enabled the author to refine the old concept of the 'salt mirror' and to suggest the new concept of the 'gypsum mirror'. The term 'salt mirror' is taken to mean the surface of the first appearance of halite in the section, independently of wheather this mineral forms a salt body or fills the cracks in country rocks. A similar definition may be applied to the concept of the 'gypsum mirror'. Both have been subdivided into two qualitatively opposite categories, the 'front of solution' and the 'front of infilling of cracks'.

It has been found that a relatively higher lithium content is confined to polyhalite deposits (up to 0.001 per cent) and the lowest content is observed in halite (2×10^{-5} per cent) and carnallite deposits. The investigation of the distribution of lithium in natural salts on the basis of the solubility of their compounds leads to the conclusion that there are three types of occurrences of lithium: water-soluble compounds, compounds which are soluble in 6 per cent hydrochloric acid and isomorphously substituted lithium in clay minerals.

The value of the coefficient of rubidium distribution (D) between crystals of carnallite and solution in the system $NaCl-KCl-MgCl_2-H_2O$ at 25 °C has been determined during the slow isothermal evaporation of a large volume of synthesized solutions of the composition of the carnallite field [92]. The average D values is 2.44 ± 0.18. On the basis of experimental and literary data the values of the Rb/K ratios have been calculated for the first carnallite crystals crystallized from seawater, as well as for secondary sylvines and carnallites formed during the dissolution of carnallite deposits and the repeated evaporation of the solution. The data obtained may be used during prospecting for potash salts, to establish the origin and formation conditions of carnallite and sylvine in saline deposits.

The problems of gas content and gas formation in various saliferous rocks and mines have been considered using the example of old potash deposits (Verkhnekamsk, Starobinsk, Stassfurt). The isotopic composition of noble gases for sylvinites from the Starobinsk deposit is as follows: ^{40}Ar rad makes up about 99 per cent by volume; $^3He/^4He = 0.5 \times 10^{-8}$. The quantitatively calculated values of the helium content in salts show good agreement with experimentally determined values. The radiogenic nature and good state of preservation of the rare gas salts confirm the primary-sedimentation origin of the old halogenic deposits. The characteristics of the chemical composition and forms of occurrence of the gases in saliferous rocks, the connection between the gas content and microelements, in particular lead, and certain geochemical indices suggest that the role of volcanism in the processes of salt formation and salt accumulation could have been underestimated.

Methods of work

The applicability of geochemical prospecting depends on the presence of sampling objects in the area under study. Those for hydrogeochemical surveys are ground-water and open water courses. The sampling grid depends on the extension of the dispersion trains. Special attention is given to those water courses that flow from gypsum caps. Sampling of streams recharged mainly by groundwater is obligatory.

When carrying out prospecting surveys at a scale of 1:50 000 it is necessary to take one sample per 0.5–2.0 km². Samples may be taken:

(a) along survey routes on the slopes of river valleys at intervals controlled by the density of sampling;

(b) from small tributaries at a uniform areal distribution of sampling points;

(c) from shallow boreholes (to sample brines that were not diluted by purched water); and

(d) from the holes drilled in order to tap salt strata.

Selection of a sampling variant depends on the local conditions, survey goals and type of deposits.

It should be pointed out that hydrogeochemical testing of the drilling mud can be performed in prospecting for potash salts, and, incidentally, in the search for other types of deposits. The following rules must be observed [121]:

(1) a drill-hole design should exclude the drilling mud dilution by water from the supra-salt aquifers in the process of drilling salt-bearing and sub-salt-bearing deposits;

(2) systematic observation of the balance of the drilling mud and registration of its loss (absorption) per metre of drilling should be carried out; and

(3) one-litre samples of the drilling mud should be taken in 1–5 m intervals of drilling and subjected to reduced chemical analysis for Br, Cl, K, Mg and Ca in the field.

Analyses for HCO_3^-, SO_4^{2-}, Cl^-, Br^-, $B_4O_7^{2-}$, Ca^{2+}, Mg^{2+}, Na^+ and K^+ are made in the laboratory. Individual and integrated methods of determining each element are used. Samples for determining individual elements are subjected to colorimetric analysis, whereas alkaline elements are determined by flame pho-tometry. Interpretation of these data results in: (a) defining the localities containing salt-bearing horizons; (b) locating natural (sedimentation and leaching) water solutions; and (c) determining the ratios of indicator elements, making it possible to distinguish mother brines from those that leach out halitic formations. It should be borne in mind that the discovered sedimentation brines are characterized by an appreciable increase in the bromine–chlorine ratio and a certain decrease in the potassium–bromine and potassium–chlorine ratios. An increase in the magnesium–chlorine ratio, and a decrease in the sodium–chlorine ratio, are additional indications. Use is also made of the standard diagram based on experimental data and observations which make it possible to gain an impression of the type of groundwater and the degree of seawater condensation [121].

Thus, the ratio ENa$/E$Cl (where E denotes the number of equivalents) in seawater amounts to 0.85. This index decreases with the extent of condensation (from the very start of halite crystallization) and at the final stage of crystallization equals 0.02. The quantity of magnesium bound with chlorine is calculated from the following formula [121]:

$$\left(E\text{Mg}^{2+} + E\text{Ca}^{2+}\right) - \left(E\text{HCO}_3^- + E\text{SO}_4^{2-}\right) = E\text{Mg}^{2+}.$$

In determining $E\text{Mg}^{2+}/E$Cl, account is taken of that part of magnesium which is bound with chlorine. The content of the latter indicates the degree of water condensation. The $(\text{Mg}+\text{Ca})/\text{Rb}$, $(\text{Mg}+\text{Ca})/\text{Br}$ and K/Br ratios are used to distinguish highly saline waters from the sedimentation highly condensed K- and Mg-rich brines, which failed to release potash salts in the process of evaporation. The presence of potash salts is indicated by the following ratios: $(\text{Mg}+\text{Ca})/\text{Rb} < 300$ and $(\text{Mg}+\text{Ca})/\text{Br} < 14$.

As a result of oil and potash exploration many new salt deposits have been found in various countries (Brazil, Thailand, Morocco, France, Portugal). The work of an exploration geologist looking for salt is usually a matter of the 'final' delineation of the deposit and its environment. The manner in which exploration is conducted is governed to a large extent by the planned method for extracting the salt.

In evaluating a salt deposit for dry mining, the geological aspects surrounding the deposit are almost as important as the deposit itself.

The amount of ground water and its hydrostatic head will have a direct influence on the method and cost of shaft sinking. If this ground water is a potable water supply for a community, the protection of the aquifer will require careful hydrological studies. Shaft sinking difficulties will materially increase if an aquifer contains a brine or significant amounts of hydrogen sulphide, ammonia, or hydrocarbon gases or if the hydrostatic pressure is very high.

In general, it may be said that the deposit should have a minimum sodium chloride content of 95 per cent or greater to be commercial. In the north-eastern part of the United States the average sodium chloride content required is 97 per cent; where salts must be competitive with either rock salt from southern salt domes or solar evaporated salt, the sodium chloride content must approach 99 per cent. Depending upon the end use of the salt, various trace elements such as copper, vanadium, boron, chromium and iron as well as calcium sulphate, ammonia and water insolubles must be considered. If we consider the overburden as the alluvium plus the rock sequences above the salt deposit, it is essential that the physical characteristics of these horizons be known as fully as possible. It is of particular importance that this information be gathered from the area in which the shaft will be sunk. These data should include the thickness and hardness of the beds, drillability, fracture pattern, friability and tendency to oxidize. Shaft sinking contractors generally require this information and occasionally make a physical examination of the cores themselves.

Physical characteristics of the rocks associated with the salt

In sedimentary layered evaporite deposits, the structural competency of the rocks both underlying and overlying the bed to be mined are of particular importance. Roof spans, pillar size, percentage of extraction, mine layout, and the roof bolting programme are some of the main features which will be dictated by the rocks surrounding the deposit. Fault systems which are common to many salt deposits can create numerous serious problems and should be defined as carefully as possible during the initial exploration programme. Domes in the southern United States have numerous faults associated with the spines that characterize their upper reaches. Vertical or near vertical fractures in them often contain brine, oil and gases.

Downdropped graben blocks are often the result of post-depositional intrusions of water along faults. The dissolving action of the water on one or more of the salt beds may cause a collapse that can take the form of a single block fault or en echelon step faults. Either of these may cause expensive realignment of the mining pattern.

Both minor and major variations in facies may occur in salt beds over a relatively short distance. In Michigan the first salt grades from a pure anhydrite on the edge of the basin near Detroit to an almost pure salt in East China Township, Michigan, and then to a heavily contaminated potash salt in the area of Midland.

The salt at Realmonte, Sicily, although metamorphosed, shows a similar gradational change from lean potash mineralization to a pure halite. The attitude and position of the salt beds or in the case of domes, the flowage lines, influence the mining system, the percentage of extraction, the depth of the mining level, the selection of a shaft versus an inclined entry, and the rock mechanics of superimposing pillars where multilevel systems are contemplated.

Where solution mining is contemplated, exploration is largely concomitant with the expansion or replacement of the present brining facilities. Usually, regional geology leads to the selection of the brine field site. On the basis of previous published information, a decision is made as to whether or not geophysical information would be useful in guiding subsequent drilling.

In cases where domes are involved, or the beds are known to have been altered by diapirism or faulting, geophysical studies may be of distinct value.

The geochemical data provided by testing of individual key wells are generalized at the stage of prospecting surveys. The indicators of mineralization are potassium anomalies in rocks, and the ratio $(Br \times 10^3)/Cl$, whose value is indicative of the stage of seawater evaporation. Zones of halogen sedimentation with potash salt horizons are located. The following tabulated data are taken into consideration: at the beginning of halite settling the $(Br \times 10^3)/Cl$ ratio is 0.1, whereas at the formation of sylvine it equals 0.4, and at carnallite precipitation it is 0.6–0.8 [121]. A decrease in the ratios may result from large sampling intervals. It can also be conditioned by the presence of the recrystallized salt horizons or by the redeposition of salts. Conversely, these ratios may increase in the mother brines contained in clay intercalations. Therefore, along with the determination of the $(Br \times 10^3)/Cl$ ratio, it is advisable to use the $(K \times 10^3)/Cl$ value, permitting selection of those ratios from the bromine–chlorine ratio maxima that deserve special attention. A cymbate increase in bromine–chlorine and potassium–chlorine ratios in halite indicates the prospects of potash salt discovery in the given cycle. Compared to the enclosing rocks, the salt-bearing horizons are characterized by negative Ti, V, Cr, Co and Ni anomalies.

Use is also made of the data on the composition of the gaseous–liquid inclusions and content of the insoluble residue in salts. The brine concentration and composition are determined, along with the temperature of precipitation of solid phases. A prospecting indicator of productive horizons is the presence of a potash-salt saturated brine. The porous bottom brines (which had existed in the basin prior to the beginning of settling of salt-free intercalations) contained in the persistent halopelitic interbeds are converted into water extracts and analysed. The degree of brine concentration is determined by analysing brine from clay intercalations, gaseous–liquid inclusions in minerals, and the bromine quantity in halite. Potassium concentrations of up to 20 g/l in the solutions of inclusions in halite are typical of potassium-free halogen formations, whereas the potential availability of potash salts is indicated by potassium concentrations of 20–40 g/l [91]. An increased content of the insoluble residue in productive potassium-bearing horizons is also a prospecting indication of potash salts.

In the process of generalization of geochemical surveys, the above-stated ratios and potassium contents in the solutions of inclusions in halite are shown on graphs and sections. The boundaries of halite, sylvine and carnallite zones are also shown on the graphs and sections, wherein completed (prospective for searching potash salts) and non-completed sedimentation cycles are identified. These data are used to correct dry (or with the help of carnallite, mud) drilling of exploratory wells.

Reconnaissance surveys involve geochemical sampling of core to detect in rocks K, Br, Cl and other elements whose haloes are identified within the exposure interval of salt-bearing horizons. These haloes are elongated along the plane of productive zones but have limited vertical extension, which should be taken into account when interpreting geochemical data. It should be emphasized that among considerable outcrops of halite deposits, potash salts occupy (in the overall mass of salt deposits) smaller volumes, forming interbeds and lenses. Such potash salt horizons may be detected by geochemical methods through a continuous point-to-point sampling with an interval of 15–20 cm. Every 1–2 m, depending on the thickness of horizons, the collected material should be combined into one sample. Special attention should be given to taking halite samples from those horizons that occur below and above clayey or carbonate–gypsum–anhydrite layers. Each salt variety is sampled individually. Upon stripping of pure halite, the sampling intervals are increased up to 5 m. In core examination, special attention is given to those intervals where among the total mass of uniform rock salt, appear clay intercalations, phenocrysts or seasonal interbeds of red or orange potash minerals, and recrystallized salts consisting of large halite individuals and skeletal forms.

The first two metres of rock salt intervals, directly overlying the clay–carbonate interbeds, are sampled at a rate of three to four samples per 1 m. If halite horizons along the section are up to 10 m or over 20 m thick, they are sampled at a rate of one sample per 1 m, but in the latter case only every second sample is analysed. The number of samples depends on the analytical results and character of Br distribution [121]. Samples with a mass of 50–100 g are taken by breaking the core along its axis with subsequent drilling or sawing of salt interbeds. Sylvine and rock-salt samples are wrapped up in hard-textured paper, and those of carnallite are put into glassware. To ensure complete extraction of the core in salt drilling, it is advisable to wash the boreholes with sodium chloride- and carnallite-saturated solutions; this guarantees the preservation of potash minerals. Attention should be given to the contamination of salts by clay material containing increased Br concentrations, since this may produce false anomalies. In the case of non-core drilling, well-logging assists in determining potassium by the radiation of radioactive isotope ^{40}C contained in potash salts. Rock and potash salts are characterized, respectively, by minimum (negative anomalies) and maximum (positive anomalies) γ-activity. γ-logging helps differentiate the rocks of the salt-bearing series, locate good quality potash salt beds, and control the completeness of core extraction.

The geochemical zonality, stipulated by a differentiated spatial distribution of elements, should be considered. Potash salts occur commonly in the upper horizons of the section of completed cycles of halogen sedimentation; these sections are notable for the maximum content of bromine in halite. Potash formations at depth may be predicted from the high values of the geochemical index $(Br \times 10^3)/Cl$ in rocks and salts. Rubidium, bromine and chlorine are the indicators of supra-ore zones. Their concentration increases with an increase in the thickness and number of productive beds, and in the horizons of potash salts. Potassium content in the solutions of inclusions in halites aids in determining the maximum depth H of the salt-forming basin [91]:

$$H = 5.7 \frac{l}{1 - b/\alpha},$$

where l is the real thickness of a seasonal salt intercalation (cm), b is the potassium concentration at the initial period of salt precipitate formation (g/l), and α is the potassium concentration at the final stage of halite precipitation (g/l).

Furthermore, the bromine–chlorine ratio is used to locate those cycles and stratigraphic horizons that may confine potash deposits. The smallest value of this ratio is found in halite at the contact with the underlying clay–carbonate and anhydrite benches.

Stages of work

The first stage (geological surveys at a scale of 1 : 200 000) involves the identification of areas whose geochemical, paleogeographic and tectonic features are favourable for carrying out detailed hydrogeochemical and γ-ray surveys. Hydrogeochemical investigations may be a certain amount of help but the structure, composition and type of salt-bearing deposits, as well as their economic value, can in every concrete case be elucidated and assessed only by studying the section with the help of drill holes. Anhydrites, magnetites, carbonates and the salt-bearing horizons in halogen formations are located from well-logging data. The stratigraphic position of salts in the cross-section of the salt-bearing series and the cyclic pattern in the structure of productive horizons are identified at this stage. The results of reconnaissance surveys are used to define concrete zones favourable for prospecting, and to evaluate the commercial significance of deposits and the quality of salts.

3.8. NATURAL SODA

The following types of natural soda deposits are distinguished: (a) volcanogenic-sedimentary; (b) hydrothermal; and (c) exogenous. The latter are of greatest significance. Natural soda occurs in contemporaneous salt lakes, and in continental formations located in the vicinity of volcanic rocks (ash, lavas), enriched in soda and potash. Soda components were dissolved by surface and subsurface waters and transported into lakes located in drainless depressions. The paragenetic associations are represented by halite, calcite, dolomite, shortite, pirssonite, sepiolite, labuntsovite, pyrite and acmite, whereas sodium, bromine, chlorine, lithium, phosphorus, arsenic, fluorine and boron are the indicator elements. The general prospecting indications of natural soda are as follows: (a) extensive development of sodium-rich rocks; (b) large intermontane depressions and hollows filled with rhythmically layered polymictic sediments; and (c) traces of intense volcanic activity.

Application of geochemical methods in the search for natural soda is based on the specific properties (enrichment in sodium, high salinity) of ground waters confined to the zone containing the above-stated formations. Being easily soluble, carbonate and bicarbonate salts actively affect the chemical composition and salinity of natural water. Minerals containing sodium carbonates and bicarbonates (and dawsonite) are not formed in non-sodic waters. Prospecting surveys should primarily be oriented towards the elucidation of the hydrochemical prerequisites and indications of soda occurrence. The subsurface sodic water salinity in soda-free formations does not exceed 55 g/l, and the content of sodium sulphate is below 400 mg/l. Sufficient amounts of the soda-type surface and subsurface waters had to inflow a salt-forming basin.

Zones in contemporaneous lakes as well as ancient soda-bearing formations, occurring among slightly saline or fresh water sediments in piedmont and intermontane depressions, are prospective for this type of raw material. The latter are generally localized in the regions of an extensive development of compact-crystalline rocks enriched in alkaline elements (or products of their destruction), and in the areas affected by recent and ancient volcanic activity. Alpine foredeeps and intermontane troughs, localized near acid magmatic rocks and filled with continental fresh water, dolomite- and dawsonite-containing sediments, may be prospective for natural soda deposits.

Hydrochemical surveys for natural soda are rather specific. Prior to field work, the objects deserving exploration are selected. Depressions resulting from tectonic dislocations and filled with rhythmically layered polymictic sediments that originated in a marine or continental brackish water basin, are identified.

The area prospective for exploratory work is surveyed. The chemical composition and salinity of subsurface sodic waters are analysed. The following types of waters are distinguished by the predominance of anions and cations: (a) hydrocarbonate–calcium; (b) hydrocarbonate–magnesium; (c) hydrocarbonate–sodium; and (d) chloride–sodium waters. Of special interest are subsurface waters whose salinity is in excess of 150 g/l. When carrying out hydrochemical investigations, E. F. Stankevich and Yu. V. Batalin propose taking into account the following indications characteristic of the anomalies associated with the development of mineralization:

(a) presence of subsurface hydrocarbonate–sodium, sodic water whose salinity exceeds 25 g/l;

(b) occurrence of subsurface sulphate–sodium water with a salt content exceeding 10 g/l;

(c) manifestation of subsurface chloride–sodium waters with an increased content of soda ($NaHCO_3 + NaCO_3 > 10$ eq per cent or 4 g/l) at a total salinity of 35–40 g/l and more;

(d) occurrence of subsurface sodic water with a total salinity of 60–70 g/l and more;

(e) regional distribution of subsurface sodic waters with a salt content of 5–10 g/l; and

(g) an increased content of sodium sulphites (> 1 g/l) in subsurface sodic waters enriched in Br, I, Cl and Na.

Special attention is given to hydrochemical testing of local springs. Optimal sites and productive horizons for core drilling are defined. The core from the holes drilled in the zones with lacustrine-continental deposits is analysed for Br, Cl, Na and Rb. In the foothill regions, continental formations that do not contain mineralization are sampled. Samples of dawsonite are taken from a depth of 0.5 m due to hypergene alteration of this mineral. Rocks are subjected to qualitative chemical analyses in order to detect soluble, non-silicate Al_2O_3 bound in alumohydrocarbonates and alumophosphates. Qualitative determination of soluble non-silicate Al_2O_3 in rocks is possible owing to an easy dissolution of these minerals in weak acetic acid. Aluminum is determined by its colour (orange or red), resulting from the reaction with alizarin. Samples characterized by the positive reaction should be subjected to laboratory analysis. It is advisable to analyse the drilling mud, as soda is dissolved in the process of drilling. Soda deposits are outlined on the basis of sodium, bromine and chlorine haloes.

3.9. GYPSUM AND ANHYDRITE

Gypsum and anhydrite rocks are composed mainly of the corresponding minerals and an insignificant admixture of calcite and dolomite. The following types of deposits are known:

(a) sedimentary which, in their turn, are sub-divided into syngenetic (formed by chemical precipitation of minerals from solutions) and epigenetic (resulting from the hydration of anhydrite) deposits;

(b) infiltrational, divided into metasomatic (substitution of limestones) and re-deposited (dissolution of the redeposited gypsum) fractions; and

(c) residual (occur together with salt deposits).

Deposits of the first group have the greatest commercial significance.

Gypsum and anhydrite can be prospected by the same geochemical methods that are employed in the search for salts. Geophysical methods (electrical prospecting) can be used to locate gypsum and anhydrite, characterized by higher resistivity as compared to the overlying loose formations (clays, loams and sandstones). Vertical electric sounding (VES) is applied to outline the deposits and to determine the depth of productive horizons, whereas electrical profiling assists in detecting karsted zones in the gypsiferous rocks, and buried sink holes.

3.10. LIMESTONES AND DOLOMITES

Pure varieties of limestones and dolomites approach in their composition calcite and dolomite, respectively. The following deposits are distinguished: (a) chemogenic; (b) organogenic; and (c) fragmental carbonate deposits. Deposits of sedimentary marine genesis have the major commercial value. Their significance depends on the conditions of occurrence of productive horizons and on the raw material quality, i.e. on the chemical composition and hardness of rocks.

Limestone and dolomite beds generally spread over vast areas, being controlled by lithological and facies conditions. The regional search for limestones is based on geological (lithological) criteria (development of carbonate formations) that make it possible to identify areas favourable for their occurrence.

Mechanical ($CaCO_3$, Ca, $Mg(CO_3)_2$) and salt ($Ca(HCO_3)_2$ and $Mg(HCO_3)_2$) haloes are formed around carbonate rocks in the process of weathering. They can be contoured by the quantity of CO_2, determined in a special device by decomposing carbonates in a 10 per cent HCl. The content of carbon dioxide in loose formations overlying carbonate rocks in commonly twice as much as above other rocks.

Geological survey is carried out in insufficiently studied areas lacking geological maps. It is aimed at detecting outcrops of carbonate rocks and at elucidating the conditions and boundaries of their occurrence. The survey routes should run across the strike of rocks at equal distances (0.3–1.0 km) from each other, primarily along the slopes of river valleys and elevations. These routes should be combined with a small number of sections (1–3) along the strike. Closed terrains should be explored by geophysical methods (vertical electrical sounding and symmetrical electrical profiling), which make it possible to determine the thickness of the overburden and to delineate the carbon rock zones. Detailed prospecting is carried out in areas with detected horizons prospective for carbonate rocks or in the vicinity of the deposits, that are already being mined.

Carbonate survey makes it possible to trace out limestone horizons and is of appreciable help in the areas with poor outcropping. Samples of eluvial–illuvial formation, with a mass of 50 g, are taken from a depth of 0.2 m. The spacing and number of observation points per unit area depend on particular conditions. The content of carbonates is determined by the gas-volume method, and their composition is elucidated by the rate of decay.

Combined surveys are aimed at studying carbonate rocks and elucidating their potential utilization. This is achieved by mineralogical–petrographic and chemical–spectral analyses. The latter aid in determining useful and harmful (magnesium oxide) admixtures. The type and scale of dolomitization should be studied in the area of its development.

The layer-by-layer characterization of sections of drill-holes, workings and exposures is undertaken to establish the geological features controlling the distribution of dolomitized rocks, and to identify the stratigraphic levels and zones of tectonic dislocations. The content of magnesium oxide is determined by analysing samples and heavy concentrates upon their obligatory dyeing. Samples are taken for chemical and thermal analysis. The latter technique is used to determine the quantity of calcite. Having established the type of dolomitization at the stage of prospecting or preliminary survey, it is necessary to study the lithological changeability of dolomitized carbonate rocks and their composition in two or three parallel profiles and in the section cutting the entire thickness of these rocks across their strike.

The sequence of prospecting surveys is as follows. The sites where limestones occur at the minimum depth are identified by the gas method along a grid of 0.5×0.05 km. Drilling and geochemical analyses of the core are performed in the prospective zones. Detailed mapping of carbonate rock horizons is made to detect the purest varieties. Rocks are analysed not only for outlining the limestones and dolomites but also for determining their quality. The chemical composition of carbonate rocks is studied with a view to determining: (a) silicate $\eta = SiO_2/(Al_2O_3 + Fe_2O_3)$, and (b) alumina $\beta = Al_2O_3/Fe_2O_3$ moduli. These parameters are used to grade carbonate rocks and to relate them to a raw material suitable for making cement or to be used in chemical and other industries.

Based on the preliminary analysis of the composition of carbonate rocks, combined layer-by-layer samples are taken at 5–10 m intervals in the workings. If limestones are to be used as flux in the chemical industry, and in cement production, the unsorted samples are analysed, respectively, for: (a) Mg, CaO, SiO_2, Al_2O_3, P and S (to lower the cost, thermal determinations may be substituted for chemical analyses); (b) CaO, MgO, SiO_2, Al_2O_3, Fe_2O_3, As, F, K, Na and H_2O (chemical analysis may be supplemented by quantitative spectral analysis); and (c) determining the coefficient of saturation (C_S) indicating what part of CaO in the clinker is bound with silica in the form of allite:

$$C_S = \frac{Ca - (1.65\ Al_2O_3 + 0.35\ Fe_2O_3 + 0.7\ SO_3)}{2.8\ SiO_2}.$$

Commonly, reduced analyses are made to determine CaO and SiO_2 (for flux production), and CaO, MgO and $SiO_2+P_2O_5$ (for making lime). Geochemical methods are combined with geophysical ones. Carbonate rocks, forming lenses and benches sufficiently thick and persistent along their strike beds, are detected by their electric resistivity on the VES curves. Electrical prospecting is most advisable in the search for limestones whose electrical resistivity drastically differs from that of dolomites and loose formations. Geophysical methods assist in tracing out the zones with limestones, dolomites, and thick loose formations. They also aid in determining the shape and dimensions of productive horizons.

3.11. CONCLUSIONS

1. The diversity of genetic types of chemical raw material deposits demands a non-uniform approach to their search by geochemical methods. The use of chlorine, fluorine and mercury as indicator elements is the most general criterion of prospecting. Haloes, substantially exceeding commercial accumulations, are identified in the deposits of magmatic, metasomatic, hydrothermal and even sedimentary types. There is commonly a direct dependence between the halo and deposit dimensions.

2. Other characteristic features are more specific. Magmatic deposits of apatite are distinguished by their petrochemical properties. The boron content in skarns is evaluated on the basis of geochemical specialization of granite intrusives. Metamorphogenic apatite deposits, formed at certain stages of metamorphism, are characterized by intensive radioactive anomalies and negative anomalies of potassium. The lithological–facies indications and rock enrichment in phosphorus and potassium are significant factors in the search for exogenous deposits of sulphur, potash salts, and phosphorites. Gradual transitions between commercial concentrations and dispersed elemental contents are observed in sedimentary formations. Around productive bodies, haloes form intermediate zones between background and anomalous concentrations. Gently dipping and horizontal occurrences of ore bodies (phosphorites) are characterized by the haloes that are elongated along the strike of the bedded deposits.

3. Contrary to the deposits of non-metallic minerals of the first group, the ore bodies of the second group of deposits are surrounded by haloes of highly contrasting elements. Phosphorus, fluorine, chlorine, mercury and bromine anomalies are distinctly detected by geochemical surveys. At the same time, haloes, formed due to the removal of sodium and iron group elements, were observed on certain deposits of mining and chemical raw materials of the hydrothermal class (fluorite), as also was the case in the first group deposits.

4. When carrying out geochemical prospecting for certain types of mining and chemical raw materials, it is advisable to use radioactive elements (U, Th and K) that have contrasting geochemical indications. The greatest differentiation in the distribution of these elements is characteristic of zones of hydrothermal alteration of rocks. Apatite deposits are reliably identified by an increased content of thorium, and a decreased content of potassium at a neutral behaviour or introduction of uranium. Phosphorite deposits are characterized by the phosphorus and uranium correlation and by the Th/U value that indicates the depth of erosional truncation of the ore bodies.

5. Remote-sensing nuclear-physical methods are applied to determine the economic value of deposits of mining chemical raw materials.

6. In the search for mining chemical raw materials, geochemical surveys should be combined with methods of thermography (sulphur, fluorite), isotopic methods (sulphur), thermal luminescence (apatite, fluorite), and decriptophonic investigations (fluorite).

CHAPTER 4

Deposits of the third group

This section is concerned with the mineralogo-geochemical methods applied in the search for the deposits of minerals whose utilization depends on: (a) the chemical composition and physical properties of the aggregate as in refractory materials (magnesite), and ceramic materials (wollastonite); and (b) the physical properties of the aggregate as in colour stones (nephrite), fillers (talc), pigments and fillers of concrete, building materials (magmatic, metamorphic and sedimentary rocks), heat insulators (vermiculite, diatomite, perlite), adsorbents, and absorbents (gaize, tripoli, clay). The chemical composition of the raw materials, relating to the second sub-group, is of secondary importance because they are valued primarily of their hardness and decorative properties.

Non-metals of this group have an exceptionally great economic significance. They are mined and used in different branches of industry on a tremendous scale. The distinctive feature of the deposits under consideration is the need for their qualitative evaluation based on the analysis of the chemical and physical properties of useful minerals and their substance composition. The geochemical differentiation of productive and enclosing rocks is generally less distinct than in ore deposits.

Minerals of this group are formed in rather diverse conditions, therefore, endogenic metamorphic, and sedimentary types of deposits are distinguished.

4.1. MAGNESITE

Magnesite ($MgCO_3$) is a monomineral rock consisting mainly of the magnesite mineral: crystalline and cryptocrystalline varieties are distinguished. There are hydrothermal–metasomatic and infiltrational types of magnesite deposits. The former are of primary commercial value. The deposits are composed of crystalline magnesite having bedded on lens-like structures. Deposits of the first type belong to the medium-temperature ones that originated at moderate depths. Magnesites resulted from the hydrothermal impact on limestones exerted by magnesium-rich chloridic solutions. Infiltrational deposits represented by amorphous magnesite are localized among serpentinous ultrabasites affected by hypergenesis in the zone of weathering. Magnesite accumulations form lens-like deposits. The parameters and commercial value of such deposits are not high.

Magnesite, when pure, contains 47.8 per cent MgO and 52.2 per cent CO_2. The pure mineral is sometimes, but rarely, found as transparent crystals resembling calcite. However, magnesite predominantly contains variable amounts of the carbonates, oxides, and silicates of iron, calcium manganese, and aluminum. Magnesite may be either crystalline or 'amorphous' (cryptocrystalline). Accessory siliceous minerals such as serpentine, quartz, or chalcedony are usually present. Calcium minerals are usually absent or in low concentrations in cryptocrystalline magnesite, in contrast to their almost invariable presence and higher concentrations in the crystalline variety.

Magnesite dissociates upon heating to form magnesia (MgO) and carbon dioxide. When heated sufficiently, magnesia develops a crystal structure identified with that of the natural mineral periclase. The mineral periclase occurs only rarely in nature and not in any known, commercially workable deposits.

Deposits of crystalline magnesite are usually found associated with dolomite, but some major deposits, such as those found in Brazil, are in limestone measures. But even in these cases, the magnesite is not in direct contact with the limestone, but is separated by a dolomitized zone. The major deposits of the world are those in Austria, Russia, Korea, Brazil, Canada, Australia, Nepal and the United States. All are situated in dolomite host measures or zones. Further, all are in regions which have had orogenic activity. Commonly, such areas also show igneous activity, and it is postulated that magnesite resulted from the action of igneous intrusions and associated solutions on the dolomite (the measures are widely viewed as being sedimentary). The intermixed rock of magnesite–dolomite is thought to be the product of the hydrothermal dolomitization of limestone.

Igneous rocks associated with magnesite vary widely in composition (pyroxenite, diabase, peridotite, amphibolite, basalt, granite and others).

The minerals associated with several magnesites include dolomite, talc, serpentine, diopside, haematite, enstatite, olivine, pyrrhotite, magnetite and scapolite. The magnesite deposits formed at a temperature in excess of $300\,^\circ$C.

The mechanism suggested for the hydrothermal emplacement of magnesite in dolomite involves a reaction with Mg-rich, CO_2-bearing solutions.

Cryptocrystalline or 'amorphous' magnesite is an alteration product of serpentine or allied magnesium rocks which have been subjected to the action of carbonate waters. The serpentine which lies near or surrounds the magnesite is itself an alteration product of the ultrabasic rocks. The mode of formation of the magnesite usually limits the amount of impurities to small amounts of iron, lime and silica.

Occurrences of this type of magnesite are fairly common throughout the world, but are usually of limited size. The action of surface water containing carbon dioxide percolating down through serpentinized fissures can convert serpentine to magnesite and other minerals. Also, waters rising through fissures could produce magnesite. Most of the deposits of this type of magnesite are found near the surface and have a limited extent in depth.

The search for deposits of any type of magnesite should be guided by a consideration of the probable origin of the magnesite. In the case of crystalline magnesite, possible locations of deposits are areas of limestone or dolomite terrain that have been subjected to folding or igneous activity. In the case of cryptocrystalline magnesite, possible locations are areas of ultrabasic rocks which show extensive alteration to serpentine. Thus, in either case, the search can be confined within the boundaries of areas that show evidence of dynamic geological activity.

Magnesite formations are generally more resistant to weathering than associated formations; consequently, magnesite deposits are characterized by bold outcrops. Outcrops are sampled by surface chipping and shallow trenching. Field differentiation between commercial and non-commercial magnesite can be aided by the use of cold, dilute hydrochloric acid; limestone will effervesce when treated with hydrochloric acid whereas magnesite will not.

Following the preliminary investigation, a diamond drilling programme may be carried out to assess the commercial potential of the deposits. The evaluation is based on the size and location of the deposit and on the quantity and distribution of unwanted minerals containing silica, lime, iron-oxide and alumina.

The favourable indications in the search for magnesite are:

(1) The presence of carbonate rocks at the contact with granitoid outcrops.

(2) Tectonic dislocations and fissured zones in carbonate and dolomitized rocks.

(3) The presence of ultrabasites with distinct manifestations of hypergenic processes and crust of weathering formation.

Deposits of the first type contain the following minerals: calcite, dolomite, ankerite, aragonite, as well as barite, apatite talc, serpentine, chlorite, quartz, graphite, pyrite, chalcopyrite, arsenopyrite, zircon, pyrochlore and goethite. Within infiltrational deposits, magnesite is associated with calcite, aragonite, opal, nontronite, gymnite and picrolite.

Magnesite fragments, that are frequently found in deluvial–eluvial formations, are regarded as a prospecting indication of the primary sources.

The following admixtures have been found in the hydrothermally altered rocks occurring in the vicinity of magnesite bodies: chlorine, potassium, sodium, lithium, barium, strontium, copper, arsenic, zirconium and mercury. Magnesites themselves contain admixtures of K_2O (0.38 per cent) and Na_2O (0.34 per cent), as well as the following (in $n \times 10^{-3}$ per cent): Cu-0.6; Zn-3; Pb-0.3; V-0.3; Sc-0.8; Li-0.2; Nb-5; Sn-0.8; and B-6.

Therefore, typomorphic indicator elements of magnesite mineralization are chlorine, alkaline elements, barium, strontium, zinc and lead. This makes it possible to apply geochemical methods in the search for magnesite deposits. Closed areas are prospected by hydrochemical methods (detection of anomalous Mg concentrations in waters) and by metallometric surveys. When carrying out the latter, it is necessary to take into consideration the enclosing rock composition, thickness of loose formations, and genetic peculiarities of deposits. Magnesite deposits in dolomites are detected by the secondary haloes of chlorine, barium, strontium, mercury and arsenic. Their occurrence is also indicated by a combination of zones of introduction and removal of the ferric group elements. Magnetometry, electrical prospecting and radiometry are employed. The results of radiometric investigations (γ-ray logging) assist in establishing reduced radioactivity of magnesites and dolomites in productive formations.

At the stage of prospecting surveys, it is recommended to investigate the primary haloes in order to identify blind and deep-seated magnesite bodies. According to universal zonality, the supra-ore zones are characterized by the development of calcite and barite and by increased contents of chlorine, arsenic, mercury and barium, whereas the sub-ore zones are notable for increasing concentration of copper, zinc and vanadium.

4.2. WOLLASTONITE

Wollastonite $Ca_3(Si_3O_9)$ relates to the silicates of the pyroxenoid group, CaSiO has the composition of 48.3 per cent CaO and 51.7 per cent SiO. However, it is seldom found in the pure state due to the ease with which it takes into solution the metasilicates of manganese, magnesium, iron and strontium. Wollastonite occurs predominantly as a contact metamorphic deposit forming between limestones and igneous rocks. Commonly associated minerals are garnet, diopside, epidote, calcite and quartz.

Wollastonite occurs in coarse-bladed masses rarely showing good crystal form. It is usually acicular or fibrous, even in the smallest of particles. Fragments of crushed wollastonite tend to be needle-shaped, imparting a high strength, and this property is the basics for many of its uses. The fibre lengths are commonly in the ratio of 7 or 8 to 1, length to diameter. Some crystals of wollastonite fluoresce under short-wave or long-wave ultraviolet light, or both; colours range from yellow-orange to pink-orange. Specimens may also show phosphorescence.

Wollastonite is chemically inert and this property makes it useful as a filler and reinforcing agent.

There are two polymorphs of calcium silicates: wollastonite, a low temperature form, and pseudowollastonite, a high temperature form.

Wollastonite is a contact metamorphic mineral, occurring within impure limestones near intrusive bodies of granite or other acidic rocks. It can also be formed by the metasomatism of calcareous sediments and by the crystallization of certain magmas.

In contact metamorphism, limestones are recrystallized by the heat of the intrusive rocks. Silica emanations from igneous rocks yield calc-silicate hornfels along with 'skarn' rocks formed by the transfer of manganese, silicon, aluminum and iron from the magma to the limestone. Skarn minerals of wollastonite, garnet, and diopside form in the adjacent limestone, usually in a well-defined zone and probably at the final stages of igneous activity.

The output of the United States, the largest producer in the world, comes from New York and California.

A typical analysis of wollastonite-bearing rocks from the Willsboro deposit consists of (per cent): SiO_2 47.7; CaO 37.8; Al_2O_3 3.1; FeO 6.6; MgO 3.6; and MnO 1.2 [122].

Diopside, garnets, mostly almandine, idocrase, tremolite, quartz and calcite are the contaminants present.

The Belka mining district (India) contains the first deposits of wollastonite ever discovered and exploited. The skarn mineralogy is represented primarily by hedenburgite, diopside, andradite, vesuvianite, calcite, quartz and bebingtonite.

In composition the wollastonite averages (per cent): CaO 48.02; SiO_2 48.77; MgO 0.06; Al_2O_3 0.66; Fe_2O_3 0.43; MnO 0.96; Na_2O 0.02; and K_2O 0.11. Thermal studies of wollastonite in the temperature range of 1100–1350 °C show an intensification of colour from grey to dark grey, which is found to be iron dependent [3].

Zoning characteristics are displayed as the minor occurrence of magnetite and skarn minerals followed by a band of wollastonite and calcite. Wollastonite formations, confined to the Grenville series of the Upper Archean (USA), and represented by wollastonite-diopside rocks containing up to 75 per cent of wollastonite, are the most prospective ones. Wollastonite results from the thermal metamorphism of sand–clay carbonate-containing rocks. The presence of garnet–diopside rocks is a favourable prospecting indication of wollastonite.

The characteristic series of wollastonite accessories is as follows: garnet, diopside, vesuvianite, hedenbergite, epidote, zoisite, phlogopite, axinite, chlorite and scapolite. Boron, fluorine, chlorine, strontium, barium, mercury, sodium, potassium, gallium, lead, copper and zinc are the indicator elements of wollastonite mineralization. The following elemental contents have been established in wollastonite (per cent): Na up to 0.15; K up to 0.07 and (in $n \times 10^{-3}$ per cent): Cu-0.5; Zn-6; Pb-0.8; Ni-0.5; Co-1.0; Cr-1.0; Mo-15; Sr-0.2; Sc-1.0; Ti-5; Be-0.4; Zr-10; La-8; and B-10. Wollastonite occurrences in skarns are mainly represented by pyroxene–garnet formations.

In searching for wollastonite deposits it should be borne in mind that their primary monomineral bodies are seldom exposed at the surface. Wollastonite formations are accompanied by bleached feldspar and propylitized rocks surrounded by strontium, barium, sodium, chlorine and mercury anomalies. The contact zones of wollastonite skarns drastically differ from the enclosing rocks. Deep-seated wollastonite bodies are indirectly indicated by the presence (at the present level of erosion) of bleached marbled rocks (calciphyres), metasomatites and skarns wherein andradite, hedenbergite, arsenopyrite and scapolite are encountered. Wollastonite deposits have not been subjected to experimental geochemical surveys.

4.3. BRUCITE

Brucite, a crystalline magnesium hydroxide, is the major component of brucite rocks which also contain carbonates and diopsides.

Brucite ($Mg[OH]_2$) has been mined for the production of magnesia, but is no longer an important source. Theoretically it contains 69.1 per cent MgO and 20.9 per cent H_2O. The mineral is often associated with limestone, but most commonly with magnesite. It has a translucent appearance and is relatively soft and lightweight.

Brucite is sometimes found associated with commercial deposits of magnesite, as in Nevada and Washington. In those cases, it would be processed along with the magnesite to produce refractory magnesia.

There are no known operations for brucite as a separate exploitable mineral. The brucite in deposits occurs as rice-like grains 1/16 to 1/8 in diameter disseminated through a limestone matrix. The process of winning the magnesia was based on controlled slaking of carefully sized and calcined ore. The system was controlled so that the discharge from the slaker consisted of an impalpable dust of hydrated lime mixed with magnesia granules. The granules were air separated, washed, and wet classified. The lime was sold directly, and the magnesia granules were further processed to refractory magnesia.

Brucite deposits are only formed upon metamorphism of magnesian magmatic (hyperbasites) and sedimentary (carbonate) rocks. Brucitites are formed as a result of contact metamorphism. The localization of brucitite is controlled by the deposition of the initial magnesites in shallow water-lagoonal dolomites. The mass sedimentation of magnesite was stimulated by the supply of Mg from old apobasitic crusts of weathering.

A magnesite–brucite transformation is determined if the post-folded intrusions are oxitic derivatives of a basaltic magma and hypabyssal and subvolcanic magmas related to a deep magma chamber. The input of H_2O by ascending magmatic fluids is indicated by the occurrence of brucitites in the intrusive roof.

The most complete zoning of the various genetic forms of brucite is formed in a subvolcanic monticellite corneous (sanidinite) facies: here, from the external to the internal parts the zones are: plate protobrucite, directly replacing magnesite, fine-flaky pseudomorphs of brucite after intermediate aqua-periclase, and the largest zone of fibrous apopericlase brucite.

Brucites occur in contact magnesium carbonate formations that resulted from the interaction of granitoids with ultramagnesian substratum rocks at a shallow depth. The latter are the source of magnesium in the process of brucite formation. P. P. Smolin [107] suggested that metamagnesian rocks are associated with the manifestation of explosive volcanism, whereas distinct assimilation of the substrance results from the interaction of intrusives with the enclosing rocks. The metamagnesian formations are represented by their transitional differences from the silicate (hornfels) to pure carbonate (marble, brucitites), and intermediate (calciphyres) ones as well as by silicate–carbonate ophicalcites and silicate serpentinites. The criteria used in prospecting for brucite depend on the conditions of its formation. Localization of brucite ores in the prospective zones depends on the nature of the primary enclosing rocks whose total magnesium content corresponds to that in dolomite–magnesite series. Monomineral brucites are formed from ancient magnesites in the process of contact metamorphism.

Brucite accessory minerals are: forsterite, humic fluorsilicates, phlogopite, diopside, edenite, spinel, augite, enstatite, cummingtonite, talc, anthophyllite, tremolite, fluoborite, crysotile, antigorite, chlorite and calcite. Forsterite is associated mainly with calciphyres, whereas diopside and edenite are associated with skarns. Fluorine (up to 0.86 per cent) and sulphur (up to 2.4 per cent) are the admixture elements in magnesian skarnoids. The former is rather characteristic of ophicalcites. Magnesian skarnoids are notable for sharp non-uniform fluctuations in the contents of alkalis at the predominance of potassium. Brucites are characterized by the low content of iron oxide (up to 0.2 per cent) and silica (tenths of a per cent). Among the indicators of brucite mineralization are strontium, barium, boron and lithium whose increased quantities are found in magnesian skarnoids. Special geochemical investigations have not practically been carried out on brucite deposits.

4.4. TALC

Talc, $Mg_6SiO_{12}(OH)_4$, is a hydrous magnesium silicate. Its contact-metasomatic, hydrothermal, and metamorphic deposits are known. The former originate near the contact of dolomites and dolomitized limestones with granitoids and are divided into two subgroups: (a) those that originated with the introduction of CO_2, and (b) those which were formed due to the introduction of SiO_2. The latter, associated with serpentinous ultrabasites, are grouped into the deposits formed: (a) due to the introduction of CO_2; and (b) owing to the introduction of SiO_2. Metamorphogenetic deposits of talcites and soap stones are associated with the introduction of MgO and CO_2 into schists and dolomites.

Talc mineralization resulted from the impact of carbonaceous and siliceous, water and chlorine-rich hydrothermal solutions genetically associated with granitoids, on magnesian–carbonate rocks and ultrabasites. Ferrugenous silicates of ultrabasites turn into talc, magnesite and magnetite under the effect of hydrothermal metasomatism. With the introduction of CO_2, metamorphic processes result in the formation of talc–carbonate and carbonate–talc rocks on ultrabasites, whereas the introduction of SiO_2 leads to the formation of talcites. Metamorphism of marls (without CO_2 introduction) gives rise to the formation of talc–chloritic rocks. Talc deposits originate within a broad temperature and pressure range, predominantly in alkaline conditions resulting from the post-magmatic activity of intrusions and regional metamorphism.

In exogenic conditions, talc is preserved but its accessories are disintegrated and altered.

The genetic features of talc have bearing on its prospecting indications. These are as follows:

(1) presence of magnesian rocks (ultrabasites, serpentinites, magnesian carbonate rocks) and granitoids;

(2) rocks that were metamorphosed at the stage of green schists and antigoritic serpentinites;

(3) tectonic dislocations in rocks accompanied by the zones of crumpling, schist formation, and outcrops of quartz veins.

Prerequisites for the application of geochemical methods in the search for talc

The paragenetic accessories of talc are those minerals whose species composition depends mainly on the nature of the enclosing rocks. Talc accumulations in ultrabasites are accompanied by magnetite, haematite, serpentine, brucite, calcite, asbestos, amphibole, chlorite, biotite, chromite, tourmaline, montmorillonite, magnesite and gypsum. Carbonate rock metamorphism results in the formation of

talc in association with diopside, tremolite, enstatite, phlogopite, scapolite, actinolite, magnetite, graphite and serpentine. In the occurrences of rocks of different
composition (schists with relics of dolomites and diabases), talc accessories are:
sericite, quartz, chlorite, actinolite, graphite and tremolite. The major accessory
minerals of talc associated with the contact–reaction rocks of the magnesian skarn
formation are: diopside, forsterite, tremolite, anthophyllite, phlogopite, graphite,
spinel and quartz. Secondary minerals are represented by serpentine and brucite.
Among the above-enumerated minerals, the most significant ones are forsterite,
serpentine, tremolite, actinolite, sericite, graphite, magnesite, anthophyllite and
diopside. The following minerals are of secondary importance: phlogopite, chlorite, carbonate, quartz, pyrite, magnetite, gypsum, tourmaline, asbestos, chromite,
haematite, scapolite and brucite.

As has been evidenced by the analysis of the mineralogical composition of talc
ores and accompanying solid phases in the surrounding rocks, their major chemical components are SiO_2, MgO, CaO, $FeO + Fe_2O_3$, Al_2O_3, K_2O, Na_2O, TiO_2,
H_2O and CO_2. The geochemical indicators of talc ores, associated with tremolite
and formed due to the alteration of carbonate rocks, are F and Na whose contents fluctuate from 0.11 to 4.65 per cent and from 0.5 to 2 per cent, respectively.
Apocarbonate talc formations contain very small quantities of admixture elements
that are mainly represented by Sr, P, F and Ba. When talc formation resulted
from ultrabasite hydrothermal metamorphism, ores contain Ni, Cr, Co, Cs, Cu
and V. Talc formations, originating from certain substratum rocks inherit some of
their components and admixture elements. Water, silica, carbon dioxide, chlorine,
fluorine, sodium, potassium and lithium, incorporated in talc-formation processes,
are transported from the outside, whereas the main component — magnesium —
is derived from the enclosing rocks. The mineral-forming solutions were partly
supplied with CO_2 by the enclosing rocks (marbles, coaly substances). All this is
responsible for the migration and redistribution of those elements that are closely
associated with magnesium, fluorine and chlorine in talc zones. It is not only
magnesium but also nickel, chromium, titanium, and phosphorus that are removed
from the enclosing rocks. The talc rocks, which originated from ultrabasites, are
characterized by the presence of cobalt (up to 0.02 per cent), whereas talc, formed
from sand–clay rocks, contains zirconium (up to 0.03 per cent).

Such petrochemical properties as the development of antigoric serpentinites,
wherein the ratio of magnesium oxide to the sum of the ferric group oxides ranges
from 7 to 8.5, are regarded as the criteria to be used in the search for talc [74].
The outcrops of acid intrusives constitute a supplementary favourable prerequisite.
Of special interest are the direct contacts of ultrabasites with granites. Being the
most accessible to solutions, these contacts are detected by the mercury–metric
survey. An increased radioactivity of granites permits the identification of the
contact positive by radiometry. Based on the data supplied, the typomorphic
indicator elements of talc mineralization are: fluorine, chlorine, potassium, sodium,
phosphorus, zirconium and elements of the iron group. Thermography may also
be employed in prospecting for talc whose loss on ignition (if it contains small
quantity of carbonates) ranges from 6 to 13 per cent.

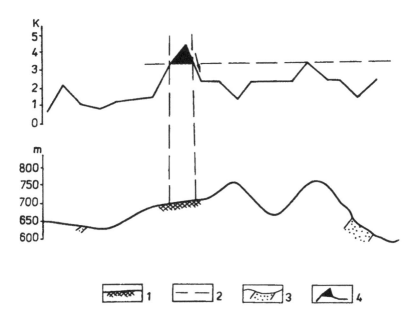

Figure 4.1. Distribution of concentration of coefficient K = (Ni × Cr × Co)/(Ba × Cu) across the strike of a talc zone I: 1 — talc body, 2 — anomalous level, 3 — quartzites, 4 — anomaly.

Certain differentiation in the composition of talc-containing and enclosing rocks favours the application of geochemical methods. Positive Cr, Ni, Cu, Co and Zn anomalies have been detected in loose formations overlying ultrabasites and in plants growing above them (Fig. (4.1)).

Methods of work

Primary talc deposits are very seldom detected at the surface. Generally, they are overlain by eluvial–deluvial formations. Prospecting at a scale of 1 : 50 000 should involve hydrochemical methods (to detect magnesium anomalies) and geobotanical methods in the presence of thick loose formations. Since talc is stable to hypergenesis, it is well preserved in eluvial–deluvial formations. Well preserved talc flakes are found in certain terrains. Talc fragments are preserved in alluvial formations and, in their appearance, distinctly, differ from rocks of other composition.

When prospecting for talc mineralization associated with ultrabasites, it should be taken into account that the latter contain increased contents of Ni, Co, Cr, V and Cu. Lithochemical surveys along exposed dispersion haloes envisage sampling of fine (< 0.5 mm) sand–clay fraction of eluvial–deluvial formations from a depth of 0.2 m. The samples are analysed for Cl, F and Ag (by special methods), for K, Na, Li and Rb (by flame photometry), and for Ni, Co, V, Cr, Zn, Y and Zr (by spectral analysis).

Talc accumulations occur commonly in the periphery of ultrabasites. It should be borne in mind that they are indicated by positive and negative haloes of the ferric group elements, as well as by increased concentrations of F, Cl, Hg and P. Where dykes of granite–porphyry occur near ultrabasites, attention should be given to the dispersions in the contents of such elements as Ba, Pb, Zn, Zr and P. Magnetic exploration is carried out to identify the zones of contact of ultrabasites as the most prospective sites for detecting talc deposits. Talc bodies are distinctly localized with the help of the emanation survey by the minimum radon concentration. In the process of prospecting, it is advisable to explore the endogenic haloes of Cl, F, Hg, K, Na, Rb and Li, indicating talc mineralization. Special attention should be given to the coincidence of the haloes formed by the introduction or removal of elements that indicate the epicentres of productive bodies. It is rather efficient to combine geochemical methods with thermography which helps detect steaming haloes and water-abundant zones in rocks. A distinct and intensive endothermal peak at 965 °C is identified in talc formations from the thermal analysis data. Zonality is an important geochemical indication. By determining zonality indices, studying the correlation of elements, and the data provided by the differential phase analysis for mercury, it is possible to predict talc mineralization at the present level of erosion or at a certain depth.

Prospecting for talc deposits is mainly based on direct field observations by geologists, geochemical methods being considered inefficient. This book demonstrates the efficiency of geochemical methods for talc prospecting. The methodology employed to achieve maximum information from the results by means of coefficients allowing the identification of potential talc areas is also explained.

4.5. JADEITE

In natural conditions, jadeite is a polymineral fine-, small-, or medium-grained aggregate, consisting of sodium-rich pyroxenes that are intimately articulated with diopside, hedenbergite and aegirine. Well formed jadeite (pyroxene) crystals are encountered very seldom. Jadeite proper (no less than 80 per cent of jadeite) and diopside–jadeite (40–60 per cent of jadeite) are distinguished. Endogenic deposits of jadeite are represented by:

(a) jadeitized veins of leucocratic granites;

(b) transformed into jadeite veins of leucocratic gabbroids; and

(c) jadeitized effusive sedimentary formations.

Eluvial–deluvial and alluvial jadeite placers are also known. Endogenic deposits result from high-pressure metamorphic transformations of the plagioclase-rich veined alumosilicate rock inclusions in ultrabasites. The transformation of gangue leucocratic granites and gabbroids occurs under the effect of soda solutions, formed during regional metamorphism under high pressure (2.5 GPa).

The general prospecting indications of jadeite deposits are:

(1) Presence of ultrabasic rocks of the dunite–harzburgite formation (harzburgites, dunites, peridotites) with an extensive development of a series of veins (plagiogranites, quartz–diorites, gabbroid bodies).

(2) Intensive serpentinization of ultrabasites in the marginal endocontact zone of rocks, presence of margins of cherry-coloured serpentinites; ultrabasites are prospective if they contain inclusions of vein bodies of granitoid composition.

(3) Development of metasomatites – albites, plagioclasites as well as xenoliths, metamorphosed spilites, and keratophyres in ultrabasites.

(4) Extensive development of soda granites, metasomatites and monomineral albites.

(5) Tectonic dislocations by shifting that contribute to the splitting of ultrabasites into separate blocks; presence of tectonites, schist-forming zones, and gliding plane.

(6) Presence in serpentinites of zonal metasomatic bodies with amphibolic, chloritic and scurfy serpentinous fringes.

Conditions for the application of geochemical methods in prospecting for jadeite

The multistage nature of mineralization is responsible for the diversity of minerals in jadeite deposits. The following three groups of minerals are discriminated:

(1) relict minerals of magmatic rocks: albite, hornblende, olivine and orthorhombic pyroxene;

(2) metamorphic minerals associated with the productive, high-temperature stage: jadeite, diopside, omphacite, chloromelanite, spinel, natrolite, albite, muscovite (fuchsite), zoisite, arfvedsonite, crossite, pectolite, grossular, magnetite and native lead; and

(3) hydrothermal, hydrothermally-metasomatic minerals superimposed over jadeite mineralization: nepheline, albite, quartz, diopside – jadeite, chloromelanite; aegirine, phlogopite, vesuvianite, zoisite, actinolite, tremolite, chlorite, alkaline amphiboles, analcite, pyrite, muscovite, crysotile, antigorisphene, pumpellyite, prehnite, stilpnomelane, zircon, talc, cancrinite and haematite.

Hydrogarnet of andradite and grossular composition, Na-actinolite, pyroxene, and magnetite are the indicator minerals of jadeite occurrences. The development of aegirine, chloromelanite, cancrinite, soda amphiboles and other sodium-rich minerals in metasomatic bodies is a mineralogical prospecting indication of jadeite mineralization. Wide distribution of minerals enriched in sodium is a favourable indication of high concentrations of this element in the mineral-forming solutions. Jadeite mineralization is indirectly indicated by the occurrence of pyroxenes of complicated and non-stable composition (aegirine, diopside, jadeite) in rocks, by the antigoritic composition of serpentinites, and by the presence of magnetite-rich tectonites of characteristic black colour. A potential discovery of green jadeite is evidenced by the presence of cherry-coloured (due to thin haematite films) serpentinites and by the occurrence of omphacites in rocks. The primary composition of the substratum rocks is appreciably changed under the effect of mineral-forming solutions. As compared to fresh harzburgites and slightly altered serpentinites, the circum-ore antigoritic tectonites contain greater amounts of hydrogarnet (by five times), Na-actinolite (by three times), pyroxene of complex composition with a jadeite component (by 100–200 times), magnetite (by 3 times), and native lead (by 7 times) [114]. The opposite tendency is characteristic of magnetite in chromium spinel, which is disintegrated and altered in the circum-ore zones.

It was found that jadeite contains admixtures of manganese, nickel, chromium, ferrous and ferric iron, whereas its gaseous–liquid inclusions include halite, carbon dioxide, and bicarbonites.

The typomorphic indicator elements of productive mineralization are: Na, K, Cr, Ti, Cl, Cu, Ni, V, As, Sb, Pb, Hg, Zr, Hf, Zn and Sn (Fig. (4.2)). Migration and redistribution of indicator elements in the jadeite-bearing zones in the process of mineral formation was accompanied by the formation of positive and negative anomalies and a sharp variation in the contents of admixtures. Owing to chromium and iron diffusion from serpentinites in the process of metasomatism, these admixtures penetrated into the crystalline lattice of minerals, thus giving green colour to jadeite and zoisite. Albites and plagioaplites are affected by natrometasomatosis. Since these minerals are localized at the contact with serpentinites, 'antagonistic' elements characteristic of granitoid (Zr, Hf, Zn, Pb, Sn, W), and basic (Ni, Cr,

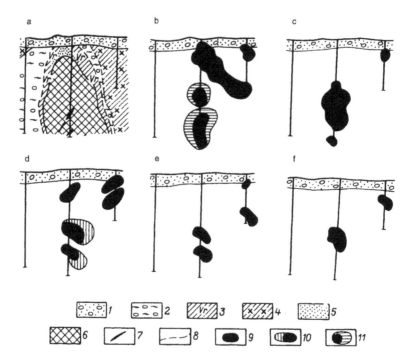

Figure 4.2. Primary haloes around jadeite body: a — rocks, b–f — haloes: b — Hg, As, Sb, Ba, c — Br, Li, Rb, d — Ni, Co, V, Cr, e — Y, Yb, Zr, f — Zn, Pb. 1 — terrigenous formations, 2 — glimmerites, 3 — metasomatic rocks with vermiculite, 4 — serpentinites, 5 — sand rocks, 6 — jadeite, 7 — gems jadeite, 8 — zones of tectonic dislocations, 9 — haloes, 10 — contact negative and positive haloes Cr, 11 — anomalies Ba, Sr.

V, Ti) rock series occur in the productive zones. There are insignificant quantities of chlorine, sodium, and mercury in the enclosing rocks outside the jadeite-bearing zones. The concentrations of these elements sharply increase in the areas prospective for jadeite. At the final stages of jadeite formation, solutions are enriched in Cr, Ni and Ca leached out from the enclosing substratum rocks.

Methods of work

The analysis and generalization of geological data, obtained in the process of prospecting and geological survey at a scale of $1:200\,000$, aid in defining the areas and zones with ultrabasites and albites. The jadeite abundance is evaluated from the petrogeochemical criteria that make it possible to identify those strata prospective for jadeite from the total mass of ultrabasite intrusives. Such factors are taken into account as the depth of formation of intrusives, enclosing rock composition, and direction of the post-magmatic processes. The depth of formation and degree of rock erosion (facies character) are determined from the totality of petrochemical indices: facies coefficient K_f and magnesium-to-iron ratio M/F. Rocks with $K_f < 0.1$ and $M/F > 10$ are not prospective for jadeite prospecting [114]. The most favourable for the formation of productive mineralization are rock strata that are characterized by a moderate facies character and have the following typical quantitative indices of jadeite abundance: $K_f = 0.13$, $M/F = 9.1 - 9.3$, and a moderate ferruginosity ($f = 0.09$). Jadeite formation is favoured by:

(a) occurrence of cherry-coloured serpentinites with insignificant concentrations of Fe^{2+} and Cr, and maximum quantities of Mn, Al, Ca and Fe^{3+};

(b) presence of dark-green pyroxenes–omphicites and chloromelanites;

(c) development of zones of metasomatites accompanied by a thick clay margin; and

(d) development (around large jadeite bodies) of tectonites characterized by negative haloes resulting from the explosion of gaseous–liquid inclusions.

Geochemical methods are most advisable at the stage of prospecting followed by expensive cutting of workings. Geochemical surveys are capable of determining the intensity of jadeite mineralization, the dimensions of the mineralized zones, and prospects for the development of mineralization at depth. Bedrocks are sampled to detect the haloes of indicator elements of jadeite formation: Cl, Na, K, Fe, Cr and Mn. Bedrock samples are taken by the dotted furrow method from the exposures, workings, and holes in the process of cutting and drilling. Five or six rock chips are taken from the sampling interval (3–5 m) at equal spacings and are combined into one sample with a mass of 150–200 g. Tectonites, dykes, and veins encountered in the profile are sampled separately. When processing the results of analyses, special attention should be given to Cl, Na and K haloes in the circum-ore zones represented by chlorite–actinolite–phlogopite metasomatites. Negative haloes, characteristic of the external zone of the circum-ore serpentinites, are located. The presence of green jadeites is indicated by black and cherry-coloured serpentinites enriched in F, Mn, Al and Ca.

Anomalies are graded on the basis of their complex nature, contrast range, high dispersion, and contact of positive and negative haloes, which indicate the location of the epicentres of jadeite-formation zones. The axis zonality of haloes is used to evaluate the present level of erosion. The ratios of linear productivities with the parameters of the supra-ore elements in the nominator and those of the sub-ore elements in the denominator constitute the criteria used in evaluating the rock strata for concealed mineralization. The level of erosional truncation of the anomalies is evaluated by comparing the above-stated parameters with those of the haloes surrounding the known productive bodies. The detected jadeite-bearing bodies are stripped by workings with the purpose of determining both the dimensions of deposits and jadeite quality (colour, uniformity or spottiness of colouring, and degree of translucency).

4.6. NEPHRITE

Nephrite is a solid, massive rock of predominantly bright green colour, composed of tremolite, actinolite, and hornblende. Endogenic and exogenic nephrite deposits are known. The former include nephrite occurrences associated with serpentinites and metamorphic rocks of basic composition, whereas the latter comprise alluvial and eluvial–deluvial formations resulting from the destruction of bedrocks. Nephrite is a product of iron–magnesium–calcium metasomatism of ultrabasites (occurring at the boundary of alumosilicate rocks with serpentinite) situated in tectonically weakened zones controlled by dykes of granite–porphyry and granites. The majority of nephrite deposits have formed at the boundary of two media (serpentinite + rodingite + albilite; serpentinite + dykes of the basic composition; serpentinite + granitoids). The mineral-forming solutions, which derived calcium from the substratum rocks and, primarily, from the dykes of basic rocks, were migrating along the faults. Some of the components (Cl, F, Hg, CO_2 and others) were transported from deep magmatic chambers. The geological prerequisites and prospecting indications of nephrite occurrences are:

(1) Outcrops of ultrabasic and basic rocks of the dunite–harzburgite formation, with serpentinization zones along the contacts with the dykes of acid rocks, accompanied by different metasomatic formations: albites, rodingites, listvenites and tremolitized rocks.

(2) The presence of elongated and persistent zones of faults (inside partly serpentinized rock strata and at their contact zones) wherein gabbro and granite dykes were emplaced and through which mineral-forming solutions were penetrating.

(3) Manifestation of zones of cataclasm of ultrabasites; development of a typical metasomatic column: chrysolite-lizarditic serpentine-recrystallized antigoritic serpentine-nephrite rodingite-aluminosilicate rocks of basic or acidic composition.

The diversity of rocks enclosing the nephrite-bearing zones is responsible for various paragenetic associations of minerals formed during metasomatic processes. These are as follows:

1. The deposits associated with ultrabasites contacting with alumino-silicate–carbonate rocks are characterized by the development of such minerals as diopside, grossular, hydrogrossular, pectolite, wollastonite, vesuvianite (calcium silicates) as well as tremolite, actinolite, ziosite, albite, quartz, garnet, carbonate, leucoxene, biotite, fuchsite, apatite, sphene, magnetite, haematite, graphite and asbestos. The deposits that contain apogabbro-rodingites are notable for an extensive occurrence of diopside and zoisite, which are characterized by a bright light-yellow colour and can be distinctly traced in the field. Hydro-grossular, pyroxene, and vesuvianite are also encountered. If serpentinites contain granite dykes, then quartz, epidote, tremolite, chrome spinellids and magnetite occur in the vicinity of nephrite-bearing bodies. When nephrite is formed at the contact of serpentine with albite, then the former is substituted by tremolite, diopside, talc and carbonate. According to their composition, rocks with rodingite and nephrite mineralization may be grouped into chlorite–lizarditic and lizarditic serpentinites devoid of antigorite.

2. Nephrite deposits associated with metamorphic rocks are characterized by an extensive development of hornblende, augite, uralite, tremolite, chlorite, epidote, zoisite, quartz, serpentine, talc, wollastonite, calcite and pyrite.

Characteristic admixture elements of nephrites are: titanium (up to 0.1 per cent), chromium (up to 1.0 per cent), nickel (0.17 per cent), calcium (0.1 per cent), strontium, barium, phosphorus, carbon dioxide and chlorine.

According to the petrochemical indications, nephrite-bearing rocks occur at a medium depth. As to their facies character, they occupy the following place in the general series of ultrabasite intrusives: barren $\rightarrow$ nephrite-bearing $\rightarrow$ jadeite-bearing $\rightarrow$ chrysoprase-bearing intrusives [114].

The M/F ratio (Hess–Sobolev number) and the facies coefficient (K_f) of ultrabasites containing nephrite mineralization are 9.9–10.4 and 0.11–0.12, respectively. The great difference in the composition of the serpentinites and surrounding alumosilicate rocks is a favourable prerequisite for the application of geochemical methods in the search for nephrite mineralization. Nephrite bodies are accompanied by primary haloes of indicator elements and by steaming haloes in the enclosing serpentinites (Fig. (4.3)). Serpentinites are characterized by increased calcium concentration at stable CO_2 concentrations, which is a peculiar feature of the tremolitization haloes. A contrasting increase in the concentrations of barium, strontium, mercury, zinc and lead has been detected. The dimensions of anomalies exceed by two to three times the thickness of the nephrite veins, which are also surrounded by positive and negative titanium anomalies. There is a sharp variance in the elemental contents of these anomalies. When approaching the productive zones, the water content in the rocks increases from 1.0 to 10.8 per cent which makes it possible to apply thermography in the search for nephrite bodies.

Studies of nephrite deposits (Jordanów, Poland) have shown that an extremely varied mineral composition of the deposits is related to leucocratic rocks occurring among serpentinites. The studies have confirmed the presence of beryllium in leucocratic rocks. A new mineral association comprising gahnite, cassiterite, columbite, fluorite, almandine, dravite, biotite and muscovite has also been found. The association yields some elements (Be, Zn, Sn, Nb, Ta, B and F) which may

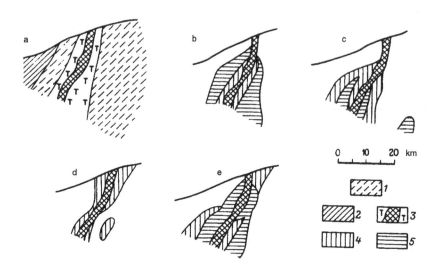

Figure 4.3. Endogenic haloes around the productive nephrite bodies — Siberia. a — rocks, b–e — haloes: b — Ti, c — Ni, Co, V, Cr, d — As, Ag, Hg, Ba, e — K, Na, Rb, Cs. 1 — serpentinites, 2 — slate band, 3 — tremolitized rocks and vein nephrite, 4–5 — haloes: 4 — positive, 5 — negative haloes.

produce contrasting geochemical anomalies, indicating the presence of leucocratic rocks and, thus, the presence of zones to which nephrite may be genetically related.

A study of the range of the East Sayan nephrite deposits has shown that not only is a narrow contact area of nephrite proper captured by hydrothermal alteration, but also most of the serpentinite manifesting the influence of fluid traces it.

The contents of incoherent titanium, zinc, strontium and barium elements in the normal hyperbasites are minimum in comparison with other rock types. No mineral has concentrations of these elements, including chromespinellids. The investigation of the elements makes them useful as indicator elements. Three stages of initial haloes can be distinguished in nephrite-bearing zones: premineral, synmineral and post-mineral. The primary cycle of the premineral stage is rock chrysotilization along the zones of cataclasis. This results in a greater ratio of Fe_2O_3/FeO than that of the primary rocks. At this stage of the process there are no incoherent elements for a hyperbasite.

The intrusion of gabbroid and plagiogranite dykes and veins of the tholeiitic series is accompanied by the formation of a 0.1–3 m zone of microantigorite serpentinite. Fibroblastic tremolite aggregates develop along the microantigorite serpentinites. In the exocontact of the serpentinites the formation of low-powered calcium haloes is noted. At the same time, more powerful titanium, zinc, strontium and barium haloes appear. The haloes of these elements, the concentration of which is 2–10 times more than that of the background elements, make it possible to pick out halo zones in serpentinites with a thickness of up to 15–20 m.

In the halo zone the contents of the indicator elements, zinc and titanium, increase first in chromespinellids, where the elements enter in an isomorphic manner, forming zinc- and titanium-bearing ferrichromites, which can be detected by a microprobe [113]. The halo power of these elements and their contents in chromespinellids from nephrites and serpentinites is correlated. The increased concentration of elements correlates with the increase in calcium in the haloes, with evidence of the distribution of finely-dispersed microfibrous tremolite.

The indicator element contents in primary haloes vary within the range of $(1-30)\times10^{-3}$ per cent by weight.

IBM polyelement geochemical maps of nephrite-bearing zones eliminate the occasional omission of informative haloes and the recognition of false anomalies.

Methods of work

Areas with a known nephrite mineralization should be subjected to follow-up prospecting. The ultrabasite areas are defined by analysing and generalizing the available geological materials, taking into account magmatic, structural–tectonic, metamorphic and petrochemical prerequisities. When carrying out regional geological surveys at a scale of $1:200\,000$ efforts should be concentrated mainly on the identification and tracing out of favourable geological structures and petrographic complexes in ultrabasites. Granitoids are located within the surveyed fields of gangue rocks in ultrabasites, as well as in the zones of serpentinization and schist formation.

The petrochemical criteria (at an M/F of ratio 10 and a facies coefficient of 0.12), combined with the geological data, may be used to discriminate the nephrite-bearing rock strata from the total mass of ultrabasitic intrusives. Particular attention should be given to surveying river valleys that cross ultrabasites and to examining fragments of nephrite, altered rocks and characteristic enclosing rocks that differ in their appearance from fragments of different composition. Nephrite and serpentine, which are similar in appearance, are discriminated by their hardness, viscosity, geochemical properties and by the results of the X-ray diffraction analysis.

Prospecting for nephrite is combined with geological survey at a scale of $1:50\,000$ and is confined to the sites with favourable magmatic and metamorphic prospecting indications. When carrying out lithogeochemical surveys along secondary dispersion haloes, samples with a mass of 200 g are taken from a representative (for the given conditions) horizon of loose formations. Detailed prospecting at a scale of $1:10\,000$ is conducted in regions with earlier identified nephrite mineralization and in the adjacent areas. Eluvial formations in semi-closed areas are sampled along the profiles oriented across the strike of the ore-bearing structures. The fine fraction of the samples $(-0.1$ mm) is ground and analysed.

Such nephrite metasomatic accessory formations as rodingites consisting of zoisite, chlorite, tremolite, albite, garnet and vesuvianite, and characterized by increased contents of calcium, phosphorus, sodium and chlorine are located by geochemical surveys (due to a small size of nephrite bodies). It should be taken into account that the transformation of gabbro, granites, and serpentinites in the process of infiltrational–diffusional metasomatosis occurs on the both sides of their contact; serpentinites are substituted by nephrite, whereas gabbroids are substituted by amphibolitized rocks. In this case, haloes of Li, Zn, Be, P, W, B, Sn and Na are formed that are not characteristic of ultrabasites but which are associated with granites and contain increased contents of the ferric group elements, calcium, magnesium and chlorine.

Primary haloes are examined in the open regions with a sufficient degree of bedrock exposure allowing to locate and outline anomalies at a preset scale. Rocks are sampled in the profiles running across the strike of mineralized zones and spaced at 50–100 m. When studying primary haloes, samples are taken by the dotted furrow method. Five to seven rock chips (sized 3–4 cm) are hammered from the sampling interval (5–10 m) at approximately equal distances (from 0.5 to 2 m) and are combined in each interval into one sample with a mass of 200 g. Since endogenous haloes are formed above nephrite veins, geochemical methods may be used to predict mineralization at depth. Mercury, chlorine, barium and strontium are indicators of the supra-ore zones. In determining the level of erosional truncation of geochemical anomalies, it is advisable to use the parameters of multiplicative haloes as having more contrast zonality. The located nephrite bodies are sampled. Lump samples sized $50 \times 30 \times 30$ mm are taken from each nephrite body to determine its quality.

4.7. LAZURITE

Semi-precious or jewellery, dark-blue, blue, or dark blue-violet lazurite is a polymineral aggregate consisting of isometric irregular grains of lazurite proper and intimately accreted calcite, dolomite, diopside, phlogopite and sodalite.

Commercial lazurite accumulations in association with silica magnesial skarns are localized in the zones of silicitization of silicate rocks (granites, aplites, pegmatites and seldom gneisses) among dolomitic marbles and calciphyres. Lazurite was formed as a result of reaction–metasomatic processes. Silicate substratum rocks (pegmatites, aplites, granites) were the source of silica and aluminum, whereas calcium and sulphur were derived from dolomitic marbles. The redistribution of elements (Al, Mg, Si) between silicate rocks and contacting dolomitic marbles occurred under the impact of ascending post-magmatic solutions, that brought Cl, Na, F and Hg from the outside, and owing to the diffusion of pore solutions. Lazuritization of feldspar rocks is most complete at a low content of silica (relative to alumina) in the mineral-forming solutions and at a high chemical potential of sodium.

Prospecting indications of lazurite deposits are:

(1) terrigenous–carbonate formations (dolomites, marbles) contacting with boudinated vein bodies of acid magmatic rocks that are localized in the zones of the regressive stage of ultrametamorphism;

(2) presence of boudinated thin veins of feldspar rocks, as well as sulphur in the horizons of dolomitic and calcite–dolomitic marbles;

(3) presence of boudinage-structures, anticlinal structures, small folds, and layered fissure zones in dolomitic marbles;

(4) magmatic or ultrametamorphic rocks of increased alkalinity.

Indicator minerals

Lazurite is generally intimately accreted with calcite, diopside and phlogopite, and associated with feldspars, hauyne, sodolite and pyrite. Furthermore, dolomite, forsterite, phlogopite, spinel, hastingsite, aegirine, scapolite, nepheline, zeolites, anhydrite, serpentine, graphite, scapolite (marialite), pyrrhotite and marcasite have been encountered in lazurite deposits of the magnesian-skarn formation. Lazurite deposits associated with calcareous skarns are characterized by a less complicated mineral composition that is commonly represented by calcite, pyrite, and wollastonite. Conditions of lazurite formation favoured the origination of both sulphides (pyrite, pyrrhotite) and sulphates (anhydrite). The mineralogical prospecting indications of lazurite mineralization are as follows: (a) presence of pyrite and sulphur inclusions in dolomitic and calcite-dolomitic marbles; and (b) development (in the enclosing carbonate formations) of potash–feldspar, pegmatites and aplites that were affected by boudinage with the resultant formation (in tectonic fractures) of scapolite, hauyne, nepheline, hastingsite, aegirine and calcitic plots. A lazurite occurrence is indicated by the presence in the boudins of feldspar rocks with the signs of an intensive desilication and dolomitization that are accompanied by a decrease in the content of quartz, by substitution of diopside for potash feldspar, and by the presence of coarse-crystalline calcite. An insignificant manifestation of spinel–forsterite and phlogopite parageneses among the metasomatic formations in dolomitic–calcitic marbles is also a favourable indication.

The specific features of the mineral composition and results of chemical analyses indicate that the main typomorphic indicator elements of lazurite mineralization are sulphur, chlorine, mercury barium, strontium, potassium, sodium, lithium, rubidium, beryllium, boron, fluorine, arsenic, antimony, copper, nickel and cobalt. The bulk of the components needed for lazurite formation (Ca, Al, Si, S) were derived from the enclosing rocks, whereas alkaline (Na, K, Li), haloid and volatile compounds (Cl, F, Hg, As) were transported by the solutions from the outside. Sodium and chlorine could partly be derived from the enclosing substratum rocks.

Lazurite zones, surrounded by altered marbles, are characterized by an anomalously high content of strontium which, together with barium, is one of the effective geochemical indicators. Dolomitic marbles and unaltered gneisses occurring outside the deposits contain, respectively, 49×10^{-4} per cent and 45×10^{-4} per cent of strontium, whose concentrations in forsteritic calciphyres increase by 3–7 and in pyroxenic by skarn 3–4 times [47]. Lazurite mineralization contributes to a considerable enrichment of rocks in strontium (from 0.11 to 0.22 per cent) which by 20–45 and 5–14 times exceeds its content in phlogopitic marbles and in skarns and calciphyres, respectively. Formation of skarns at the magmatic stage results in the introduction of barium whose content in skarns exceeds that in the substratum rocks by 1.5–1.7 times (dolomitic marbles contain 0.0017 per cent of Ba). The barium content in the rocks that enclose lazurite mineralization fluctuates appreciably and reaches 0.7 per cent. Depending on the nature of the substratum rocks, the content of Be in lazuritic rocks ranges from 2 to 3 g/tonne. However, the greatest concentrations of this element are found in varieties enriched in lazurite and pyroxene [47]. The boron content (g/tonne) is as follows: in marbles 12, skarns 15, granites 200, lazuritic calciphyres 450, and in unaltered pyroxenic skarns 750 [47]. Chlorine is a typical indicator element of lazurite mineralization. Dark-coloured isotropic phases of lazurite contain much greater contents of this element.

Methods of work

Geological survey at a scale of 1 : 200 000 envisages a small volume of geochemical prospecting restricted to sampling of marble horizons with the purpose of detecting indicator elements of productive mineralization and dolomite differences with signs of lazurite occurrence.

Prospecting for lazurite (scales 1 : 50 000 – 1 : 25 000) is carried out in the regions where a favourable situation for the formation of lazurite mineralization was evidenced by previous surveys. Alluvial formations in river valleys are explored and poorly exposed areas are subjected to areal lithogeochemical survey. In the first case, efforts are concentrated mainly on finding in alluvium the fragments, boulders and pebbles of skarns, and alumosilicate rocks with a certain dissemination of lazurite. In the second case, the data of geochemical prospecting are generalized to identify complex barium, strontium, boron, chlorine, sulphur, arsenic and mercury anomalies that indicate concealed lazurite mineralization. In the process of mining operations, attention is given to tracing out the horizons of dolomitic marbles and zones of their contact with vein granites and pegmatites, which are subjected to geochemical sampling to predict the occurrence of lazurite. Special attention should be given to the change in Sr, Be, B and Hg contents in the marbles affected by dedolomitization. Geochemical methods are combined with magnetometry and electrical prospecting. The borehole survey on a grid of 50×2 m (with the help of a portable drill) is carried out to discover and contour the lazurite-bearing zones.

The showings of lazurite are prospected and evaluated by geochemical methods allowing to specify the dimensions of mineralized zones and the scale of their manifestation, the affinity of deposits to certain formations, as well as the potentialities of lazurite mineralization at depth. The borehole core is subjected to geochemical testing in order to trace and contour productive bodies and to search for blind lazurite accumulations. Spectral and special analyses of samples are made to determine the following elements: Ba, Cr, B, Be, Cu, Zn, Pb, Cl, F, Hg, Na and K. Special attention is given to the contacts of positive and negative anomalies of Sr, Na and the ferric group elements that indicate most reliably the epicentres of productive mineralization. It should be noted that the highest strontium concentrations (up to 0.7 per cent) have been detected in recrystallized and calcitized marbles and altered calciphyres, i.e. in rocks that form external haloes around lazurite formations. Negative anomalies are identified directly in lazuritic rocks of the inner zones. The intensity of development of lazurite mineralization may be judged by the variance of the contents of indicator elements and their quantity (especially of S and Cl).

Lazurite mineralization can be prospected effectively by the mercurimetric survey. Our investigations, carried out in the Pamirs area, indicate that the mercury content in altered rocks increases with the approach to the productive beds in the following succession ($n \times 10^{-6}$ per cent): 0.2, 3.0 and 4.3 at 3, 2.5 and 0.2 m from the deposit, respectively, whereas it reaches 36×10^{-6} per cent in the lazurite deposit. The following temperature forms of mercury have been detected in lazurites and in the surrounding altered rocks: 120–200, 230, 260 and 280 °C. A combination of several forms of mercury, considerable areas of thermal peaks of sublimation, and the presence of the peak at 280 °C constitute the most reliable prospecting indication of productive lazurite bodies.

4.8. FELDSPARS AND PEGMATITES

Granite pegmatites are the major source of feldspar raw material. The differentiated coarse-grained granite pegmatites are especially favourable for obtaining feldspar raw material of the highest grades. Not infrequently, such pegmatites contain feldspar zones of considerable length and thickness. Apart from pegmatites that are the only source of ceramic raw material (feldspar, pegmatite, quartz), feldspars are extracted from micaceous pegmatites whose geochemical prospecting criteria were considered in the foregoing sections. In addition to feldspar and quartz, granite pegmatites may contain tourmaline, muscovite, beryl, garnet and biotite. Alkaline pegmatites are composed of microcline, nepheline, aegirine, biotite, apatite and calcite, and are associated genetically with nepheline-syenites. Desiliconized granite pegmatites (albites) occur in ultrabasic rocks. The main prospecting indication of the most significant feldspar deposits (differentiated granite pegmatites) is the mineralogical one including: occurrences of pegmatites, areas of muscovite deposits and finds of fragments of feldspars and tourmaline in eluvium, deluvium and in river valleys. Prospecting surveys for feldspar and pegmatites are incorporated in small-scale geological mapping. They involve inspection and evaluation of pegmatites in the zones of development of pegmatitic veins by determining in them the contents of iron, potassium, sodium and calcium. Increased contents of fluorine, rubidium, yttrium, and ytterbium have been detected in the vicinity of chambered pegmatites. A high content of these elements (fluorine forms the most contrasting halo) is commonly detected at the sites with altered circum-pegmatite rocks. The shape of haloes reflects commonly that of the altered pegmatites.

Feldspar and pegmatite, suitable for ceramic industry, should contain less than 9.5 per cent of Fe_2O_3; no less than 11 and 8 per cent of $K_2O + Na_2O$ (in feldspar and pegmatite, respectively); less than 1–2 per cent of CaO; and less than 10 and 30 per cent of SiO_2 in feldspar and pegmatite, respectively.

Sodded, poorly exposed areas are subjected to lithogeochemical survey along secondary dispersion haloes. Prospecting surveys require stripping of the overburden and making exploratory pits and trenches. Geophysical prospecting has to specify the geological structure of the site being investigated, to identify zones of tectonic dislocations, and to determine contours and contacts of pegmatitic bodies in plan and at depth.

Geochemical exploration is mandatory in the process of detailed prospecting at the scales 1 : 10 000 and 1 : 5000. Taking into account that the length of pegmatitic bodies along their strike ranges from 10 to 200 m and their depth exceeds 5 m, the grid of geochemical sampling, aimed at contouring pegmatites from the surface, must be 20 × 10 m.

It is also advisable to conduct spectrometric testing of bedrocks in order to identify the primary halo of pegmatitic bodies. To detect pegmatites in the areas where loose eluvial–deluvial formations are more than 5-m-thick, it is necessary to take the drill-hole core samples and analyse them for potassium, sodium, lithium, rubidium, cesium, boron and fluorine. Heavy concentrates are also analysed during detailed prospecting. The material for washing and obtaining a heavy concentrate is taken from workings at the drill-hole core. Heavy concentrates are subjected to mineralogical and spectral analysis. Detailed investigations involve special sampling of bedrocks and outcrops of pegmatites, identification of their coarse- and fine-grained varieties, analysis of the substance composition of pegmatites, establishment of the pattern of feldspar and quartz distribution, and, finally, determination of the raw material quality. Samples should be taken from every pegmatite variety.

Feldspar obtained from pegmatites, is very deficient. The feldspar and pegmatite raw material is generally collected manually. Feldspar may be substituted by porcelain stone, or, close in composition, sericite–quartz, or effusive rocks containing Fe < 0.5 per cent and K ≫ 0.5 per cent. The content of dyeing elements (iron, manganese, chromium and titanium) in porcelain stone should not exceed 0.4 per cent.

4.9. VERMICULITE

Vermiculite, in the broad understanding of this term (vermiculite–biotite and vermiculite–phlogopite mixed-layered varieties of micas proper) results from the hydration in the process of weathering of the Mg–Fe micas, containing up to 15 per cent of Fe^{2+} and up to 1 per cent of F.

Vermiculite is the name used for micaceous minerals with a ferromagnesian aluminium silicate composition and the unique property of exfoliating. Commercially, the exfoliated products are also called vermiculite, or more exactly, expanded vermiculite. In its exfoliated state vermiculite serves many markets. Chief among them are construction, agriculture and general industry.

Vermiculite, in a natural state, has the characteristic mica habit and splits readily into thin laminae, which are flexible but inelastic. In a mineralogical sense, vermiculite is a hydrated silicate with no exact chemical composition. When heated quickly to an elevated temperature, vermiculite expands by exfoliating at right angles to the cleavage, into wormlike pieces. The characteristic of expansion, the basis for the commercial use of the mineral, is the result of the mechanical separation of the layers by the rapid conversion of contained water to steam. The increase in the bulk volume of commercial grades is 8 to 12 times, but individual flakes may expand as much as 30 times.

The exfoliation of the vermiculite crystal results in large pores being formed between groups of platelets. Thus, exfoliation can make available a large increase in volume without significantly changing the surface area of the platelets. This characteristic is important in the chemical applications of vermiculite. Vermiculite is not a single mineral species, but families of related minerals. Additionally, commercial interest in the end product has resulted in the name vermiculite being applied to the very voluminous products of the calcination of baneritized (bleached) phlogopite.

This has also resulted in many names for the specific vermiculite found in a particular deposit. This difficulty arises from the nature of its crystal chemistry.

Vermiculites consist of talc-like layers in which the deficiency of the positive charge is compensated by the interlayer cations. Vermiculites are also closely related to biotite and phlogopite. These minerals and vermiculite often occur in the same deposits and they are sometimes intimately mixed within a crystal or across a single crystal face.

Table (4.1) shows the chemical analyses typical of the major deposits, and the discovery of the location of vermiculites reveals the general similarity of the occurrences.

Vermiculites may form by several processes. Weathering and plant action in soil will form vermiculite directly from biotite. Vermiculite is always formed merely as a result of the imposition of old crusts of weathering upon previously formed phlogopites and medium-iron biotite.

The main types of vermiculite deposits are in ultrabasite alkaline, alkaline and basic intrusive–metamorphic complexes.

Magnesian vermiculite predominates in the deposits, and calcium and sodium vermiculites occur more rarely. Vermiculite occurs at the base of eluvial terrains. In the upper parts of the section in alkaline environments it is replaced by montmorillonite, serpentines (lizardite, antigorite) and sepiolite; in an acid environment it is transformed to kaolinite. At increasing depth, vermiculite is followed by initial mica through a zone of hydromicas and ordered and disordered mixed-layered minerals or a mixture of mica and vermiculite [117]. The characteristics and trends of the transformations are determined by the composition of the initial mica and the geochemistry of the weathering environment. Hydromicas have intermediate compositions and characteristics between phlogopite and vermiculite.

Table 4.1.

Chemical analyses of vermiculite, per cent [80]

Region	SiO_2	MgO	Al_2O_3	Fe_2O_3	FeO	K_2O	CaO	H_2O	Total
Palabora, South Africa	39.37	23.37	12.08	5.45	1.17	—	1.46	11.38	94.28
West Chestes, Palabora	33.82	19.17	15.82	7.51	1.12	0.10	0.56	20.25	98.35
Libby, Montana	38.64	22.68	14.94	9.29	—	7.84	1.23	5.29	99.91
Enores	38.66	20.04	17.36	8.45	—	5.25	0.75	8.71	99.21

Low- and medium-iron micas are easily changed to vermiculite in an alkaline environment. In an acid environment, mica is often replaced by kaolinite without the intermediate vermiculite stage. These results may be used to locate and assess vermiculite deposits and to determine the vermiculite content in ores.

In general, mineralogically interesting vermiculite deposits possess features that are derived from either a basic pegmatite, alkali pyroxenite, or carbonatite complex. Most of the larger occurrences display this carbonatite origin. This association explains the thermal problem that troubled many of the original theorists in this field since the temperatures of carbonatite systems may be quite low and the aqueous content very high.

Prospecting and exploration

Vermiculite deposits are usually covered by soil and vegetation because the minerals are soft and have a considerable degree of natural water retention properties. Outcrops containing a lot of good quality vermiculite are rare, but the rock types associated with vermiculite have distinctive physical features. It is not uncommon to find outcrops of glimmerites (biotite), alkali pyroxenites, commonly with magnetite, and dykes of carbonatite or a particular type of syenite. All of these features may be devoid of vermiculite, and the initial field evidence has in the past commonly been the presence of vermiculite flakes in the soil and stream deposits. The increase in interest in carbonatite complexes for other minerals has broadened their use as a possibly favourable indicator in vermiculite exploration.

Significant occurrences of vermiculite are almost invariably found associated with ultrabasic rocks, commonly pyroxenites. These rocks may be intruded by numerous pegmatites, pegmatoid masses, carbonatites, and syenites. A combination of these features may be indicative of an area favourable for detailed examination.

Vermiculite commonly occurs in small pods of randomly oriented flakes in pyroxenite or in associated gneiss. These vermiculite pods are usually the higher grade deposits that are commonly surrounded by large zones of disseminated vermiculite. To prove the presence of an adequate volume in deposits, which may be very irregular in outline, the ore bodies are usually delineated by trenching and drilling. In some instances trenching by tractor has been sufficient; however, in most instances rotary drilling is used. Because of the friable nature of most vermiculite, diamond core drilling has not been as effective for sampling as larger-hole rotary drilling. Particle size is an important factor, and special attention should be paid to bit design to avoid the destruction of the larger flakes [80].

The primary requirement of a commercial deposit revolves around the exfoliation or expansion characteristic of the vermiculite. The vermiculite must contain very little, if any, associated biotite or phlogopite and it must be predominantly in the coarser flake sizes required by the markets. High grade deposits containing more than 50 per cent of vermiculite exist.

Vermiculite is characterized by paragenesis, with Mg, Al and Fe hydrosilicates predominating in the upper parts of profile, whereas the bulk of vermiculite is located below. The most prospective for vermiculite are moderately preserved crusts of weathering with comparatively thin uppermost clay zones and accumulations of unhydrated micas in the underlying rocks. These accumulations belong to the following types of formations:

(1) ultrabasic alkaline rocks and carbonatites;

(2) dunite–harzburgite ultrabasites in association with late- and post-orogenic granitoid or alkaline intrusions in the zones of potash metasomatosis along linear tectonic dislocations;

(3) granitized gneiss-amphibolite complexes in the zones of manifestation of the potassium stage of ultrametamorphism near the interlayer tectonic dislocations and in the nodes of their intersection with later dislocations;

(4) gneiss–biotite complexes in the periphery of alkaline and nepheline syenite bodies;

(5) melanocratic biotite gneisses and schists formed as result of epidote–amphibolite, amphibolite, and less frequently, granulitic facies of metamorphism (at generally weak granitization and migmatization), ensuring the preservation of the originally high content of biotite in them.

Since the above-stated structural types of vermiculite deposits have unfortunately not been practically subjected to geochemical prospecting, we shall emphasize only the main possibilities of using them for this purpose.

The possible utilization of such indicator elements as F, K, OH group and the Fe/Mg ratio in the search for vermiculite has been proved experimentally. Increased concentrations of K and Mg and low (< 0.15) Fe/Mg values should be expected in the majority of minerals and primarily in micas in the vicinity of vermiculite deposits and above them. The use of decreptophonic methods is rather prospective because they will evidently make it possible to evaluate the degree of hydration of Fe/Mg micas. The content of fluorine in the rocks enclosing the vermiculite bodies is below 0.05 per cent, whereas its content in vermiculite samples taken from large exploited deposits reaches 0.7 per cent. An analysis of samples taken from the eluvium of these deposits indicates a distinct increase in the fluorine content in the supra-ore zones whose dimensions make it possible to recommend a combined application of fluorometric survey and decreptophonic investigations in prospecting for vermiculite at the $1:10\,000$ and larger scales.

4.10. VEIN QUARTZ AND QUARTZITES

Certain clear varieties of vein quartz, with minimum gaseous–liquid and mineral inclusions are used as substitutes for rocks crystal in the production of transparent quartz raw material. For instance, granulated transparent vein quartz, an inexpensive raw material, can be used to produce different kinds of glass. Quartz should comply with the following major requirement: its high chemical purity should not permit harmful admixtures in excess of thousandths to hundred thousandths of a per cent depending on the purpose of glass production. The following genetic types of quartz veins are distinguished: (a) pegmatitic; (b) hydrothermal; and (c) metamorphogenetic. Chambered pegmatites, which are the source of rock crystal and vein quartz, from differentiated bodies with zones of small-grained granite. The high-purity quartz in voidless micaceous and ceramic pegmatites is separated into blocks. The mineralogical and geochemical methods of prospecting for this type of raw material have been described in the foregoing chapters and therefore, are not considered here.

The hydrothermal deposits of quartz raw material comprise ore-free, rock-crystal-bearing (void), and monomineral (voidless) veins. By its composition, quartz in the rock-crystal-bearing veins is close to rock crystal and hence can be used for manufacturing optical and transparent quartz glass. The maximum dimensions of the blocks of transparent vein quartz do not usually exceed 150 cm^2 along the plane of splitting. The geochemical criteria of rock crystal prospecting can successfully be used in detecting transparent vein quartz. Vein quartz of metamorphogenetic genesis results from the transformation and partial recrystallization of quartz contained in the hydrothermal veins. Granulated vein quartz is a uniformly grained aggregate made up of transparent and semi-transparent grains divided by polygonal cracks, the majority of which are healed. Veins consist mainly of quartz (up to 99 per cent of volume), whereas such admixtures as feldspar, muscovite, epidote, zoisite, apatite and rutile account for less than 1 per cent. Harmful admixtures are localized in the cracks of a jointing, in selvage margins, and in the intergrain spaces. As to its chemical purity, granulated quartz is not inferior to rock crystal. Deposits and shows of granulated quartz are localized among ancient metamorphic formations that have been repeatedly affected by regional metamophism from the green schist to amphibolitic stages. Granulated quartz was formed in a zone of deep faults where favourable preconditions were created for the recrystallization of vein quartz (high pressures, migration of solutions), and for the alteration of structural and mineralogical properties of granulated quartz that is sometimes used in the production of quartz and optical glass. The latter depends also on the degree of purification of quartz and on the perfection of the technology of its processing.

Quartzites and quartzy sandstones that are used to manufacture quartz glass have resulted from the regional metamorphism of sedimentary rocks. Their reserves are enormous. Quartzites are used to manufacture dinas (a refractory with a silica content of 90–93 per cent). Rocks should contain no less than 93–98 per cent SiO$_2$ and 0.5–3.0 per cent mineral admixtures.

All kinds of admixtures in quartz reduce the quality of fused quartz glass. Hard and dislocated admixtures produce the most adverse affect. Gaseous–liquid inclusions containing various chemical compounds infringe glass homogeneity. Their effect on the glass depends on the method of fusing. When quartz is used for manufacturing optical glass, gaseous–liquid inclusions do not significantly affect its quality. Fused milk-white, recrystallized and semi-transparent vein quartz of the rock-crystal-bearing veins is characterized by high chemical purity and can be used for manufacturing multicomponent optical glass and, upon deep separation, for fusing transparent quartz glass.

The geological prerequisites and prospecting indications of relatively widespread, huge deposits of vein quartz and quartzites are conditioned mainly by their formational affiliation. The direct prospecting indications are outcrops of vein quartz and sandstones, and finds of rocks fragments in dispersion haloes. Geo-chemical exploration is aimed primarily at evaluating the quality of vein quartz and quartzites, on the basis of the composition and quantity of admixtures: solid mineral and gaseous–liquid inclusions, dislocations, colloidal compounds, structural and non-structural elements.

The γ-ray method can be used to gain an idea on the composition of admixtures in quartz. Three-millimetre-thick plates are sawn from vein quartz, polished on both sides and then subjected to γ-radiation at a rate of 7 A/kg. The content of the structural admixture of aluminum is evaluated by comparing the density of the resultant colouring with standard samples. The size of the granules and the degree of uniformity of grains in the aggregates are determined in thin sections and in sections soaked in hydrofluoric acid. This method aids in identifying homogeneous and inhomogeneous granulated quartz.

Quartz samples are subjected to a high-precision chemical and spectral analysis because the quartz raw material used to manufacture multicomponent optical glass must not contain admixtures of F and Ti, Co, Ni and V in excess of hundredths (F, Ti) tens of thousandths (Co, Ni, V) of a per cent and less. When veins of water-milk-white and transparent quartz are discovered more than 2 m thick and with a small amount of admixtures, these veins are sampled to determine the light-transmitting capacity of the quartz and to subject it to technological testing. The relative content of transparent grains of the granulated quartz is determined. When producing 'grist' from the granulated quartz of three grain-size classes: (0.2–0.4 mm; 0.2–1.2 mm; and 0.2–2.0 mm) it is possible (depending on the methods of fusing) to use up to 30 per cent of the vein mass.

4.11. SAND AND GRAVEL

Sands and gravel are loose or slightly cemented rocks consisting mainly of quartz and a small admixture of feldspars, micas, magnetite, and ilmenite. The following deposits are known: (a) marine; (b) lacustrine; (c) alluvial; and (d) fluvioglacial. The physical and mechanical properties, as well as the mineral composition of these rocks, depend on the conditions of their formation and on the bedrock composition. Sands and gravel form generally irregular sheet-like and lenticular deposits, 1–30 m thick extended over a considerable length. Taking into account the appreciable spatial distribution of these useful minerals, their prospecting is based mainly on the analysis of the geological structure of the area being explored.

Poorly exposed areas are prospected by geophysical methods. The distinct differentiation of the electric properties of the above-stated rocks favours the application of electrical prospecting for contouring and determining the depth of sand and gravel occurrences. Sands and gravel are identified in the cross-section by their electric properties among such enclosing rocks as clays, limestones and dolomites. Geophysical methods have proven their high efficiency in the search for sand–gravel rocks in the deposits of the Mordovia (Russia), wherein the useful strata have not only been delineated in plan but also the depth from the roof to the bottom of a productive bed has been determined.

Gravel in a sand–gravel mixture can be detected by photographic method involving the count (with the aid of a grid stencil) of the number of intersections that indicate the components being determined in scale photographs of natural exposures. In the process of prospecting surveys, the quality of raw material is determined by granulometric analysis.

4.12. CLAYS

Clays are rather diverse in origin, conditions of occurrence, and substance composition. Bentonite, refractory, and kaolin varieties are distinguished. All of them are actually an earthy rock consisting mainly of alumosilicates (pyrophyllite, sericite, kaolinite, halloysite and montmorillonite) with an admixture of quartz, feldspar and mica.

A heavy mineral study reveals the presence of about 20 mineral species (deposits in Egypt), the most important of which are opaques, zircon, rutile, tourmaline and staurolite [49]. Pre-existing sedimentary rocks and the basement complex are postulated as the sources of the heavy minerals, as indicated by the presence of grains of different forms (zircon), and varieties (tourmaline) of the same mineral species.

DTA studies of kaolins from India have indicated that endothermic and exothermic peaks appeared in the ranges of 560–590 °C and 960–985 °C, respectively.

Thorium analysis and geochemistry can be very useful in interpreting the geological history of kaolin deposits. In the large kaolin deposits in South Carolina, thorium has been shown to be immobile when crystalline rocks are altered by phreatic water, whereas thorium is mobile when rocks and sediments are leached by vadose water [21]. The Cretaceaous kaolins have an average of 17.6 ppm of thorium and the Tertiary kaolins have an average of 12.1 ppm.

The chemical composition of bentonites from Bohemia is represented by iron-rich montmorillonite and several admixture minerals, especially kaolinite, calcite, anatase, siderite, ankerite, biotite, oxihydroxide and iron oxide. The relatively high contents of Zr, Ce and La indicate the presence of underlying gneissic materials. Bentonites with a dominant or subordinate proportion of sedimentary mud exhibit very high ppm values of Mn (1.000–5.000), P ($\gg$ 5.000) and also V ($\gg$ 300) [49]. The data obtained have been used successfully in geological prospecting in the above-mentioned area.

Hydrothermal and katagenetic bentonites from South Bulgaria are of practical importance. The main clay mineral is alkali-earth montmorillonite. Besides this, kaolinite and baidellite, accompanied by amethyst, ankerite, barite, calcite and sulphide minerals are also found in the clays.

The major method of prospecting for deposits of clays depends on their specific features. Clay rocks are detected most reliably by geological methods. Bentonite clays consist of Na- and Ca-montmorillonite (60–90 per cent), kaolinite (3 per cent), and hydromicas (10–30 per cent). The following deposits of these clays are distinguished: (a) volcanogenic-sedimentary; (b) hydrothermal; and (c) sedimentary. The first type of deposit results from the transformation of volcanic ash, tuff, or other pyroclastic material. Hydrothermal or post-volcanic deposits are formed due to the alteration of pyroclastic or volcanic rocks of basic or medium composition under the effect of hot gases, water vapour, and hydrothermal solutions. Sedimentary deposits are formed through the redeposition and alteration of the washout products of crusts of weathering of intrusive and metamorphic rocks in coastal–marine and lacustrine conditions. The formation of montmorillonitic zones in the crusts of weathering depends on the bedrock composition.

Favourable prospecting criteria for detecting bentonite clays are as follows: areas of active volcanic and tectonic activity, platform territories, crushing zones with crusts of weathering, and salt lakes.

The leading clay mineral of bentonites is Na-montmorillonite (up to 80 per cent of Na), with an admixture of cristobalite, carbonates, gypsum and quartz. The indicator elements are fluorine, lithium, sodium, potassium, arsenic, antimony and mercury.

Prospecting surveys at a scale of 1 : 50 000 are concentrated mainly on detecting the outcrops of bentonitic bodies in exposures and on finding 'vysypkas'* in deluvium, and small rounded fragments in streams. The main purpose of geochemical prospecting at this stage is to determine the composition of clay formations.

Affiliation of clays to bentonites should be determined by granulometric analysis, making it possible to establish the composition of minerals in fractions. Clay minerals are identified by defraction and thermal analyses and are subdivided into alkaline and alkaline-earth mineral, based on the results of cation-exchange analyses. The quantity of montmorillonite in polymineral mixtures can be determined by luminescence-adsorption analysis using organic dyes and luminophores. According to this procedure, a sample of clay in suspension is titrated by a solution of dyes and luminophores of known concentration. It should be taken into account that the montmorillonite exchange capacity (EC) of cations reaches 100 mmol per 100 g of clay, which makes it possible to determine the montmorillonite percentage in the sample by the EC value. Due to the near-surface alteration of clays, it is necessary to sample fresh, slightly altered rocks in deep horizons below the ground water table. When evaluating the raw material quality and predicting new target areas, it is advisable to consider the spreading of magmatic rocks whose alteration leads to the formation of bentonite clays. The effusives of basic composition are substituted by poor-in-iron, light alkaline-earth clay varieties, whereas the ultrabasic rocks are substituted by nontronites.

Refractory clays are predominantly genetically associated with continental sedimentary formations. These are classified into primary (eluvial, leached out) and secondary (lagoon-lacustrine and alluvial) refractory clays. Halloysite refractory clays originate from weathering of glassy volcanic rocks. They are represented mainly by vein-like bodies in glassy andesitic flows. Refractory clays of the sedimentary type, formed in the crusts of weathering, are widespread, and hence are prospected for by conventional geological methods that are supplemented by geochemical and geophysical surveys. The former can specifically be used to determine the quality of clays by thermal and chemical analyses. Kaolin deposits are

---

*Russian term denoting rock fragments scattered around and showing the presence of a bedrock.

of sedimentary and hydrothermal origin. The former consist of kaolinite, quartz, mica, potash feldspars and montmorillonite, whereas the latter are formed due to the alteration of volcanic rocks.

Primary kaolins, formed as a result of weathering of metamorphic and sedimentary (mainly feldspar) rocks, and secondary ones, formed in the process of redeposition of the clay substance of primary kaolins, are distinguished. Eluvial kaolins contain relict (quartz, orthoclase, muscovite, paragonite, biotite and rutile) and newly formed (halloysite, hydromicas and montmorillonite) minerals. High content of the highest grade kaolin is detected where substratum rocks are represented by granites or light-coloured micaceous-feldspar formations poor in iron. Kaolins that substituted granites have a persistent granular composition (10–30 per cent of fractions < 0.005 m and 40–70 per cent of fractions > 0.56 mm). Greisenized granites, altered in the process of pneumatolysis and enriched in lithium, that improves ceramic properties of kaolin, are especially favourable. The indicator minerals and elements of kaolins of this type are: zinnwaldite, tourmaline, potassium, sodium, fluorine, lithium, rare earth elements, titanium and phosphorus. The degrees of concentration and dispersion of elements in concrete deposits fluctuate depending on the facies and physico-chemical conditions of kaolin formation, composition of bedrocks that are the source of the raw material, and on the character of contemporaneous processes of weathering.

Kaolin deposits are generally flanged by syngenetic haloes. The linearly-elongated isometric haloes repeat the shape of kaolinic deposits. The secondary dispersion haloes are formed above kaolin occurrences exposed at the present level of erosion by the denudation processes. Such deposits are characterized by residual (exposed, buried and impoverished) haloes. On the whole, prospecting for kaolin deposits along exposed and buried secondary dispersion haloes may yield positive results in favourable geological and landscape-geochemical conditions.

Lithogeochemical survey, aimed at detecting the secondary haloes of potassium, sodium, lithium, phosphorus, fluorine, and titanium, is advisable under conditions of poor outcropping. If an overburden is up to 5 m thick, samples should be taken at a depth of 30–40 cm from the illuvial horizon and from the horizon transitional to the bedrock. Fractions of 1 mm and less are analysed. Taking into account that the dimensions of kaolin deposits fluctuate in the range 400–500 m, the sampling grid of lithochemical survey should be not less than $(500 \times 250 \times 50)$ m and (250×25) m (scale $1:50\,000$). The spectral analysis for a wide range of elements should be supplemented by fluorescence analysis and flame photometry. An optimal prospecting grid is calculated from the formula

$$p = 1 - e^{-S/m},$$

where S is the area of an anticipated halo in the units of the observational network register; m is the number of points needed to locate the halo; and p is the probability of tapping the halo.

To increase the efficiency and reliability of geochemical prospecting, the results of lithochemical survey should be represented as maps of isolines of integrated polyelemental indices of the following type: $v_1 = (\text{Al} \times \text{Fe} \times \text{Ti})/\text{Si}$;

$v_2 = (K \times Na \times Li)$. This is a compact representation of geochemical information on the occurrences of kaolin.

The quality of the raw material is determined by the granulometric analysis. The main commercial product is elutriated kaolin, with fractions sized 20 μm. The optimal size for a commercial fraction is determined individually for each deposit. The bedrock composition is taken into consideration to determine tentatively the quality of kaolins. The percentage of coarse fractions diminishes in kaolins that resulted from weathering of gneisses, schists, and veined formations as compared to those formed as a result of substitution of granites.

Vertical mineralogical zonality is taken into account in predicting and evaluating kaolin deposits. An insignificant erosional truncation of productive bodies is evidenced by the presence of hydromicas and montmorillonite in the upper part of the profile of weathering when quartz-porphyries are substituted by kaolins.

Palygorskite clays are finely dispersed rocks composed mainly (up to 70–80 per cent) of this clay mineral. The following deposits are distinguished: (a) endogenic of the hydrothermal–epithermal type; (b) hypergene (crusts of weathering); and (c) chemogenic-sedimentary. The latter are formed in the coastal-marine and lacustrine conditions due to the deposition from natural ooze micas at the stage of diagenesis.

4.13. PERLITE, PUMICE AND VOLCANIC GLASS

The majority of perlite deposits are associated with lava flows in volcanic areas. Perlite occurs together with obsidian (up to 2 per cent H_2O) and retinite (5 per cent H_2O). The stated rocks, characterized by valuable physical and mechanical properties (softened glass, foam-like light mass at 900–1200 °C), are used extensively as heat insulators and substitutes in the production of wall materials. High-quality perlite has a porous structure and swells even on slight heating.

Pumice, a naturally swelled volcanic glass, is formed as a result of the explosive-scorification process. Generally, it occurs as lenticular or laminated bodies among lavas. Deposits of pumice are localized in the areas of recent volcanic activity. It is mined directly on the slopes of volcanoes. Pumice is used for wood grinding and polishing, as a filler of concrete, and as ballast in road construction.

Obsidian (volcanic glass of acid composition) is used for manufacturing various articles. There are homogeneous, laminated, and brecciated varieties of obsidian, occurring as flows, lenses and intercalations ranging from 1 cm to 50 m in thickness and interbedded with effusive rocks, liparites, and perlites.

Prospecting for deposits of perlites, pumice and volcanic glasses is rarely carried out as an independent stage of geological prospecting surveys. These rocks are commonly detected during geological survey or on the basis of the materials of prospecting for other useful minerals. However, based on the general geological prerequisites, prospecting should be aimed at locating the sites with the above-stated rocks whose quality and possible commercial utilization are determined by geochemical methods. Rock varieties are sampled for spectral and thermographic analyses.

The varieties of water in perlites are determined as being responsible for different temperatures of glass melting. The difference in the melting of obsidian glasses in volcanic complexes depends on the velocity of cooling of the glass-like melt in the upper part of an intrusive dome. The character of glass melting is determined from the form of volatiles contained in the glass.

4.14. SILICEOUS ROCKS

Siliceous rocks comprise diatomite, tripoli and gaize whose composition is dominated by opal. Diatomite and tripoli are represented by a finely porous, light sedimentary rock with opal silicon dioxide as the main component (75–82 per cent) and quartz, feldspar and clay as admixtures. Tripoli is almost devoid of diatoms, being composed of small opal grains. Gaize, containing 90–97 per cent SiO_2, differs from tripoli by a greater degree of cementation and uneven fracture. The stated rocks are used to produce soda-lime glass, hard artificial rubble (thermolite), and as active additives in the production of facing materials and cement.

Siliceous rock deposits were mainly formed biochemically in marine conditions and in lakes. There is a close spatial relationship between siliceous rocks and volcanic formations — the source of silica. Sedimentation of the products of volcanic activity occurred in a chemical way. Wide spatial occurrence of siliceous rocks makes it possible to prospect them mainly by conventional geological methods (formational analysis, route survey), whereas geophysical methods are applied in conditions of poor outcropping or where the thickness of the Quaternary formations exceeds 5 m.

4.15. MAGMATIC, METAMORPHIC AND SEDIMENTARY ROCKS

Igneous, metamorphic and sedimentary rocks are used as a building material: dimension stone (blocks) and stone of mass production. The areas of utilization of building stone depend on a number of properties — hardness, porosity and volume mass — that result from the structure and mineral composition of rocks.

Igneous rocks (granites, basalts) are used to produce rubble, as concrete filler, in the glass industry (nepheline syenites), and as a raw material in metallurgy (dunites).

Certain igneous rocks are used as industrial and semi-precious stones: well-polishable khibinite, welded tuff, porphyries, labradorites, porphyrites and light-coloured ornamental serpentine. Sedimentary rocks are used as rubble and in industrial materials. Many metamorphic formations are utilized as facing materials (jaspers, marbles, jaspilites).

The above-stated relatively widespread kinds of igneous, sedimentary, and metamorphic rocks form sheet and lenticular deposits, as well as large batholiths. They are prospected for by conventional geological methods. Prospecting is generally preceded by a regional geological survey aimed at identifying the target areas that are worthy of detailed exploration. Prospecting is accomplished by route surveys.

In areas with good outcropping, prospecting is supplemented by making single workings or boreholes. In conditions of poor outcropping, holes are drilled with a spacing of 1–2 km. Geophysical methods and drilling are employed if the thickness of overburden exceeds 2 m. One or two workings per 1 km of route are quite sufficient to evaluate the prospected area.

The positions of rock masses, horizons, strata and beds are determined at the stage of prospecting. Rock outcrops are evaluated both quantitatively and qualitatively. The latter is based on the geochemical criteria. Dunites and olivines with a high MgO content and small losses on ignition are characterized by most valuable properties. Potential utilization of dunites as a raw material for metallurgy is evaluated by the loss on ignition which, in its turn, is inversely proportional to density. The raw material is evaluated directly in the field by determining its density with the aid of hydrostatic weighing. This makes it possible to more rationally and urgently carry out mining and drilling operations on slightly serpentinized ultrabasites without making expensive chemical analyses. The physical and mechanical properties of rocks are studied to determine the ways of their utilization. The quality of stone raw material is determined by analysing samples taken from characteristic varieties.

Prospecting and evaluation work is carried out on perspective occurrences of building and industrial (or semi-precious) stones, that is in the areas defined at the first stage of prospecting. Exploration of intrusive rock deposits involves the determination of petrographic varieties of stone, depth of weathering, as well as investigation of inclusions and foreign formations. Rocks are contoured by geophysical methods.

4.16. CONCLUSIONS

1. Non-metallic useful minerals of the third group are rather diverse as to the conditions of formation, composition of enclosing rocks, mineral composition, parameters and forms of commercial bodies.

2. An individual, specific approach to geochemical exploration of each genetic type of deposits is stipulated by considerable differences in the conditions of their formation. Geochemical methods have proven to be most efficient in the search for vermiculite (anomalies of fluorine and alkaline elements), nephrite (chlorine and fluorine haloes), and in the presence of positive and negative anomalies.

3. Such minerals as magnesite, talc, wollastonite and vein quartz generally do not create the concentrations of elements in ore bodies that produce geochemical haloes. In such cases, it is necessary to use indirect methods based on direct and inverse correlations between the contents of the sought useful minerals and geochemical anomalies because formation of ore bodies of these minerals was accompanied by the migration and redistribution of the admixture elements in the circum-ore zones. Prospecting along the secondary lithogeochemical haloes is based on the hypergene destruction of deposits and their primary haloes.

4. Deposits of building materials, of raw materials for glass production, of clay materials, diatomite and perlite are generally widespread. In conditions of

good outcropping, they are commonly reliably detected by conventional geological methods. It should be noted, however, that visual observations fail to provide complete information on the mineral and chemical composition of rocks and on their physical properties. The situation is quite different in sodded areas devoid of natural exposures, and especially in the presence of a thick crust of weathering. In such conditions, route surveying of a territory fails to provide objective data making it possible to form an opinion on the prospectiveness of all the areas under investigation with respect to finding non-metalliferous useful minerals. In this case, geophysical methods are used in combination with mineralogical–geochemical investigations.

5. It is important for the deposits of the third group to evaluate the quality of raw materials and to determine their physical properties.

6. Prognostic exploration for non-metals is based on the direct, indirect, and combined principles of interpreting geophysical data. The direct principle implies direct investigation of standard deposits; the indirect one envisages a study of the correlations between the geological and geochemical characteristics; and the combined principle is based on the integration of the first and second principles (e.g. in prospecting for kaolins).

CHAPTER 5

New types of non-metallic deposits

All the industrial minerals are classified on the basis of industrial requirements into five groups of raw materials: (1) crystals, (2) chemical, (3) ceramic and refractory, (4) binding, and (5) aggregate. Some groups of natural minerals and rocks used in many branches of industry, such as limestone, evaporite clay, sand and others, cannot be placed into a definite group.

The main requirements for crystals are quality, size and purity. There are piezoelectric crystals, mica, asbestos, graphite, gemstones and others.

The chemical groups are closest to metallic ores in the requirements of industry, as the main index for quality is the percentage of an element. The distinction between them is the use of the element not in a pure form, but in some compound. Boron, potassium, phosphorus, fluorine, barium, magnesium, as well as common salt are placed into this group. The absence or small content of detrimental colouring impurities (iron, chromium, titanium) or flusing agents (alkaline elements) is the most important requirement for ceramics and refractory materials. Kaolin, quartzite, feldspar and its substitutes belong in this group.

Binding raw materials must initially contain the necessary combination of components that will result in various cementing substances after processing. Aggregate raw materials include construction and moulding materials and sorbents, the utilization of which is determined by physical and mechanical properties.

This classification of industrial minerals makes it possible to predict and evaluate new types of deposits.

Nowadays natural zeolites are widely used in industry and agriculture. For this reason they are considered to be new types of raw materials. In many countries zeolites are being extracted on a large-scale.

Brucite is a new high-magnesian industrial mineral. Brucite is characterized by the absence of deleterious impurities (iron, sulphur, chlorine, alkalis) and high chemical reactivity at the crude and calcined stages. Brucite may be used to produce chemical magnesian products of high quality.

A new type of cryolite deposit has been discovered in Russia. It is represented by fault-line alkali feldspar metasomatites. The metasomatites are localized in regional faults; they are not connected with magmatic rocks; they are developed by the replacement of cataclastic alumino–silicate crystalline schists and gneisses; they form elongated transversely zoned bodies [93]. These metasomatites are similar to alkali granites in their chemical-petrographic and mineral-geochemical properties: they are poor in Ca and Mg, and rich in Na, K, F and a specific complex of trace elements.

Cryolite is observed in riebeckite–arfvedsonite and aegirine–arfvedsonite rocks. The content of cryolite is 12–80 per cent. The deposits of this type are large-scale. Kalsilite-bearing intrusive rocks are fairly useful minerals first found in the Baykal region. The kalsilite and other alkaline rocks form kalsilite–nepheline syenite para-genetic associations which belong to the group of feldspathoid formations. According to the predominant role of nepheline, a potassium series of alkali-aluminiferous mineragenetic formations is established: nepheline, nepheline–alkali–feldspathic, kalsilite–nepheline–alkali–feldspathic, kalsilite–alkali–feldspathic, and kalsilite. The last two are represented by kalsilite–potassium syenites, synnyrites and kalsilite pyroxenite–yakutites, respectively. Synnyrites and yakutites are new types of rocks and they form an independent class of complex ultra-potassium alumino–silicate raw materials.

The useful minerals of synnyrites, kalsilite and potassium form the basis of the rock and the average content of the main components by mass of rock is 17.5 per cent K_2O; 22.5 per cent Al_2O_3; 1 per cent Na_2O and 55 per cent SiO_2.

Synnyrites form various ore manifestations (according to scale) including unique deposits (according to reserves). They are found in the Synnir, Sakun, Yakashin and Murun Massifs (eastern Siberia), Khibina and Selawik Lake (Alaska).

Only apomagnesitic monomineralic brucitites (Russia, USA) and apodolomitic brucite-calcitic marble (Canada) deposits are being brought into operation.

Yakutites are a new type of rock of kalsilite–pyroxene composition in a syenitic massif (Murunsky, east Siberia). Metasomatic varieties of a similar composition occur along with the intrusive-magmatic rocks. Yakutites are a potential source of potassium and alumina.

The tausonite ores are a potential commercial source of natural titanate strontium ($SrTiO_3$ — tausonite). They are represented by kalsilite–pyroxenic rocks, lamprophyllite, vadeite and sphene [126].

Halleflinta, quartz porphyry, rapakivi granite and syenite represent promising new sources of feldspar raw material.

Halleflinta is a light-grey microgranular quartz–albite rock with a small amount of scurfy mica (biotite, muscovite, sericite). It occurs as a thick pipe-like deposit surrounded by the Archean volcanogenic and sedimentary rocks of western Karelia. It is a concomitant mineral mined in iron ore deposits. After removing mica from halleflinta by the flotation method, extremely pure quartz–feldspar concentrates can be obtained.

Quartz porphyry occurs as a stock among tuffaceous shales. It is a light-pink rock composed of a micro-granular quartz–microcline aggregate with bluish quartz grains. The K_2O/Na_2O ratio is in excess of 10 in feldspar concentrate.

High-quality K–feldspar concentrates may be produced from magnophyric non-ovoid varieties of rapakivi. Leucocratic syenite consists of albite, nepheline and microcline in varying proportions. Furthermore, aegirine, arfvedsonite, biotite, muscovite, titanomagnetite, ilmenite, apatite and other accessories are present.

Iron-bearing minerals are readily removed by dry and wet electromagnetic separation. Extremely pure nepheline–feldspar concentrates have been obtained from leucocratic syenite.

Porcelain stones (Primoryl, Russia) are a complex ceramic raw material. They consist of quartz, feldspar, kaolin, pyrophyllite and sericite.

Hydrothermally and metasomatically altered rocks, like the secondary quartzites, which are distributed in the porcelain stones of Primoryl, develop after acid and intermediate volcanic rocks. The following deposits are distinguished on the basis of the prevailing main mineral: 1) kaolin (dickite), 2) K–feldspar, 3) pyrophyllite, 4) hydromica (sericitolite).

Porcelain stones are a complex high quality raw material for fine ceramics.

Water-bearing glasses are a new type of deposit. Basalt and rhyolite glasses are the most common types of volcanic rocks. Basaltic glasses are mostly represented by a porous type. In rhyolite rocks massive as well as porous glasses are found. Basaltic glass contains up to 2–3 per cent H_2O and a little hydroxyl. Acid glass contains 0.1–9 per cent H_2O and OH-group. It expands intensively at 1000–1150 °C and forms artificial pumice with a 70–350 kg/m bulk density. Rhyolite glass is used as a natural pumice and expanded perlite.

Recently, in Russia and Ukraine the geologists of an industrial association have discovered numerous new types of gem deposits: charoite, scapolite, clinohumite, chrome diopside, jadeite, apocarbonate nephrite, etc.

It is proposed that raw materials which, with regard to the anticipated progress of technology, will become usable before the end of this century should be regarded as having potential. They may be exemplified by anhydrite as a substitute for gypsum added to cement clinker in the production of high-strength quick-setting cement, and cellular concrete; magnesite added to special ceramic substances; and fluorine obtained from phosphates instead of from fluorite [66].

Potential industrial minerals and rocks are unconventional raw materials, most of which would be usable only in an age of abundant energy.

Substitute raw material are also called the raw materials of the future.

Abundant and cheap energy would permit the beneficiation of substitute materials to the required quality (sulphur from gypsum and seawater, or all magnesium from seawater).

Charoitic rocks used in jewellery consist mainly of charoite (up 70–95 per cent). They are spatially and genetically connected with carbonatites (east Siberia), for which charoite is a typical mineral.

The hydrothermal–metasomatic nature of the charoite paragenesis, including charoite and associated minerals (tinaksite, canasite, miserite, fedorite, etc.), has been proved by various methods. The formation of charoite metasomatites is related to the calc-alkali metasomatism of productive meso- and melanocratic fenites, shonkinites and/or alkaline minettes. The metasomatic column includes circum-ore metasomatites (dark, intermediate in colour and composition, and light orthoclase zone), charoite ore (lilac zone) and a quartz zone (the axial part of the column). The following series of differential mobility of the components has been revealed (from inert to increasing mobility): SiO_2, $(K_2O + Na_2O)$, $(CaO + TiO_2)$, Al_2O_3, Fe_2O_3, $(FeO + MgO)$. Concentrated aqueous solutions of calcium and potassium chlorides have been found in fluid inclusions with quartz associated with high-quality charoite.

CHAPTER 6

The role of analytical techniques and instrumentation in geochemical exploration

6.1. DECOMPOSITION TECHNIQUES

Sample decomposition is the first step after pulverization in the procedure of geochemical analysis. Its aim is to convert the solid sample into a solution suitable for elemental determination by modern atomic spectrometry or other instrumental methods. The complexity of geological materials makes it necessary to choose a sample decomposition technique that is compatible with the specific objective of the analysis. When selecting a decomposition technique, consideration should be given to the chemical and mineralogical characteristics of the sample, the elements to be determined, the requirement for precision and accuracy, sample throughput, the technical capability of the personnel, and the constraints of time [17].

Acid digestion is a convenient way to decompose geological samples for analysis (hydrofluoric acid — HF; hydrochloric acid — HCl; hydrobromic acid — HBr; nitric acid — HNO_3; sulphuric acid — H_2SO_4; and phosphoric acid — H_3PO_4).

Sample decomposition by fusion relies on chemical and thermal energy to break up the original mineral phases and convert them into different solid forms which are more readily dissolved by acids or water. The fluxes used for sample decomposition are most often salts and alkali metals. They may be alkali or acid in reaction and oxidizing or reducing in nature. Sodium carbonate and sodium hydroxide have long been used to fuse rock and mineral samples. This is to show, in a very brief form, the use of sample decomposition techniques with a combination of acids and/or fluxes in either open or closed vessels [17].

Automation will increase the efficiency and speed of sample decomposition which may eventually allow this step in geochemical analysis.

This developing technology with automation and robotic technology is sure to affect the future concept of sample decomposition and the physical appearance of the modern chemical laboratory. Automatic systems that integrate the components of the decomposition process such as sample grinding and weighing, reagent or flux dispensing, decomposition, and dilution before determination, are being developed.

6.2. NEW ANALYTICAL TECHNIQUES AND LABORATORY

The analysis of geochemical materials involves four main stages. These are the collection and preparation of the sample materials, decomposition, the separation of the sought element and the determination of concentration.

Sample collection and preparation techniques vary with the material. Drying and sieving in the field reduces the size of sample to be transported to the laboratory, resulting in lower mailing or carrier charges.

Decomposition of the sample involves the breakdown of the solid phases to release the elements in extractable form. The determinative stage uses either instrumental or comparative techniques. Those used for geochemical work are listed below: (1) optical determinations include colorimetry, paper chromatography, spot tests, visible fluorescence, and turbidimetry; (2) radiation determinations include X-ray fluorescence, emission and flame spectroscopy, radiometry, atomic absorption, and thermoluminescence; (3) electrical measurements, such as specific ion activity, and polarography.

Some of these methods, such as X-ray fluorescence and emission spectroscopy, bypass all but the sample preparation stage. Specific ion determinations of natural waters require only separation of the sought element.

The choice of the analytical method to be used is mainly dependent on the economics and logistics of the operation, the nature of the sample material, and the element to be determined.

Colorimetric analysis is usually cheaper than some instrumental techniques. This is because cheap labour can be utilized, whereas instruments require a more skilled operator and involve initial capital plus current servicing costs. The demand for more sensitivity and accuracy in the analysis of geological materials in recent years has been concomitant with the development of atomic absorption analysis. Techniques have been developed which are rapid, simple, and sensitive for most of the more common ore elements, and so satisfy the requirements for trace element analysis in geochemical exploration programme. The advantage of atomic absorption spectrophotometry over other instrumental techniques is primarily its simplicity. The only requirements are that the element to be determined is in solution. The method is not beset by such problems as mass absorption, and spectral or chemical interference.

An extensive range of highly sophisticated, automated instruments, utilizing a wide variety of analytical techniques, are employed to determine the concentrations of a large number of elements in rock, mineral, ore, soil and stream-sediment samples. The techniques used include atomic absorption spectroscopy, inductively coupled plasma–optical emission spectroscopy, X-ray fluorescence spectroscopy, wet-chemical procedures, mass spectrometry, gas and liquid chromatography, assaying, physical chemistry and radiometric analysis by β- and/or γ-ray counting. The laboratory also determines the composition and concentration of gas in soil and performs fluid-inclusion studies. High-quality mineral identifications using XRD-systems, a macroprobe, microprobes and scanning-electron microscope are undertaken for the staff.

'Dtransu' is a new advection-dispersion finite element analysis code which uses the Eulerian–Lagrangian (EL) method. 'Dtransu' can simulate the behaviour of ground water contaminants more exactly than other existing methods. An application of 'Dtransu' to synthetic data is presented in comparison with that of a conventional method.

Methods of radioactivation analysis have been developed.

In a modern sense these methods utilize certain nuclear properties of the isotopes of the elements sought in the sample. Most methods utilize thermal neutrons as the bombarding particles (neutron activation analysis). Such methods have an extremely high sensitivity in the parts per billion range for elements such as uranium, etc. They have been used extensively in basic geochemical research for many years and are now finding increasing use in geochemical prospecting.

The development of gas, partition, paper, adsorption, and ion exchange chromatography and other precise methods of trace analysis of hydrocarbon compounds and various gaseous inorganic substances is noteworthy. These are probably the most significant developments as regards the rapid analysis of traces of hydrocarbons and associated gaseous inorganic compounds in prospecting surveys of non-metallic deposits, utilizing rocks, soils, waters, and lake and marine sediments.

Radioactive methods have been developed for the detection of the radioelements, particularly uranium, thorium and their daughter elements in apatite.

Airborne spectrometers have reached a high degree of sophistication, being used for total gamma counting and specific quantitative analyses of elements such as potassium (^{40}K), uranium (as ^{214}Bi), and thorium (as ^{208}Th). Today radiometric surveys are carborne, airborne, or shipborne.

Atomic absorption spectroscopy is a particularly rapid and accurate method of determining many of the elements in the materials of the earth in the parts per million range. For this advance we owe a great debt to the substained research in instrumentation and methods at CSIRO in Australia.

Flame emission spectroscopy is similar to atomic absorption spectroscopy, but the sample element is raised to a high energy state for the measurement of radiation. In atomic absorption spectroscopy, the flame dissociates the element of interest from its chemical bonds. In flame emission spectroscopy, the excited electrons in orbit give off energy as they return to their former state. The flame emission method is plagued with 'spectral' interference.

Direct reading emission spectroscopy uses electrical energy to bring a solid or a liquid to the vapour phase in which it will emit light. The equipment is complex, and the technique requires considerable skill.

The decade of the 1980s witnessed great advances in analytical techniques and instrumentation: inductively coupled plasma–atomic emission spectrometry (ICP-AES) moved from the research to the production-oriented laboratory, the concept of ICP mass spectrometry (ICP-MS) grew into a commercial instrument to find early application in geochemistry [39]. The process of analysis became fully and intelligently automated with the proliferation of low cost microcomputers. Efficient multi-element packages were developed in instrumental neutron activation analysis (INAA), and interference in graphite furnace atomic absorption spectrometry (GFAAS) was largely mitigated by improved atomizer design and matrix modification.

The analytical laboratory is eyes and ears of any mineralogical operation. It is the key to new methods of exploring deposits. And, whether it is rocks or minerals, the final seal of quality depends on the results of analysis (assay, classical chemistry, atomic absorption spectrometry, X-ray fluorescence, inductively coupled plasma–optical emission spectroscopy). Methods tend to become redundant in terms of cost effectiveness, accuracy and precision.

The field of exploration geochemistry has benefited and is continuing to benefit from these advances in technology. A much broader range of elements can be determined at modest cost, there is an improved detection limit, and the capability to analyse accurately a wide spectrum of sample matrices.

6.3. INSTRUMENTAL NEUTRON ACTIVATION IN GEOANALYSIS

Instrumental neutron activation analysis (INAA) is an analytical technique which is dependent primarily on measuring γ-radiation induced in the sample by irradiation with neutrons. The primary source of neutrons for irradiation is usually a nuclear reactor. The method is highly selective and extremely sensitive for a wide range of elements.

The process of activating samples is inherently simple. The samples are encapsulated and placed into or near the core of a neutron source.

A major advantage of INAA is the fact that the sample does not have to be put into solution, which is particularly beneficial for some geological matrices where complete dissolution is difficult to achieve. INAA is a complementary analytical technique. A different suite of elements is measured using INAA than with other techniques.

A major application in the field of geoscience research involves the determination of REEs and other elements like Sc, Hf, Ta, Th and U, which are important in petrogenetic studies as well as in the study of rock-water interactions and various types of alteration associated with deposits. Geological materials containing refractory minerals such as chromite, garnet, sphene, and zircon are commonly regarded as difficult samples to put into solution.

A suite of short-lived isotopes extends the range of elements being determined by INAA. Typical detection limits for rocks are: Al — 0.0001 per cent, Mg — 0.05 per cent, Na — 0.01 per cent, Cl — 100 ppm, Mn — 0.1 ppm, U – 0.1 ppm and V — 0.5 ppm.

The major factor influencing the level of detection for these elements is the amount of Al and Mn present in the sample. The high sensitivity of Al and Mn results in large amounts of induced radioactivity which may result in increased detection limits. The detection limits for short-lived isotopes are generally more matrix-dependent and hence more variable than for long-lived isotopes.

Water samples can also be analyses by INAA. Water can be analysed for U to 0.2 ppb by evaporation of 100 ml samples.

INAA provides a means of rapidly determining as many as 36 elements simultaneously on virtually all types of geological matrices including vegetation using the long-lived isotopes [39].

The cost of analysis by INAA is usually competitive or much lower than with other analytical techniques. Advances in computer technology will continue to reduce the capital cost of multidetector computer systems. There are some elements which are in such low concentrations in geological material that they can only be measured after radiochemical separation if INAA is used.

Radiochemical separations have several advantages over other analytical techniques, in particular the lack of contamination and the ability to add a carrier to monitor and correct for losses. Radiochemical separation schemes are described for the following elements, or groups of elements: As, Bi, Tl, Cd and Hg; Cl and Br, Re, Rh, Ag and In, Rb, Cs, Cr and P; Te; Se and Mo; Zr and Hf; and the REEs. Therefore, radiochemical neutron activation analysis (RNAA) is used to measure sub μg g^{-1} (parts per million) concentrations of many elements in preference to other techniques. In general, there are far fewer examples of current applications in RNAA than for instrumental techniques.

Radiochemical NAA is simply a separation procedure applied to an activated sample. There are three basic stages: irradiation, separation and counting. The process of activation with neutrons is exactly the same as for INAA. RNAA can make use of all the conventional separation procedures available to the chemist including solvent extraction, ion exchange chromatography and precipitation. If γ-ray analysis is used, the counting process will also be the same as for INAA. The only different technique that is encountered in RNAA is the use of beta counting, which is rarely applicable to INAA.

The advantages of RNAA for geochemical applications can be given in three main categories [44, 88]. The first case includes those elements which may suffer losses when the sample is being prepared for analysis. The second application is where contamination is introduced during preparation for analysis. The third case is when INAA is already being used to determine certain elements and the separation stage will provide additional information on the same sample. The general applications of RNAA in geoanalysis are in those areas where the elements are present in sub − μg^{-1} (parts per million) concentrations. Any new method of separating trace elements from geological samples can be applied to RNAA and so recent developments in other areas may be untapped sources of new radiochemical methods. Therefore, in the future there may be new procedures for elements which are currently difficult to determine by other techniques. At present the most likely areas of development remain with the REEs and PGEs.

Neutron activation analysis is one of the newest, most exciting developments. It is based on the use of a stream of high energy neutrons to bombard a sample, followed by the analysis of the resulting radioactivity. It has good capability for analysing microgram quantities of many elements.

6.4. ATOMIC ABSORPTION SPECTROMETRY IN GEOCHEMICAL EXPLORATION

Atomic absorption spectrometry (AAS) is one of three analytical techniques of atomic spectroscopy, the others being atomic emission and atomic fluorescence spectroscopy.

Atomic absorption spectroscopy is based on the fact that the atoms of every element can absorb light of the same wavelength emitted by that metal. In order to be in a condition to absorb, the atoms must be chemically unbound and in their minimum energy state. In general, this is achieved by vaporizing the sample in a flame. Now used for gold and silver determinations in place of fire assaying, atomic absorption is useful for trace analysis work in geochemical campaigns. The method is quite free of the interference factors which have plagued most wet or instrumental analysis techniques.

The introduction of AAS revolutionized geochemical exploration. Rapid, low-cost, sensitive measurements of more than 30 elements in a variety of sample media assured the universal acceptance of AAS in geochemical exploration laboratories [124].

Current applications of modern AAS instruments in microelectronics, advanced optics, and computer-controlled instrument setup, sampling, and data collection could hardly have been imagined by analysts only two decades ago. Paralleling the advances in hardware, innovations in chemical methodology such as selective digestions, hydride generation, and organic separations have evolved to meet the unique requirements of geochemical exploration.

Atomic absorption occurs when an outer electron of an analyte atom changes energy levels from a neutral ground state to a higher energy level by absorbing radiant energy of a specific wavelength [124]. The AAS measurement of the absorption of energy by the analyte atoms requires a light source, a sample cell, and a specific light measurement device. The concentration of the analyte in an unknown sample is measured by comparing its atomic absorption signal with the absorption signal of standards with known analyte concentrations. Prior to AAS, the main analytical tools in geochemical exploration were E-Spec and colorimetric techniques. E-Spec has the advantage of simultaneous determination of at least 30 elements, essentially on a total basis, using a solid sample. Colorimetric techniques played a complementary role to E-Spec in determining near crustal abundance levels of important elements such as As, Bi, Hg, Mo, Sb, S and Zn.

Since its introduction, AAS has replaced most of the colorimetric methods used in geochemical exploration and has lived up to most of its heralded advantages of being rapid, sensitive, reproducible, and relatively inexpensive and interference free. Flame atomic absorption spectrophotometry (FAAS) may be used to measure Al, Ba, Ca, Cd, Co, Cr, Cu, Fe, K, Li, Mg, Mn, Mo, Na, Ni, Pb, Si, Sr, V and Zn.

Figure (6.1) shows the elements of the periodic table that can be determined in geochemical exploration samples using a variety of AAS methods.

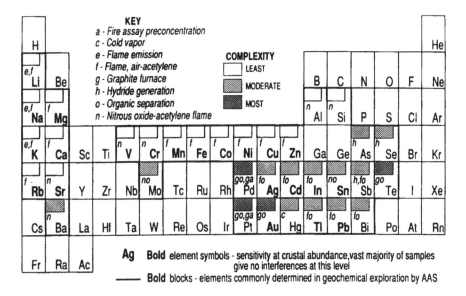

Figure 6.1. AAS methods compared in terms of complexity, sensitivity and freedom from interferences in determining crustal abundance levels of metals useful in geochemical exploration. All determinations are made from acid solutions unless noted otherwise [124].

6.5. INDUCTIVELY COUPLED PLASMA-ATOMIC EMISSION SPECTROMETRY IN EXPLORATION GEOCHEMISTRY

Inductively coupled plasma-atomic emission spectrometry (ICP-AES) is a well-established, cost-effective, multi-element technique which is routinely used for geochemical analysis in mineral exploration. Systems available range from relatively expensive simultaneous instruments capable of precisely and accurately determining more than 35 major and trace elements in less than 2 minutes, down to much cheaper but slower sequential machines which compete directly with atomic absorption spectrometers. The advantages and disadvantages of different instrumental options and analytical protocols are stressed with specific reference to the routine analysis of exploration samples. Sample preparation, and particularly sample dissolution procedures will ultimately limit the range of elements which can be accurately quantified by ICP-AES.

Analytical requirements will vary enormously depending on the specific application. Exploration work for non-metallic deposits requires extensive geochemical surveys of the soil and sediments in the target area. In such cases it is necessary rapidly to determine a comprehensive suite of elements in large batches of samples at low cost. Lithogeochemical surveys require the measurement of total sample composition, necessitating the use of procedures which ensure the complete dissolution of silicate material.

Biogeochemical surveys are being used increasingly in mineral exploration. Many plants extract metals from soils and concentrate them in their tissue at levels which, in ashed samples, are greater than those found in the associated soils, sediments and bedrocks. The analysis of plant materials such as leaves and twigs, therefore, provides a sensitive exploration tool. For all botanical samples ashing is necessary to ensure their complete decomposition. Two procedures are employed [63]: (1) wet ashing; and (2) dry ashing. Wet ashing uses oxidizing acid mixtures to destroy organic matter. Dry ashing is rapid and inexpensive since it merely involves the oxidation of samples in an oven at temperatures of 450–600 °C.

The analysis of natural waters provides another potential tool in exploration geochemistry. Natural waters are close to neutral (6.5–8.0 pH) and therefore carry only very low levels of dissolved transition metals.

The elemental capabitities of ICP-AES in geoanalysis may be summarised as follows: (1) all major elements may be determined with precision and accuracy in the solutions prepared; and (2) trace elements may be routinely determined in mixed acid digestions employing $HF/HClO_4$.

The technique displays excellent sensitivity for many low atomic number elements (B, Be, Li, P and S), the alkali earths (Ba, Ca, Mg and Sr), refractories (Al, Ti and Zr) and the rare earth elements, Sc and Y. Performance for the first-row transition metals is also generally good. The number of elements which can be determined varies greatly depending on sample type and preparation procedures used. Spectral interference commonly limits the range of trace elements which can be measured, but this can be usefully extended using a simple chemical separation technique such as cation-exchange chromatography prior to the determination of the REEs, hydride generation for As, Bi, Ge, Sb, Se, Tl and Sn, or solvent extraction for many trace metals [48].

Comparisons with other instrumental methods indicate that ICP-AES is particularly versatile and compares very favourably with AAS, INAA and XRF for the determination of major and trace elements in a wide range of matrices. It is ideally complemented by ICP-MS, which allows the determination of ultra-trace levels of heavier elements without the need for chemical separation procedures and can be employed on the same samples prepared for ICP-AES. The use of ICP-AES is now widespread in exploration geochemistry. Considerable research effort continues to be expended on expanding the capabilities of ICP-AES to include solid samples by using laser ablation microanalysis (direct multi-element analysis of solid samples). The laser Lab (VG system) incorporates a pulsed Nd–YAG laser (at 1064 nm), which is focussed into the target sample, located in a sample cell through which a continuous stream of argon gas is passed. A series of laser shots are fired at the sample, vaporizing the analyte material to form a microparticulate cloud which is carried by the argon flow into the plasma, where it is ionised.

Targeting is achieved by means of three stepper motors which allow the sample to be positioned at the appropriate place beneath the beam of the laser. This can be accomplished both manually and under computer control.

Once focussed, the system may be set up either to sample from the targeted point only, or to perform rasters across the surface, firing at each point.

During bulk analysis the average value for the overall composition of a sample is obtained, inhomogeneities present in the sample being averaged out. In order to minimise the errors in such an analysis it is necessary to sample over a large area.

The laser Lab has been designed to meet such requirements. The beam may be defocussed to increase the sampling area, and the rastering capability enables the user to perform accurate multiple analyses while ensuring that the same size of sample is analysed each time.

One of the most important aspects of solid sampling is the ability to select a specific feature in a sample for individual analysis. This has been found to be invaluable in the geological fields where samples are often heterogeneous and important information would be lost by bulk sampling. Feature analysis is also important in other application areas, even if theoretically.

The VG Laser Lab is capable of focussing down to spot sizes of 20 μm. As the spot size, and hence the amount of material ablated is reduced, there is a corresponding reduction in the signal produced. However, the sensitivity of the Plasma Quad/Laser Lab combination is such that excellent sensitivity and precision is maintained during feature analysis.

The Time Resolved Analysis (TRA) feature of the Plasma Quad PQ 2 plus may be used in conjunction with the Laser Lab to enable rapid, sensitive homogeneity measurements to be performed. Using TRA, variations in the multi-element signal with respect to time can be measured. By rastering across the surface of a sample at a constant rate, the variation in elemental concentrations for all elements in the sample may be determined with respect to the position on the sample (chromite). Such three-dimensional surface mapping may be achieved in under 2 minutes, enabling homogeneity assessments to be performed as often as required. In addition to bulk sampling, the VG Laser Lab offers the capability of analysing individual minerals or phases within a sample, for both trace metal concentration and isotopic screening information.

To summarise, VG Laser — direct multi-element analysis of solid samples has the following main features: (1) very rapid analysis, typically 60 seconds for all elements; (2) multi-element analysis from ppb to per cent; (3) analysis of conducting and non-conducting materials; and (4) direct analysis, minimal sample preparation.

6.6. THE TECHNIQUE FOR ELEMENTAL ANALYSIS (ICP-MS)

The acknowledged benefits of ICP-MS — sensitivity, speed and flexibility — have resulted in the rapid acceptance of the technique for environmental analysis.

The VG PQe is a combination of innovative design and engineering, specifically developed to meet the needs of environmental analysis laboratories for cost effective performance.

The detection power of the VG PQe is such that it can measure all environmentally significant elements and more — to well below the current legislated detection limit. Each sample is analysed for all required elements in under 90 seconds. The VG PQe is fully automated. The powerful 32 bit microcomputer and multi-tasking operating system, controls all instrument functions, ensuring simple operation.

Inductively coupled plasma mass spectrometry (ICP-MS) is the most up-to-date option for the analysis of environmental samples. ICP-MS offers parts per trillion limits of detection for most elements. This excellent detection capability combined with a high sample throughput and eight orders of linear dynamic range make it uniquely suited to the requirements of the routine environmental laboratory.

Non-routine samples may be rapidly and easily screened, and a complete unknown may be characterized for major, minor, trace and ultra-trace elements automatically without operator intervention in under three minutes.

The programmed access design of the optimal autosampler ensures that analysis proceeds unattended. The use of Laser Ablation allows the user to perform direct ICP-MS analysis on solid samples without pretreatment.

Excellent precision and accuracy have been obtained using laser ablation for a wide variety of environmental samples. The excellent solution detection limits available using ICP-MS may be improved even further with an electrothermal vaporisation, laser ablation.

The following products are available for mass spectrometry in environmental analysis — (VG products): 1) NS Plasma Quad (ICP-MS); 2) VG Plasma Quad with flow injection accessory; 3) Laser Ablation ICP-MS; 4) VG Microtherm ETV; 5) VG survey; 6) VG Petra Sentinel.

The Inorganic Division of VG Instruments embraces five operating companies, each specialising in a sector of mass spectrometry. Together they provide a unique capability for the elemental, isotopic and chemical analysis of solids, liquids and gases.

Rapid and precise methods have been developed to analyse various volatile elements such as mercury, sulphur compounds, helium and radon in rocks, soil gases, waters, and the atmosphere. The instruments adapted for such work include the gas chromatograph, atomic absorption spectrometer, mass spectrometer, and a number of variations of alpha counters.

6.7. INDUCTIVELY COUPLED PLASMA MASS SPECTROMETRY IN GEOANALYSIS

Inductively coupled plasma mass spectrometry (ICP-MS) has had a profound effect in the discipline of geoanalysis since its commercial availability in 1983. In particular, the ability to determine precious metals, rare earth elements (REEs) and refractory elements such as Hf, Ta and W at their natural levels in geological materials has dramatically improved through the application of this highly sensitive, multi-element technique. Its capability to measure individual isotopes has been directed mainly towards the Re–Os chronometer pair in ore genesis studies, and to a lesser extent to the quantification of Pb isotope ratios.

The reasons for such widespread enthusiasm for this technique and its adoption lie in the advantages it enjoys over its chief rival, ICP atomic emission spectrometry (ICP-AES). These comprise: superior sensitivity, generally by several orders of magnitude; simpler spectra; and the ability to obtain isotopic information. Commercial instruments use a radio-frequency quadrupole mass filter, as such devices are relatively inexpensive and compact and provide adequate resolution over the desired mass range. The analyte signal depends on the number of analyte ions in the plasma and on their potentially mass-dependent transport to the detector, which operates on a rapid scanning principle.

As with every analytical technique, there are disadvantages. Like ICP-AES and AAS, ICP-MS is at present solution-based and samples must be dissolved.

ICP-MS is ideal for the analysis of waters and selective leaches, areas of current interest in exploration and environmental geochemistry.

The detection limits of elements are in the range $0.01-0.1$ ppb (nq ml^{-1}), generally two to three orders of magnitude better than ICP-AES and competitive with GF-AAS. This sensitivity, coupled with the capability of multi-element analysis, has been used to advantage with elements whose determination posed problems using existing techniques.

There are three areas where ICP-MS has had a major impact on analysis: (a) what is known as the refractory elements (Mo, W, Hf and Ta); (b) REEs; and (c) precious metals. The ability of ICP-MS to measure isotope ratios efficiently is proving to be one of its greatest strengths and the field of geochemistry is gradually exploiting this capability.

The elements best served by each technique are indicated in Fig. (6.2).

Figure 6.2. Periodic table illustrating the complementary roles of ICP-MS and ICP-AES in the analysis of geological materials [39].

6.8. X-RAY FLUORESCENCE SPECTROMETRY

X-ray fluorescence (XRF) spectrometric analysis is a technique that is widely used to determine the major and trace elements in a wide variety of geological materials. With the exception of instrumental neutron activation analysis (INAA), XRF is one of the few analytical methods in which determinations are normally made on solid samples, so avoiding the dissolution stage that is a prerequisite for most other atomic spectroscopy techniques. XRF has a justifiable reputation for the high precision of analytical measurements, second only to mass spectrometry techniques. In consequence, XRF is often the preferred technique for determining the major elements in rocks (Na, Mg, Al, P, K, Ca, Ti, Mn and Fe), where relative uncertainties of less than 0.2 to 0.4 are required to ensure the confident summation of the major elements to 100 per cent. Furthermore, XRF is capable of determining, often to ppm detection limits, a wide range of trace elements, some of which cannot readily be measured by competitive techniques. Modern computer-controlled XRF instrumentation benefits from full spectrometer automation and on-line data processing facilities for quantifying the results. Matrix types of samples collected as part of exploration programmes may be more variable and may include chromitites, carbonates, sulphides and various oxide ores. The elements determinable by XRF may range from Na to U in the Periodic Table. For many elements determination may by made over a concentration range that extends from high wt. per cent down to low ppm levels.

Solid samples are the preferred form of sample presentation, preparation techniques being particularly simple and cheap for the determination of trace elements in powder pellets. The technique is nondestructive; the preparations have relatively long shelf lives and can be kept as archive materials. Instrumentation and data processing can be fully automated, reducing the incidence of operator error, and permitting unattended routine operation and high sample throughput rates. XRF is one of the few instrumental methods of determining important trace elements and is particularly effective for Rb, Sr, Y, Zr, Nb, Pb and Th.

X-ray fluorescence generates X-rays characteristic of the sample when the sample absorbs rays critically shorter in wavelength than the emitted radiation. It is good for the routine analysis of many elements. Its chief drawback in geochemistry is that the system is not very good for trace analysis work.

X-ray microfluorescence (omicron) analyser combines the excellent elemental sensitivity of X-ray fluorescence (XRF) with spatial resolution in the range of SEM (scanning electron microscope) microanalysis. A colour television monitor provides real-time imaging of samples during analysis. X-ray microfluorescence means: (1) non-destructive elemental analysis; (2) no need for sample conductivity; (3) elemental sensitivity to a few ppm; (4) rapid analysis of the elements sodium through uranium, often in less than a minute; and (5) automated, unattended analysis of hundreds of points per sample.

Mineral inclusions in geological samples are often analysed by microanalysis using an SEM, Omicron can also perform microfluorescence analysis at multiple points on a sample, providing an average composition and information on compositional heterogeneity.

Energy-dispersive X-ray fluorescence spectrometry (EDXRF) is a method for the qualitative and quantitative analysis of elemental composition, which is steadily gaining popularity among rock analysts. In contrast to traditional forms of chemical analysis, EDXRF is based on the real-time generation, detection and measurement of X-rays emitted by the elements in a sample. The analytical process is non-destructive, requires little or no sample preparation and can be performed in minutes.

The EDXRF technique finds useful application wherever the elemental composition of a solid or liquid geological material needs to be determined.

The EDX-771 system of X-ray fluorescence spectrometers is useful for identifying unknown minerals and determining elemental compositions. The EDX-771 features a PC-based processor, standardizing the system with current trends in laboratory computing. The sample chamber of the EDX-771 can accommodate samples ranging from microgram-size particles to large, irregularly shaped parts.

X-ray fluorescence spectrometer for element analysis Spectro X-Lab is a newly developed instrument for element analysis in research laboratories. It combines all the advantages of energy-dispersive X-ray fluorescence analysis with detection limits which are comparable to those of wavelength dispersive RFA. There is a demand for analyses in environmental protection and geochemical prospecting. Energy dispersive X-ray fluorescence analysis is an ideal procedure for the analysis of samples with variable compositions, because all elements from 7–11 (Na) to 92 (U) and the total concentration range are analysed simultaneously and sample preparation is very easy.

Sensitivity ranges characteristic in trace analyses which up to now were possible exclusively for other analytical methods, can now be achieved with the Spectro X-Lab. The detection limits are within the range of 0.3 ppm of many elements. In comparison to the traditional energy dispersive X-ray fluorescence analysis this means a lowering to factor 10 as well as comparable net peak intensities with shorter measuring times. These advantages are achieved by exciting the sample with polychromatic, polarized X-ray radiation. The process leads to an improved peak-background-ratio in the spectrum and the measuring line can work with more usable signals.

This instrument makes use of the special advantages of energy dispersive X-ray fluorescence analysis coupled with a special way of exciting the fluorescence radiation. This allows a considerable reduction in measuring time, and the sensitivities of detection are significantly improved. X-Lab can be applied in geological prospecting in a wide concentration range.

6.9. THE NEW GENERATION OF MICROANALYSERS

The division is being equipped with automated electron microprobes and scanning electron microscopes to analyse a wide variety of rocks and minerals, and to compile maps showing elemental concentrations. Programmes are available by means of which a grid of points overlying a sample, showing complex element distribution patterns can be analysed quantitatively and automatically. The result can be represented graphically in a variety of forms. The usual qualitative concentration profiles are easily generated using standard microprobe facilities. The ability to carry out a complete quantitative analysis along a line of closely spaced points is an advance on this technique. For many applications in mineralogical analysis, digital X-ray mapping is widely accepted as one of the most efficient methods of determining the spatial distribution of elements in a sample. Until recently, however, the excessive time required to complete a digital X-ray map — rarely less than 30 minutes, more often an hour or longer — severely restricted the widespread use of the technique [50].

The CSIRO division of mineral products has automated a scanning electron microscope (SEM) and developed a fast search programme for rare phases in minerals, to be located at sizes down to 1 micron. This automated location facility (ALF) in conjunction with a series of leaching tests, permits the content of precious elements to be determined in polished sections of diamond drill core sections and crushed ore [68].

Kevex Delta Class analysers include numerous capabilities found in no other system. The key element of the Delta Class in this: no other system integrates high-performance elemental and image analysis so thoroughly into a family of completely upgradeable systems.

At the heart of the system is a firmware based X-ray and image analyser with distributed microprocessor control of all hardware functions. The Delta Class analyser is the world's only system which may be upgraded from its simplest to its most complex configuration without removing or replacing a single component. Upgrading at any time from one level to the next requires only the augmentation of the existing hardware and software; the display and other hardware components need not be replaced.

Standard with every Delta Class analyser is a high-speed, bidirectional printer/ graphic plotter. Optionally available is a complete range of multi-pen plotter, colour video printers for report generation.

The key to quality in X-ray microanalysis is superlative performance in the X-ray detector.

Feature analysis is the technique of extracting morphological and chemical information from samples during and after the acquisition of digital images.

Automated image analysis can be integrated with Kevex Sesame for combined SEM stage automation and image analysis.

In addition, this capability can be implemented on any automatable model of SEM.

Mineralogical instruments include X-ray diffractometry (XRD), petrological microscopes, scanning electron microscope (SEM) equipped with EDS microanalysis facility and an automated image analysis system linked to the SEM and petrological microscope. The concentration of minerals to assist mineralogical investigation is often necessary and for this purpose a modern superpanner or magnetic separator are used. Heavy liquid separations of minerals can be carried out at varying SG cutoffs. The detailed mineralogical investigation of residues and tailings to identify the mode of occurrence of unrecovered minerals can provide vital information for the prospecting of deposits.

6.10. HOMOGENEOUS SAMPLE FOR MINERAL ANALYSIS AND AUTOMATED FIELD IDENTIFICATION OF ALTERATION MINERALS

Labtechnics' vibratory (Australia) mill stands alone in its ability to prepare up to a 5 kg batch of a more representative sample for mineral analysis. Optimum pulverising and mixing efficiency is assured by Labtechnics' internationally patented, time-tested pulverising bowl. A conventional ring and roller type bowl option is available. Maximum performance power in the bowl is provided by an integral, minimum-maintenance electro-mechanical drive.

The fully-automated methods of image analysis fall into two major categories [50]: (a) area-measuring, and (b) line-measuring. These categories are sub-divided according to the basic image-forming method used. These methods include: (1) optical illumination, (2) electron-beam illumination, and (3) X-ray signal response to electron beam illumination.

All the fully-automatic area-measuring systems use television-type scanning techniques and most of them rely on optical signals to distinguish the various minerals from one another. In some instruments (including the Quantimet) it is possible to use electron optical signals (derived from a scanning-electron microscope) to distinguish the various minerals.

Television-type systems that employ electron-beam illumination use back-scattered primary electron signals, secondary electron signals, absorbed current signals, or various combinations of these to distinguish the various minerals in a specimen.

In automatic line measuring systems a suitably prepared specimen is traversed in a series of steps under a small beam of light or under a focussed beam of electrons. The signal derived from the specimen is used to distinguish the mineral at that point. The results can be used as the basis of a linear measuring system for determining random intercept lengths, grain sizes, grain shapes, particle compositions, etc.

All image analysers provide statistical data based on zero, one or two-dimensional measurements: these measurements must be converted into the equivalent three-dimensional values. This can occasionally be done by comparison with standard materials but, more usually, it can be done by mathematical transformation of the raw data.

A five-year project of 'Automated field identification of alteration minerals based upon reflective spectroscopy' by MMAJ has been in progress since 1989. The goal of this project is to develop a new field spectroradiometer linked with an 'expert system' for the *in situ* identification of minerals commonly found in hydrothermal alteration zones. The procedure to create an 'expert system' is summarized as follows [13, 50]:

(1) Spectral measurements of minerals. Spectral reflections of thirty nine standard minerals were measured by an FTIR spectroradiometer with high spectral resolution of 0.3 nm to confirm the precise wavelength of absorption features. A grating spectroradiometer of 5 nm was also used to obtain spectral data in order to evaluate the effect of spectral resolution.

(2) Data base creation. The extracted absorption features, including wavelengths position, normalized depth, sharpness, symmetry and the basic score

of each absorption according to its importance in the identification of the particular mineral, were loaded into the data base.

(3) Matching between a target mineral and the database minerals. The absorption features of a target mineral are compared with those of the standard minerals in the database by calculating a few types of score values and correlation coefficients. The mineral obtaining the highest score will be chosen as the identified mineral in the target.

(4) Examination of identified minerals. This 'expert system' currently has a 96 per cent rate of success in identifying one mineral based upon a database containing 39 pure minerals.

6.11. THE EXPERIENCE OF THE APPLICATION OF FULL AND PARTIAL PHASE CHEMICAL ANALYSIS IN GEOCHEMICAL RESEARCH

The possibility of the application of phase chemical analysis for exploration purposes has been established.

At the experimental and methodical stages phase analysis can be used to reveal the mineralogical and geochemical nature of trace elements in haloes. On chrysoprase deposits the following mineral forms of nickel are found: silicates with Fe-form oxide sulphides as a water-soluble compound; for copper on malachite deposits, secondary sulphides, native copper carbonates, halogen oxide silicates, phosphates with Fe-form oxide primary sulphides as water-soluble compounds. For these purposes we use solutions of chloric acid, sodium pyrophosphate, NH_4 acetate, and NH_4 oxalate.

The study described elements of the application of selective chemical extractions on metal-bearing minerals and soils to geochemical exploration. Specifically, the study aims to detect anomalous soil in malachite mineralized zones. The weathering products of the mineralization are mainly malachite, amorphous iron oxides and haematite. The soil samples were submitted to an extraction procedure using the following reagents in sequence: NH_4 acetate, hydroxylamine hydrochloride, NH_4 oxalate and finally strong acids.

The determination of different forms of distribution in the halo zone near the deposit is important for the location of the anomaly. In addition, the results of full phase analysis lead to the well-grounded selection of extracts used on a large scale in geochemical research. The use of partial phase analysis, intensifying the useful geochemical 'signals' at the exploration stage, makes it possible to find informative forms of minerals and metalloorganic compounds.

The use of geochemical methods of prospecting for non-metallic deposits leads to the discovery of 'elementary' geochemical anomalies and their interpretation. A substantiated assessment of such anomalies requires isotopic and geochemical evidence. The latter makes it possible to determine the type and to predict the prospects of even unimportant 'elementary' anomalies and mineral occurrences from the isotopic age and isotopic criteria of the sources of ore material.

The isotopic composition of elements is the most typical (invariant) characteristic of the components of fluid systems and so is less liable to the influence of accidental factors. Changes in isotopic composition reflect the qualitatively different conditions of the fluid system participating in tectonic movements of the relevant section of the Earth's crust.

The isotopic characteristics of gases are not only a sensitive indicator of geochemical processes since they also reveal variations in the isotopic composition of carbon in carbon dioxide and hydrocarbon, and also helium and argon [13].

CHAPTER 7

New progressive technologies for mineral exploration

The introduction of deep-seated (more than 300–400 m) mineral deposits into economic production has required new geochemical technologies which involve multi-level investigations, the 'space-air-ground-borehole', combined with drilling, mining, the determination of ores and petrophysical investigations.

The combination of various methods which are different in nature, selecting the most informative methods for the given region and amount of work, makes it possible to minimise the amount of investigations carried out and to gain a better understanding of the target from the geological point of view. In some case this permits a reconsideration of the concepts, prospects and amount of work being planned. The capabilities of hydrolithochemistry and the complex of airborne geophysics have been seen in a new light as regards the assessment of the prospects for minerals in considerably large areas. This has permitted regional and local ore-controlling zones to be detected and follow-up ground surveys to be carried out within promising structures with various concentrations of ore.

The proposed technology features the employment of three-dimensional seismic data, modifications of geoelectrochemical methods (MPF, CHIM, KSPK), the mise-a-la-masse method which measures the magnetic field, TEM and the natural electromagnetic field, acoustic and radiowave methods, a combination of geophysical investigations in boreholes, including borehole magnetic data, density and acoustic logging and determinations.

Due to the development of new economically beneficial and ecologically safe production technologies (including hydraulic methods of mining) the role of geophysics, which provides information on the state of worked out areas, physico-mechanical and other properties of rocks and ores, and stress in rocks within massifs, becomes promising for the geometrical determination of ore targets, and the geophysical determination of ores. The introduction of geologica–geophysical technologies into mining provides a sharp reduction in the proportion of heavy mining operations and the number of holes on a grid in core drilling.

7.1. GEOLOGICAL MAPPING USING AIRBORNE GEOPHYSICS — NEW FRONTIERS

Airborne geophysics, principally airborne magnetics and radiometrics, is becoming far more widely used in mineral exploration and resource assessment as a tool for detailed geological mapping, relative to its traditional application of target detection and regional studies [95].

Image processing has played a dominant role in widening the areas of application of airborne geophysics, although advances in the acquisition, processing and navigation technologies have all had some impact. The ease with which the eye can resolve both fine and coarse information from an appropriately stretched image reduces both the level of skill and effort required to interpret the data. Image processing is now used to measure the effectiveness of developments in acquisition, processing and navigation.

New frontiers in airborne geophysics which, when developed, will give the technology even wider application are:

(1) Data and image processing. Filtering and levelling techniques prior to image processing are especially useful in isolating and enhancing fine structure in the data. Significant geological features often have only subtle expression, and quite specific processing is required to allow the interpreter to interrogate the data for such features. Good communication between the geophysicists processing the data and the geoscientists interpreting the data is important.

(2) Data acquisition. Lowering the interference levels in the data, both in geophysical measurements and aircraft position, increases the area of application of airborne geophysics.

(3) Data integration and digital data bases. The image processor, through judicious use of graphics and raster planes, can superimpose sets of data which are conventionally overlain on a light table. Superimposing Landsat TNI data with aeromagnetics is a powerful display tool for structural analysis; locating geochemical and potassium radiometric responses in the vicinity of strike faults defined by magnetics is important in the exploration of deposits. Display and data base technologies have evolved to the point where the integration of digital data bases can be done on a routine basis. Recent advances in airborne electromagnetics, including the development of digital systems with large bandwidth, offer the potential for a new airborne mapping technique that resolves silicate lithologies through differential weathering in the regolith.

7.2. GEOCHEMICAL SURVEYS

Helicopter-supported regional surveys of areas are undertaken by means of stream-sediment geochemistry using a sampling density of one sample per square kilometre. In South Africa twenty elements are quantitatively determined for each sample. These surveys are initially carried out in selected areas with a higher resource potential, but this programme will eventually be extended to embrace the whole country. To date, about one tenth of the country has been covered. Together with the routine analysis of chip samples from water boreholes drilled country-wide by the Department of Water Affairs, these analytical result constitute an extensive geochemical data base. The interpretation of these geochemical results forms the basis for a variety of detailed studies aimed at identifying possible exploration targets.

Research aimed at the development of geochemical exploration techniques, specifically adapted to South African conditions, constitutes an integral part of the activities of this section.

7.3. IMAGING SPECTROMETRY AND RADAR TECHNIQUES

Imaging spectrometry for Earth observations has been developed over the last decade in order to provide more definitive and quantitative information about the surface of the Earth from remote sensing. Space-based multispectral scanners, the USA's Landsat and France's SPOT, have shown many of the advantages of the perspective from space. National Aeronautics is designed to acquire 24 km wide, 30 m pixel images in 192 spectral bands simultaneously in the 0.4–2.45 μm wavelength region. By using pointing mirrors, it will be able to sample any place on Earth, except the poles, every 2 days [35].

Spectral features for minerals are associated with an electronic transition in transition elements. Vibrational features, generally overtone bending-stretching vibrations, are exhibited by mineral-bearing $Al-OH$, $Mg-OH$, W_2 and H_2O constituents. These minerals have features in some cases as small as 10 nm full-width, one-half-maximum values that require 5 nm sampling for a proper description. The majority of mineral features can be described completely using 10 nm sampling. A significant degradation in identification, particularly in magnesium-bearing minerals, as well as in limestone and dolomite, is seen when the sampling interval in increased to 20 nm.

Syntitic Aperture radar (SAR) data have been used increasingly in support of geological mapping and in studies related to the investigation of geochemical exploration. The radar data were flown in support of (1) geological mapping, and (2) structural geological investigations searching for mineral deposits. The results of the airborne SAR campaign are encouraging for a number of multidisciplinary geological programmes and have led to the increased use of radar imagery [106].

In areas with a mix of exposed soil and rocks, combinations of the Landsat Thematic Mapper (TM) bands and the stretched and filtered C-band radar image were useful.

Radar images also provide valuable information on landforms and the roughness of the terrain.

SAR, TM and other geological data have been integrated in multidisciplinary geological mapping programmes.

Other techniques combining SAR and geophysical data have been used for structural interpretation in support of exploration for non-metallic minerals exploration. A combination of SAR, TM and magnetic vertical gradient data created an enhanced image that was used to produce lineament density maps and in support of diamond deposits.

Several techniques have been developed that may result in automated landform mapping. A new technology (sensor capabilities, 35 and 90 GHz systems and lightweight electronics) may be used for new applications in the exploration of deposits [106].

7.4. ATMOSPHERIC GEOCHEMICAL MEASUREMENT

The atmospheric layer close to the Earth's surface carries extensive geochemical information relating to the composition of the underlying terrain. During active mixing, there is an upward flux of both gaseous and particulate material from the surface. Gaseous forms of atmospheric geochemical interest include mercury vapour, halogen vapours, sulphur compounds and radon. These gases diffuse rapidly and need to be measured very close to the surface to be of value. In general, experiments into the atmospheric geochemical measurement of trace gases have been mainly confined to mercury and radon, and have not led to the development of techniques that have been widely applied. This is due to the inherent problems relating to sensitivity requirements and the effects of rapid dilution.

Atmospheric particles can be of both inorganic and organic orgin, much of the latter arising from vegetation. It has been established both in the laboratory and in the field that there can be movement of elements through vegetation to leaf surfaces, followed by the dispersion of these elements into the atmosphere as particulate material. Particulates derived from vegetation are, therefore related geochemically to the composition of the underlying soils. In practice, material arising from both vegetation and residual soil surfaces can be utilized for geo-chemical exploration purposes.

In the case of atmospheric particulates, comparatively large fluxes of material rise into the atmosphere when mixing conditions are good, and with appropriate instrumentation it has been shown that it is feasible to carry out atmospheric geochemical surveys using this material. If it is desired, however, to collect particulate material that relates closely to the underlying terrain, it is important to separate the material carried in parcels of rapidly rising air from particulates associated with neutral or sinking conditions. It is also advantageous to use coarse particulate fractions in size ranges larger than 30 microns to minimize the effects of lateral migration and the re-entrainment of particles that have been previously translocated by wind [8].

Rapid and precise methods of analysing various types of both organic and in-organic particulates in the atmosphere have been developed (techniques utilizing aircraft as sampling vehicles). The methods provide a broad scanning technique of a terrain and under favourable circumstances may indicate the location of mineral deposits. Several types of airborne equipment have been developed for carrying out systematic atmospheric geochemical surveys. The spatial resolution of these systems varies according to design between 100 m and several kilometres. Ana-lytical methods employed have included conventional emission spectroscopy, laser vaporization coupled with emission spectroscopy, X-ray fluorescence spectroscopy, and fission trace-etch counting. The spectroscopic measurements provide analyses for 20 or more elements, including all the base metals, while fission track-etch methods give exceptionally high sensivity and specificity for uranium alone, and are intensitive to radon and bismuth 214.

Atmospheric airborne geochemical prospecting appears to offer important potential as a complementary tool to a variety of airborne geophysical methods [8]. In the case of phosphorite exploration, it can provide information that is unaffected by surface disequilibrium effects and the fine-size aerosol interference that can considerably modify airborne γ-ray spectrometer results.

7.5. GAS SURVEYING APPLICATION IN PROSPECTING FOR NON-METALLIC MINERALS

Natural gases of differing composition and origin are widely distributed within the Earth. They are either syngenetic with the ore process, represent a product of subsurface chemical reaction, or reflect the constant and general process of the degassing of the Earth.

The extraordinarily wide range of gas distribution makes it possible to use them for a variety of different geochemical tasks. However, the importance of further research into mobile gases of deep origin must be stressed. The migration of elements in the gaseous phase is a widely distributed phenomenon. The phase itself constitutes an important component in the formation of endogenous deposits, and also remains active in the post-ore formation period.

In the supergene conditions of the parts of ore bodies immediately below the surface, gaseous components connected with the oxidation of ores may also be formed. Thus, gases of different origin and composition accompany deposits during the various stages of their evolution. Such gases are either preserved in the gaseous and gaseous–liquid micro-inclusions of rocks and ores, or migrate along joints related to ore-controlling fault systems. Endogenous and exogenous gas aureoles around non-metallic deposits can therefore be used as highly sensitive geochemical indicators, and in structural mapping they can be used to classify fracture systems of different ages, and also to investigate the origin and depth of bedding. The same methods may also be employed to outline those parts of an ore body which have the highest economic potential. It is also possible to determine the nature and trend of superimposed processes of ore formation. Ores containing barite are characterized by mercury vapour, and phosphorite and apatite ores, by radon.

Gases generated at depth are mobile and closely related to systems of deep tectonic faulting. In the variety of their forms, aureole intensity and abundance they greatly predominate over gases of subsurface origin. Among such gases carbon dioxide, hydrogen, helium, methane, nitrogen and mercury are the most typical, and they are quantitatively proportioned by their abundance in the Earth's crust.

A variety of sampling methods are employed when mobile gases are being investigated. Boreholes and probes can be operated within the limits of the aeration zone (i.e. boring to 0.5–1.5 m and probes down to 5–10 m), and springs, wells and boreholes may be used to provide access to the zone of saturation. In the first case, carbon dioxide, hydrogen, methane, mercury, radon and, rarely, helium are sampled for analysis, whereas in the second case, helium and, less usually, radon or methane are employed.

Gas surveys make it possible to recognize fracture systems with deep tectonic penetration. The results of such surveys can be used to locate endogenous mineralization.

These aspects show that geochemical studies of natural gases constitute an important branch of geological science.

The method of gas surveying is now effectively used for detecting non-metallic deposits and revealing and tracing fracture structures in a number of regions with different geological, geochemical and environmental conditions. Geological work has ascertained that the method of gas surveying can be effectively used in geochemical exploration and prospecting for graphite and apatite and solving a number of other geological problems [2, 13, 68].

The composition of natural gas, characteristic of large geological structures which differ in their tectonics and the history of their geological development, changes in places where the invasion of veins and hydrothermal solutions and the formation of endogenous deposits has taken place. Gases genetically connected with the formation and evolution of non-metallic deposits are superposed on the regional gas background here.

Work which aimed to apply the method of gas surveying in prospecting for graphite, apatite, boron and other deposits provided numerous facts showing the prospects for its application in revealing and tracing zones of breaks in continuity and in detecting blind bodies overlapped by a thick cover of sedimentary rocks.

This is determined by the fact that the perceptible migration of natural gas takes place along fissure zones and zones of higher jointing, and this leads to the formation of intensive gas haloes in near-surface sediments.

Carbon dioxide, hydrogen, methane, mercury, helium and heavy hydrocarbon gases were used as gas indicator components. These components are confined to fissure zones and also connected with the thermal influence of intrusive bodies on the organic material of the enclosing rocks. Two types of zones of higher rock permeability were fixed by abnormal concentrations of gas indicator components in the underground air: linear and areal.

Zones of the linear type correspond to permeable breaks in continuity, displaced under loose sedimentary rocks. Zones of higher rock permeability of the areal type are due to the presence of areas of jointed rocks, above large intrusive bodies.

The revelation, by the method of gas surveying, of zones of breaks in continuity and higher rock permeability of the areal type developed over large intrusive bodies with which apatite mineralization may be connected, makes it possible to choose effectively areas for detailed exploration and to increase the effectiveness of prospecting.

The investigations produced the conclusion that deep fractured zones are characterized mainly by contrasting anomalies of hydrogen, methane and mercury vapours, increasing the background by 5–10 times and more, and less contrasting anomalies of radon, carbon dioxide and other rarer gases. The extent of anomalies, depending on the thickness of zones, varies from 0.5–1.5 km. The anomalous gas spectrum above different deep fracture zones remains different. The most contrasting and complex gas anomalies develop above the zones at the joints of different tectonic blocks. Complexes of various ages are characterized by gas fields of free and sorbed gases, different in intensity and structure [29].

A natural field of mobile helium is connected with durable deep fractures and reflects the permeability of the Earth's crust. The correspondence of endogenous ore deposition and deep fractures enables the data obtained by the helium survey to be used to locate areas favourable for exploration for non-metallic deposits. The practical significance of the helium method lies in deep fracture mapping and forecasting the most suitable sites for ore deposits. The use of helium surveying is also advisable in combination with the methods of abyssal geophysics.

The He-geochemical criteria for phosphates, uranium and other deposits have been developed. Helium leak detectors and simple sampling equipment is used to measure He in surface and near-surface environment. A novel method of sampling the He content of organic lake bottom sediments has been developed [9, 24]. A ping pong ball is immersed in freshly collected lake sediment contained in a sealable glass jar. When the air in the ball has equilibrated with the gas in the sediment the He content of the ball is measured.

The bottle has proven effective and inexpensive for the collection and storage of water samples for He analyses. He is useful as a tracer for deep tectonic features in inhabited areas; the He trend coincides with U, Rn and CH_4. The gas phase in combination with lithogeochemical criteria may serve as an additional physico-chemical indicator of the depth of formation and the difference between productive and non-productive ore formations, including the estimation of kimberlite diathermancy for typomorphic metallization.

7.6. APPLICATION OF RESISTIVITY TOMOGRAPHY TO MINERAL EXPLORATION

Resistivity tomography, one the geotomographical techniques, has been developed intensively from the late 1980s to 1992. The Technical Development Group of the Mining Agency of Japan has been conducting research into the application of resistivity tomography to mineral exploration. The research consists of the following stages [42]:

(1) The data acquisition system of resistivity tomography was developed on the basis of the DC electrical prospecting method in order to acquire substantial good quality potential data in a reasonable time. The system consists of the Mc-OHM-21 main manufactured by Oyo Corp., which can be programmed to measure simultaneously three potential as well as injected current wave forms, and a geoelectric scanner, which switches the electrodes both on the surface and in each borehole array.

(2) By using the above system, resistivity tomography data was obtained in a test field in the mining area. The adjacent alternation zones remained in the test field.

(3) Field data were analysed using the alphacenters inversion techniques, which involves two-dimensional auto-analysis. The analysed section was marked off using the boreholes and the surface. The reconstructed resistivity image was compared with a detailed geological section based on several exploration drillings and resistivity sections created from conventional ground electrical prospecting.

The method of resistivity tomography provides a more accurate description of a geological section than that produced using conventional methods of ground survey.

7.7. THE METHOD OF THE APPLICATION OF ARTIFICIAL SORBENTS

The method is based on the application of artificial sorbents and successfully combines the possibilities of hydrogeochemical and sorption methods of prospecting. It makes it possible to fix the mobile form of many sought elements, and to estimate it more confidently when using ion-exchange resin kationites and anionites.

In order to accumulate a significant quantity of migrating elements and to obtain them in the form of substances with a definite chemical composition, synthetic ion exchange resins as well as other sorbents were used as their concentrators. Batches of them were buried in soil about 1 m deep 50–100 m apart, in the region of ore prospecting and kept there for a longer period of time [8].

The positive characteristics of the method are also the possibilities of applying a greater number of elements, the analysis of which, without preliminary enrichment, proves to be complicated and labour-consuming in the process of ordinary sampling. The method proved to be effective enough under all testing conditions. In a dry climate there appeared to be enough moisture in the soil at a depth 1 m to fix the aureoles of a vein body lying at a depth of 3–10 m. The concentration of some metals on the sorbent (Cu, Zn, etc.) is several times higher than the background. High concentrations of some metals (0, n per cent) were obtained on the sorbents submerged in springs and mountain streams in a region with F-deposits. After being buried for some months in the soil (peat-bog soil and gley quartz sands) under the conditions of a swamped fluvioglacial and alluvial plain, the sorbents quite clearly revealed the secondary aureole of the ore body hidden at a depth of 17–30 m. The method may be used in prospecting on the flanks of known deposits as well as in testing geochemical anomalies discovered by means of other prospecting methods.

Recently, a number of scientific and research institutes have analysed the emissions of some elements from geological structures into the soil or atmospheric air respectively. The elements in a molecular form can penetrate through even a very thick overburden, or through sea water. The method of the molecular form of elements can be applied even in localities where traditional geochemical methods cannot be used (villages, large settlements, etc.). The concentration of the emitted elements is able to reveal hidden mineral deposits and buried geological fault structures.

The relics of laterites are widespread in exploration areas. The pisolitic and stony varieties enriched with iron and manganese oxides are good sorbents for Ni, Cu, Zn and Co. Such laterites can also be a subject for prospecting.

7.8. FLUID INCLUSIONS AND AREALS OF FLUID INCLUSIONS IN GEOCHEMICAL PROSPECTING

Because fluid inclusions represent trapped portions of the fluids responsible for hydrothermal mineralization, they have enormous potential as a tool in mineral exploration. Furthermore, hydrothermal activity may result in the development of fluid inclusion aureoles that are several orders of magnitude larger than the area of mineralization [2, 82]. Although this potential has been appreciated for several decades, and some promising results have been obtained (e.g. see papers [11, 22, 82]), attempts to use fluid inclusions as a direct geochemical sampling medium for mineral exploration have been neglected.

In fact, in many mineral deposits, a direct correlation has been observed between the abundance of fluid inclusions (as deduced by simple counting methods or decrepitometry) and the areas of mineralization [40]. The analysis of individual fluid inclusions by electron and laser microprobe techniques, and the observation of productive solid phases (both daughter and captive phases) in some fluid inclusions, have shown that certain hydrothermal fluids were highly enriched with the ore components.

Inductively coupled plasma atomic emission spectrometry (ICPAES) is capable of the routine, semi-quantitative and qualitative analysis of fluid inclusions because of its rapidity, sensitivity and multi-element capabilities. We have already established that the ICPAES-method, in combination with decrepitometry, can provide important information regarding the bulk fluid inclusion chemistry of a sample, and is particularly good for the detection of various trace elements — subject to certain limitations.

The dimensions of fluid inclusions vary within wide limits, from macro- and microcavities to single defect-vacancies in the crystal lattice. The contents of inclusions of ultramicroscopic size, equivalent to vacancies in the lattice, can differ considerably from the composition of the microinclusions and, consequently, from the primary mother fluid. This effect takes place because of the differential capture of the medium elements by electrostatic power. The method of vacuum gross analysis shows the constant presence of hydrogen in the gas mixture extracted from minerals; separately analysed inclusions from the same minerals do not contain hydrogen. It is shown that the gas mixture from inclusions of usual size is not enriched with hydrogen. It is concentrated because of the isolation of hydrogen from the intergranular and structural pores in the metal of the grating device and the ultramicrocavities in the mineral. All minerals, including minerals of surface origin, contain hydrogen of this type.

Decrepitation anomalies 50–60 times wider than the thickness of the ore body were observed in the neighbourhood of quartz veins. The anomaly consists of a decreptominimum that occurs in the selvage — near the zone of the substance transition and the decreptomaximum. An anomaly intensity is of great significance for prospecting while its structural peculiarities can be used for estimation.

Most endogenous non-metallic deposits are accompanied by wide areals of fluid inclusions in the minerals of the country rocks. They are genetically connected with the hydrothermal process and contain important information about its physical and chemical characteristics. Their investigation is very useful for geochemical prospecting. Many express methods have recently been proposed for investigating the different properties of fluid inclusions. The data obtained by these methods after statistical manipulation along with the methods of geological mapping give important information about the spatial anisotropism of hydrothermal activity. The comparison of the relative differences in actual regions makes it possible to discover the typomorphic peculiarities connected with different newly-forming minerals and to distinguish the most promising sections for different non-metallic types.

The amalgamation of different data on fluid inclusions permits us to judge the intensity and main temperature of the hydrothermal characteristics of mineral-forming solutions. The acquisition of this information and its utilization, together with traditional special geological, geophysical and metallometric maps, opens up new possibilities for the improvement of geological prospecting. Because fluid inclusions represent trapped portions of the fluids responsible for hydrothermal mineralization, they have enormous potential as a tool in mineral exploration. Furthermore, hydrothermal activity may result in the development of fluid inclusion aureoles that are several orders of magnitude large than the area of mineralization [22]. Although this potential has been appreciated for several decades, and some promising results have been obtained, attempts to use fluid inclusions as a direct geochemical sampling medium for mineral exploration have been neglected. The results of a study of the chemical characteristics of fluid inclusions in quartz suggest that certain major and minor elements in the inclusion fluids could be used successfully in exploration for mica and rock crystal [82].

7.9. GEOELECTROCHEMICAL METHODS OF PROSPECTING

The need to solve geological problems associated with prospecting results in the application of new methods. In the last decade new technical means and methods of prospecting and evaluating mineral deposits based on the behaviour, transformation and distribution in ore bodies and host rocks of various forms of occurrence of elements termed geoelectrochemical have been developed.

The first group of methods, based on the extraction (including electrical excitation) and analysis of elements in easily mobile forms of occurrence, within which they are able to migrate for large distances, makes it possible to reveal and trace deep-seated and overlapped ore objects. This type of method includes: methods of prospecting based on metalloorganic compounds (MPF), the thermomagnetic geochemical method (TMGM), and the method of diffuse extraction of metals under the action of electrical current (CHIM). These methods may be used at all the stages of geological prospecting work, beginning with regional geologico-geophysical investigations of vast, poorly studied terrains. It is advisable to use MPF and TMGM as the more efficient methods at the early stages of geological surveying work at a scale of 1 : 50 000, together with general prospecting. The CHIM method, as a more tiresome but more informative method, is to be used at the stages of prospecting and evaluating work (1 : 10 000–1 : 50 000). The application of these methods is more efficient during prospecting for deposits in closed regions with a thickness of overlapping deposits of up to 200 m and with a depth of ore bedding of up to 800 m.

The application of a complex of geoelectrochemical methods makes it possible to accelerate the study of territories and to increase the geological and economic efficiency of geological surveying and detailed prospecting and evaluating work. The experiment shows that prospecting work can be carried out more than twice as quickly, while the bulk of drilling in separate areas can be reduced by 40–50 per cent using new methods [2, 13, 40]. The distant jet migration of chemical elements in the crust can be used as a basis for regional exploration for deep-lying mineral deposits.

A new type of dispersion haloes of chemical elements from vein bodies and other geological targets has been revealed. These haloes take the form of jets stretching from the sources of the elements to the surface. They are grouped in zones of hundreds of metre in size. The jets feature an increased or decreased content of individual elements, as well as various physical properties of the medium (specific electrical resistivity, magnetic susceptibility, etc.) that differ from those of the host rocks within the jets. Chemical elements move from depths of hundreds of metre and a few kilometre at great speeds. Some of the elements cross the earth-air interface and enter the atmosphere.

The discovery of these haloes has become a basis for the development of new geoelectrochemical methods of regional exploration (MPF, TMGM, CHIM and others) capable of detecting from the surface deep mineral deposits, including those under clay and other overburden tens and hundreds of metre thick.

The geoelectrochemical methods of regional and detailed exploration have been developed for the detection from the surface of deep-seated (hundred of metre, a few kilometre) mineral deposits and the distant characteristic determination of the content and reserves of useful components. The methods are based on the application of natural and man-made induced electrochemical processes in the earth.

The discovery of the distant jet-line migration phenomenon of chemical elements has allowed the MPF, TMGM, CHIM and some other methods to be developed to detect deep-lying targets by selectively recording the elements in mobile forms of occurrence. Humate–fulvate (MPF), ferri–manganese (TMGM), ionic (CHIM) and other forms of occurrence of the elements are used.

The active extraction of electrochemical reactions on the surface of minerals and the selective recording of the processes in a potential curve form made it possible to develop the KSPK and BSPK methods. Parameters determined from the polarization curve-potentials and limiting current strength of the electrochemical reactions are used for the distant determination of minerals and their sizes within ore bodies.

7.10. APPLICATIONS TO MINERAL PROSPECTING

Radiational mineralogy

Radiation mineralogy is a branch of mineralogy that deals with the problem of the formation and change of minerals affected by natural and artificial radiation. The research includes [61]:

- a) studies of peculiarities of minerals that surround radioactive minerals;
- b) the effect of radiation (ionizing, reactor) on minerals;
- c) a study of radiation defects in minerals by the methods of electron-para-magnetic, nuclear-magnetic, and nuclear-quadrupole-resonance and infrared spectroscopy.

A study of the changes in the physical properties of crystals in the field of ioniz-ing radiation and radiation defects, which can serve as indicators of the migration of radioactive elements in the Earth's crust, as well as revealing phases resistant to radiation of high intensity, are important aspects in radiation mineralogy. The cause of the variation in the degree of alteration of minerals affected by ionizing radiation is closely related to the problem of the metamict state of certain mineral phases.

The internal structure of natural minerals is distinctly revealed by γ-irradiation, disclosing the history of the growth and development of individual grains. As it follows from the patterns that are produced, an inhomogeneous internal structure is established in natural crystals. The modifications in the internal structure (zonality, sectorial and mosaic structure, traces of dissolution-regeneration) correlate with the content of structural impurities and their mode of occurrence. The patterns of originally homogeneous and transparent minerals as revealed by γ-irradiation are the reflections of these properties. Irradiation and etching supplement each other and allow the visualization of different elements of the real structure of minerals, and can decipher changes in physico-chemical parameters during the growth of minerals. A zonal and sectorial distribution of colour was traced in numerous crystals of quartz from hydrothermal veins in the Urals, Aldan and Pamir. Zonality, which is most distinctly manifested in the smoky-citrine varieties, is the basic characteristic feature of these crystals.

The difference in colour and its rhythmical zonality reflect the change in the composition of mineral-forming solutions with time. With a gradual change of the smoky and citrine colour, the crystals are characterized by high quality. Besides quartz, the zonal-sectorial structure was established in diamond, beryl, spinels and calcite. A study of mineral zonality allows one to judge the conditions of change in the state of mineral systems. Dissolution and regeneration zones are established in absolutely homogeneous transparent minerals after irradiation. From two to seven zones are distinguished in such minerals. Some of the dissolved crystals that have been regenerated are coloured more intensely than the layers that have grown simultaneously.

The factors that determine dissolution regeneration are as diverse as changes in thermodynamic parameters (PT conditions), the degree of oversaturation of mineral-forming solutions and tectonic movements. The geochemical heterogeneity of minerals as established by their colour differences should be taken into account when typomorphic impurities in crystals are defined, as well as when doing age determinations. Non-uniform sectorial occurrence of impurities and variations in the crystal-lattice parameters connected with impurities lead to an incoherence of crystals with each other. The incoherence creates stresses producing fissures round them, and evokes considerable changes in the composition and lattice parameters, which become visible after irradiation, but cannot be defined by chemical analyses.

Minerals subjected to irradiation preserve a 'memory' of changes due to the ionization processes. 'Memory' acquired during mineral irradiation depends on the absorbed dose and type of radiation.

Thermal erasure of paramagnetic damage occurs in minerals following superposed annealing. The activity from radiation sources under natural conditions changes the properties of the surrounding minerals. For example, radiation–chemical reactions of iron oxidation are reflected in the iron-containing silicates. Spectroscopic peculiarities of the irradiated silicates and the anomalies in their properties are signs of the presence of radioactive zoning. The anatomy of each mineral should be considered not only as a process of growth but also as a process of subsequent transformation.

Typomorphic characteristics of minerals (colour) are used for the genetic correlation of rocks, and the study of the history of formation [61]. The smoky colour density in quartz decreases in rocks of varying origin in the following order: magmatic–greisen–pegmatite–hydrothermal. Ionizing radiation together with a study of the physical properties of the minerals, helps to determine their relative temperature of formation. Since the anomalous pleochroism is not observed in the smoky crystal of quartz grown above 400 °C, it is recommended that this effect should be used to determine the upper-temperature crystallization boundary of natural quartz. An anomalous pleochroism can be most precisely estimated by measuring the intensities in the groups of the EPR spectrum lines and the ratios between maximum (I_{max}) and minimum (I_{min}) intensities.

It has been established that if the crystallization temperature is increased, a change in the ratio of OH(Al−Li) concentrations and OH(Al) defects occurs, which is estimated from the ratio of the areas of the named absorption bands of infrared spectra. An increase in temperature of minerogenesis results in a decrease of the ratios of the areas of OH(Al−Li) bands to that of OH(Al) bands.

Changes in the acidity–alkalinity of solutions during crystal growth can be judged from the concentration of hydrogen defects in them. The maximum concentration of hydrogen is established in citrine and radiation-resistant crystals of quartz, which proves that the crystals have been formed in less alkaline systems than the smoky ones were. An objective criterion is suggested to determine the acid-alkali conditions of quartz formation. The criterion is based on the study of the absorption bands of infrared spectra when determining the ratio between the areas of OH(Si) bands and the total area of the bands of OH defects as well as the ratio between OH(Al−Li) and OH(Al) defects. These parameters decrease with the increase in alkalinity of the minerogenetic environment.

The method of γ-irradiation, which permits fixation of the radiation colour character, helps to determine the simultaneity of spatially separated formations and thus to judge the relative synchronism of minerals. Uniform zonality, and correct alternation of the smoky and citrine zones, may be interpreted as indicative of their simultaneous crystallization. A study of the γ-irradiated quartz in a complex by infrared spectroscopy makes it possible to estimate relative rates of crystal growth, the ratio of the areas of the absorption bands $OH_{dif.} : OH_{tot}$ being used for this purpose. High growth rates promote the formation of individuals of which the transmission infrared spectra contain a wide diffuse absorption band in the range of 3400 cm^{-1} and more intensive absorption bands at 3588 cm^{-1}, 3430 cm^{-1} and 3300 cm^{-1}.

There are various possibilities for using the methods of radiation mineralogy when searching for mineral deposits:

a) the detection of latent ore bodies by colour zonality;

b) the determination of accumulations of radioactive ores from the defects in surrounding minerals; and

c) the sorting of the veins by the ore content.

When irradiated, the non-ore-bearing quartz becomes uniformly smoky. A non-uniform, variegated, jet-like, usually smoky-citrine colour is observed in non-productive varieties. Gold-bearing quartz veins are characterized by a high intensity of aluminium centres in connection with quartz formation at high alkaline activity (sodium) in mineral-forming solutions. The dependence of the colour of fluorite on its radioactivity has been recorded. The black-violet colour of fluorites that metasomatically develop in carbonate rocks is of particular significance for the search of bertrandite–phenacite deposits. The density of colour in the minerals decreases as the mineral is removed from the radioactive mineralization. A study of the defects in the mineral phases allows contouring of the deposits and finding the radiation haloes. The presence of radiation-oxidized ions of Fe^{3+} in biotite, hornblende, riebeckite and chlorite is an indicator of their variations when interacting with radiation of the ore zones and serves as a search criterion for uranium mineralization.

The economic aspects of a deposit are to a considerable extent determined by the depth of its weathering zone. Ionizing radiation and infrared spectroscopy help to obtain rather objective information on the level of weathering in the ore-bodies. Hydrogen is more mobile than the alkali elements; acid components and lithium as high-volatiles are accumulated in the upper horizons of the deposits. The level of the weathering of ore-bodies can be determined from the formula

$$h = \frac{a - b}{g} \times 100,$$

where h is the value of the depth of weathering, a and b are the maximum and minimum contents of OH(Al–Li), and O–Al/Li defects and g is the geochemical gradient, i.e. the change of infrared spectra absorption parameters of quartz at a depth of 100 m. When the gradient of change in the mineral absorption-band areas for 100 m depth interval of the ore zone for a specific section has been calculated, one can determine vertical intervals with the analyzed section belonging to them. A high concentration of hydrogen in the upper zones promotes the formation of radiation-resistant and optical crystals [61]. The defectless zones of the crystals correspond to the period of favourable growth when thermodynamic and physico-chemical parameters were stable. Such a parameter of quality, e.g. the uniformity of properties (uniform colour in the whole volume) or the growth-stability index, is also of importance. The crystals of Iceland spar should be γ-irradiated before sawing and processing, allowing the zones with high concentrations of defects to be found. The irradiation is used to modify the structure of mineral phases, and this allows minerals with improved characteristics — high strength stability, radiation resistance — to be attained. Irradiation changes the physico-chemical properties of minerals, and both the concentration of electron vacancies and energy heterogeneities increase. Thus, the radiation considerably affects the mineral concentration. The irradiation of minerals and pulps permits raising the selection quality of similar minerals by their flotation properties (pyrochlore–zircon, apatite–calcite, dolomite–calcite) and the intensification of the flotation process, improving the separation of heterogeneous ores. The flotation of copper–lead ores after preliminary irradiation results in an increase between 4 per cent and 10 per cent in the extraction of lead and copper. The radiation colour in quartz after irradiation can be used for the classification of quartz, in which the amount of structural impurities varies, to prevent the mixing of raw materials that are different in uniformity and colour.

Infrared spectrophotometry has traditionally been a technique for the chemist, since it is sensitive to molecular bonding interactions. It has also found much application in the qualitative identification of mineral phases. The relatively recent development of Fourier-transform instruments, and of mathematical methods for the treatment of infrared spectral data, has resulted in the development of a powerful technique for the quantitative analysis of minerals (quartz, zircon).

The dynamics of hornblende oxidation and dehydroxylation during the process of thermal and radiation treatment in air and under vacuum have been studied by the methods of NGR, PMR and IR-spectroscopy. It is shown that dehydroxylation proceeds as a process parallel to oxidation. The effects of the thermal and radiation influence expressed much more feebly under vacuum than in air. Radiation oxidation causes strong distortions of the coordination polyhedra of Fe^{3+} ions, and a change in the optical and magnetic properties. These effects are of a reversible character: annealing leads to the reduction of the radiationally-oxided Fe^{3+} ions to Fe^{2+}, to magnetic susceptibility and the reduction of the refractive index to initial values. It is shown that radiation treatment has a catalytic effect upon the processes of the thermal decomposition of minerals.

Thus the use of the methods of radiation mineralogy in the analysis of geological phenomena indicates their great potential in the solving of scientific and practical problems.

— An all-round study of the inhomogeneity of mineral phases, zone-sectorial structure, anomalous pleochroism, phenomena of dissolution-regeneration, and fixed irradiation, define a possibility of obtaining new data on the conditions of minerogenesis.

— A more precise identification of minerals is performed.

— Radiation defects that appear in minerals under irradiation, reveal not only the formation conditions but also the existence of crystals, including the superposed processes (annealing, diffusion and neutralization of the colouration centres).

— Palaeotemperature conditions of formation of the ore can be reconstructed in the radiation centres in minerals, accounting for the effects of thermal destruction ('rubbing out') of paramagnetic centres (blind intrusions, deep fractures, volcanic orifices).

— The vertical evolution of hydrothermal solutions is established from the results of mineral irradiation and colour generation. It is connected with a decrease in temperature and local pulsation determined by tectonic factors. The length of the unopened or annihilated part of the ore zone, the level of intrusion of the vein bodies, can be determined in such a way.

— It is established that the minerals, which concentrated different defects in the field of radioactive irradiation, are the indicators of migration and redistribution of radioactive elements in the concrete zones of geological space.

The number of minerals used in industry can be increased by the applications of irradiation.

In summary the advantage of the method of γ-irradiation is that actual information on mineral composition can be obtained. Locality, sensitivity, amount of information, and documentality are the important features of this method.

Raman spectroscopy

The development in recent years of Raman microprobe techniques must probably be viewed as a major advance in microanalysis. With the successful application of Raman spectroscopy to the analysis of microscopic samples, it has become possible to obtain direct molecular information from samples which hitherto have only been subjected to a determination of the elemental composition.

Raman spectroscopy is ideally suited as a highly specific and sensitive technique of investigating molecular vibrational phenomena in all phases of minerals. Its uses for the determination of molecular structure and its broad range of applications to chemical analysis have been described [41]. Applied to microanalysis, the technique becomes a powerful tool for the characterization and identification of the principal molecular constituents.

Thus, the vibrational Raman spectra observed from microsamples serve as a unique fingerprint of the chemical species in such samples. In addition, these spectra often contain information on the local molecular environment and the structural coordination of the crystalline phase in which the species resides. Specific examples are presented from the study of: (1) fluid inclusions in minerals, and (2) trace organic pollutants. The spectroscopic measurements performed with Raman microprobes are based upon the excitation and detection of the normal or spontaneous Raman effect. The effect is an inelastic scattering process which, in its simplest form, involves the interaction of a monochromatic (laser) beam of visible light with the molecules of the sample. This interaction results in the appearance of scattered light at altered frequencies. These are the frequency shifts seen in the Raman spectrum of the sample. These frequency shifts are identified with the frequencies of molecular vibrations in the sample.

Various factors determine the success of a Raman microprobe measurement and the amount of useful analytical information obtained from the sample.

Track-analyses

Methods have been developed for estimating radioactive elements such as uranium, thorium, and radon by fission and alpha-track analyses. A novel in-field method for the determination of radon by alpha-sensitive dielectric film under the trade name 'Track Etch' has been developed by the Terradex Corporation (California). This method utilizes cups that are placed in a shallow hole in the ground and left for a period of 3 weeks or more; they are then recovered and the film returned to a central laboratory for analysis.

Another novel radon gas detector developed by Alpha Nuclear of Caledon East, Ontario utilizes a solid state detector (silicon diffused junction) coupled to an electronic integrating readout metering device. This reusable instrument is placed in a hole in the ground and left for one or two days, after which the radon concentration and the number of hours of integration can be read from a visual numeric display. The distribution of uranium in thin sections from a mineralized fault zone was investigated by the fission-track method. The fission-track method is a quick simple and inexpensive way to determine the location and abundance of uranium, and, in some cases, thorium, in uncovered thin sections. The method can be used to determine accurate concentrations ranging from several ppb up to several per cent. The accuracy for the determination of thorium is quite poor when $Th/U < 1$, but improves to about 25 per cent when $Th/U > 3$.

The uranium and thorium balance in a rock unit may be obtained by combining data from whole-rock γ-ray spectrometric analyses with data from fission-track measurements on populations of separated mineral grains. The intensity of fluoritization generally controls whole-rock radioactivity.

As with almost all geochemical techniques, the fission-track method is most valuable when its results are combined with chemical data obtained by other methods, e.g. electron-microprobe examinations that disclose the distributions of rare-earth elements, Zr and Ti in accessory minerals and reveal the chemical nature of uranium-rich pigmentary material.

7.11. NEW GEOCHEMICAL EXPLORATION METHOD FOR BLIND ORE DEPOSITS

Methods of prospecting for deposits (including weakly eroded ones) by their primary geochemical haloes developed by the authors have been highly appraised by many experts. The high efficiency of these methods has proved very success-ful, resulting in the discovery of many blind ore bodies and deposits both in the Ukraine and Canada.

But in cases when the ore body (including blind ones) is overlapped by younger rocks (buried and blind-buried ore bodies) the primary haloes cannot be used as prospecting indicators, since in the case of buried and blind-buried deposits their primary haloes are buried as well. The new atmoelectrogeochemical method can be used in such cases.

The atmoelectrogeochemical method is based on the phenomenon discovered by S. Grigorian of anomalies of chemical elements (typical for buried deposits) developed within the periterrestrial atmosphere above deeply buried deposits [37]. The atmoelectrogeochemical anomalies were detected by means of special ionic receivers.

It has been established, that the method of primary haloes is not effective in prospecting for blind ore bodies. The investigations of recent years have shown that the mechanism of mineralization unambiguously defines the characteristic features of primary geochemical haloes, which in F-deposits are characterized by an almost complete absence of supra-ore haloes. This circumstance restricts the possibilities of the application of primary haloes in the exploration of blind deposits. This is the reason why the problem of the development of efficient methods for the exploration of blind deposits remains an urgent priority. One of the most effective ways of solving this problem seems to be the application of the atmoelectrogeochemical method.

Deposits which are difficult to discover include: weakly eroded and blind de-posits; those overlapped by younger sediments; and deposits of new, previously unknown genetic types (non-traditional).

Some deposits are emplaced in landscapes which restrict the application of effective prospecting methods.

The main way to increase the effectiveness of prospecting for the above-mentioned deposits is to extend our knowledge of the ore-forming process and the regularity in the distribution of the deposits. A systems approach to the study of the deposits provides the opportunity to use various geological data and reach a high level of reliability.

Ore-bearing systems are complex abstractions approximating reality. They can be used to arrange and coordinate various geological data in order to localize ore bodies and estimate their possible quantitative and qualitative indices.

The areas of the generation, transportation and localization of ore material are distinguished in ore-bearing systems. The factors preserving the ore bodies from destruction by erosion make up a specific information block.

The generation area characterizes the geological process, the development of which is accompanied by the mobilization of ore material. Usually this is the area of magma formation, metamorphism and sedimentation.

The transportation area represents a part of geological space between the areas of generation and localization (ore deposition area). The degree of discreteness divides ore-bearing systems into 3 groups.

The localization area is the ore body itself with its geochemical, physical and mineralogical aureole.

The prospecting strategy for difficult-to-discovered deposits based on the idea of ore-bearing systems is illustrated by P-deposits.

Appendix

GLOSSARY [2]

Alluvium: deposited river sediment.

Anatexis: the partial or complete melting of crystal rocks in response to rising temperatures, with or without mobilization.

Anomaly: a deviation from the normal chemical pattern or background.

Background: the normal range of concentrations for an element, or elements, in an area.

Biogeochemical province: a region characterized by specific flora and fauna which have adapted to abnormalities in the geochemistry of the province.

Biogeochemical prospecting: collection of plant tissues as samples for analysis in geochemical exploration.

Cold-extractable metal (exMe): that part of total concentration of an element in a sample which can be extracted by cold aqueous reagents.

Colluvium: debris eroded from hillsides and transported by mechanical forces (i.e. landslide, and gravity slump material). It may also contain some alluvium.

Concentrator organisms: flora or fauna which selectively concentrate elements in their tissues in which this attribute is hereditary.

Diagenesis: the processes of burial and compaction whereby an unconsolidated sediment is converted into a compact rock, without chemical change.

Dispersion: see geochemical dispersion.

Epigenetic: one or more component is introduced from an external source.

Geobotanical survey: the recognition of morphological varieties of plants (indicator species, dwarf, or oddly colored forms), which only occur where a certain metal or metals are concentrated, to locate ore.

Geochemical dispersion: redistribution of the elements by physical, chemical and mechanical agencies.

Geochemical environment: the sum-total of all the physical, chemical and mechanical forces active within an area.

Geochemical landscape: the overall chemical pattern of a region including both background and anomalous areas.

Geochemical relief: variations within the geochemical landscape.

Glacial till: debris transported and deposited by glaciers.

Hydrolysis: the chemical reaction of soluble elements and rock materials with water to produce insoluble hydrous oxides and silicates (clays).

Hypogene: relates to processes and conditions within the earth, i.e. within the primary environment.

Hydrogen ion concentration (pH): a measure of the acidity of ions in solution using either a millivolt-type meter, or pH indicator papers.

Indicator element: a mobile ore mineral used to locate mineralization.

Indicator plants: species which can only exist in areas of high concentration of a specific trace element.

Individual: a single measurement (variate) from within a population (synonymous with observation).

Isograd: a contour of equal metal concentration used to depict element distribution on a geochemical map.

Mean (μ): arithmetic mean value of a population expressed as:

$$\mu = \frac{\sum X}{n},$$

where X = individual values, n = number of individuals.

Metallogenic province: a region of the earth's crust containing a characteristic mineral assemblage or assemblages.

Metamorphism: the structural and chemical alteration of solid rocks in response to changing temperatures and pressures, below the surficial zone.

Metasomatism: the simultaneous solution and replacement of minerals in a rock body by materials introduced from an external source.

Mobility: the mobility of an element in the case with which it can be moved within a specific environment.

Pathfinder: a mobile associating with the ore minerals of a polymetallic deposit and used to locate it.

Population: a set of measurements or observations made on each member of a group.

Primary environment: extends from the lowest levels of the earth where rocks can exist up to the base of the near-surface (surficial) zone affected by percolating groundwater. It is characterized by relatively high temperatures and pressures.

Probability: is the proportion (number) of times an observation with a specific value will occur when a very large number of random selections are made, and is equal to the frequency of occurrence of that value within the population.

Redox potential (Eh): the oxidation-reduction potential of ions in solution measured with a millivolt-type meter.

Resistates: minerals highly resistant to chemical weathering which are transported as elastic grains, such as sands and gravels.

Secondary environment: occupies the surficial zone of the earth's crust affected by percolating groundwater and is characterized by relatively low temperatures and pressures.

Spread or deviation: difference between the mean and each individual value (i.e. the deviation from the mean).

Standard deviation (σ): square root of the variance (σ^2) expressed as

$$\sigma = \sqrt{\frac{\sum(X - \mu)^2}{n - 1}},$$

(symbols as for mean).

Supergene: relates to processes and conditions at, or near, the surface of the earth, i.e. within the secondary environment.

Syngenetic: contemporaneous formation of all components.

Threshold: the upper limit of background values.

Total metal (Me): the total concentration of an element present in a geochemical sample.

Variance (σ^2): the mean value of the squares of the deviations of a population. It is a measure of the variation of all the observations.

Chemical elements occurring in nature

H	Hydrogen	1	Ga	Gallium	31	Eu	Europium	63
He	Helium	2	Ge	Germanium	32	Gd	Gadolinium	64
Li	Lithium	3	As	Arsenic	33	Tb	Terbium	65
Be	Beryllium	4	Se	Selenium	34	Dy	Dysprosium	66
B	Boron	5	Br	Bromine	35	Ho	Holmium	67
C	Carbon	6	Kr	Krypton	36	Er	Erbium	68
N	Nitrogen	7	Rb	Rubidium	37	Tm	Thulium	69
O	Oxygen	8	Sr	Strontium	38	Yb	Ytterbium	70
F	Fluorine	9	Y	Yttrium	39	Lu	Lutetium	71
Ne	Neon	10	Zr	Zirconium	40	Hf	Hafnium	72
Na	Sodium	11	Nb	Niobium	41	Ta	Tantalum	73
Mg	Magnesium	12	Mo	Molybdenum	42	W	Tungsten	74
Al	Aluminum	13	Ru	Ruthenium	44	Re	Rhenium	75
Si	Silicon	14	Rh	Rhodium	45	Os	Osmium	76
P	Phosphorus	15	Pb	Palladium	46	Ir	Iridium	77
S	Sulphur	16	Ag	Silver	47	Pt	Platinum	78
Cl	Chlorine	17	Cd	Cadmium	48	Au	Gold	79
Ar	Argon	18	In	Indium	49	Hg	Mercury	80
K	Potassium	19	Sn	Tin	50	Tl	Thallium	81
Ca	Calcium	20	Sb	Antimony	51	Pb	Lead	82
Sc	Scandium	21	Te	Tellurium	52	Bi	Bismuth	83
Ti	Titanium	22	I	Iodine	53	Po	Polonium	84
V	Vanadium	23	Xe	Xenon	54	Rn	Radon	86
Cr	Chromium	24	Cs	Cesium	55	Ra	Radium	88
Mn	Manganese	25	Ba	Barium	56	Ac	Actinium	89
Fe	Iron	26	La	Lanthanum	57	Th	Thorium	90
Co	Cobalt	27	Ce	Cerium	58	Pa	Protactinium	91
Ni	Nickel	28	Pr	Praseodymium	59	U	Uranium	92
Cu	Copper	29	Nd	Neodymium	60			
Zn	Zinc	30	Sm	Samarium	62			

Comparison of units of concentration

Powers of 10	Percentage (%)	Parts per million (ppm)	Parts per billion (ppb)
10^2	100	1 000 000	1 000 000 000
10^1	10	100 000	100 000 000
10^0	1	10 000	10 000 000
10^{-1}	0.1	1 000	1 000 000
10^{-2}	0.01	100	100 000
10^{-3}	0.001	10	10 000
10^{-4}	0.0001	1	1 000
10^{-5}	0.00001	0.1	100
10^{-6}	0.000001	0.01	10
10^{-7}	0.0000001	0.001	1

Parts per million (ppm) = grams/1 000 000 (10^6) grams (= 1 metric ton).
Parts per billion (ppb) = grams/1 000 000 000 (10^9) grams
Ounces (Troy) per ton = 1 oz/ton = 34 ppm.
Penny weights per ton = 1 dwt/ton = 1.7 ppm.
Micrograms per liter = 1 γ/liter = 0.001 ppm or 1 ppb.

Glossary	Analytical methods
AAS	Atomic absorption spectrometry
ANOVA	Analysis of variance
GFAAS	Graphite furnace AAS
GRM	Geological reference material
HG	Hydrine generation
ICP	Inductively coupled plasma
ICP-AES	Inductively coupled plasma atomic emission spectrometry
ICP-MS	Inductively coupled plasma mass spectrometry
INAA	Instrumental neutron activation analysis
IP	Ionisation potential
PGE	Platinum group element
QT	Quartz tube
REE	Rare earth element
RNAA	Radiochemical neutron activation analysis
RSD	Relative standard deviation
SD	Standard deviation
SRM	Standard reference material
TDS	Total dissolved solids (or salts)
XRF	X-ray fluoresence

References

1. *Airborne Gamma-Ray Spectrometric Method of Prospecting for Ore Deposits*. Nedra, Leningrad (1977).

2. Andrews-Jones, D. A. The application of geochemical techniques to mineral exploration. *Geol. Surv. Bull.*, **1252** (4), 29 (1968).

3. Andrews, R. W. *Wollastonite*. Monograph. Her Majesty's Stationery Office, Great Britain (1970).

4. Anon. *A World Survey of Phosphate Deposits*. 3rd edn, British Sulphur Corp., London (1971).

5. Baburin, L. M., Dyomin, B. G., Levitsky, V. V. and Khrenov, P. M. The estimation of endogenic metallization by means of primary lithogeochemical and gas haloes. In: *The Second Int. Symp. Methods of Applied Geoch. Irkutsk*, Vol. I, pp. 61–62 (1981).

6. Bakhtiyarova, Z. V. and Lashnev, I. M. A technique for comparing peridotites of certain intrusions in connection with determining commercial content of asbestos. *Trans. Tyumen' Indust. Inst.*, **11**, 135–146 (1971).

7. Ball, T. K., Booth, S. J., Nickless, E. F. P. and Smith, R. T. Geochemical prospecting for baryte and celestine using a portable radioisotope fluorescence analyser. *J. Geoch. Explor.*, **11** (3), 277–284 (1979).

8. Barringer, A. R. The application of atmospheric particulate geochemistry in mineral exploration. *Geol. Survey of Canada, Economic Geology Report*, **31**, 363–364 (1979).

9. Beck, L. S. and Gingrich, J. E. Track etch orientation survey in the Cluff Lake area of northern Saskatchewan. *Canad. Min. Metall. Bull.*, **69**, 104–109 (1976).

10. Beiseyev, O. B. *Geological and physico-chemical conditions of amphibole asbestos deposit formation*. Abstracts presented for 27th IGC, Vol. VII, pp. 238–239 (1984).

11. Beus, A. A. and Grigorian, S. V. *Geochemical Exploration Methods of Mineral Deposits*. Applied Publishing, Wilmette, IL, USA (1975).

12. Borodin, L. S., Lapin, A. V. and Pyatenko, I. K. *Petrology and Geochemistry of Dikes in Alkaline-Ultrabasic Rocks and Kimberlites*. Nauka, Moscow (1976).

13. Boyle, R. M. Geochemistry overview. *Geol. Survey of Canada, Economic Geology Report*, **31**, 25–31 (1979).

14. Bradshaw, P. M. D. Conceptual models in exploration geochemistry. *J. Geoch. Explor.*, **4** (1), 207 (1975).

15. Bugrov, V. A. Choice of sampling fractions in the solution of prospecting tasks using secondary aureoles and dispersion trains. In: *Int. Symp. of Geoch. Prosp. Prague*, pp. 113–114 (1990).

16. Burenkov, E. K. and Zorin, A. M. Primary haloes in exploration of sedimentary deposits. In: *The Second Int. Symp. Methods of Applied Geoch. Irkutsk*, Vol. II, pp. 190–191 (1981).

17. Chao, T. T. and Sanzolone, R. F. Decomposition techniques. *J. Geoch. Explor.*, **44**, 65–105 (1992).

18. Chernyshov, A. V. Datolite is a perspective mineral for boron. Abstracts presented for 27th IGC, Vol. VII, pp. 253–254 (1984).

19. Cox, D. P. and Singer, D. A. Mineral deposit models. *U.S. Geol. Surv. Bull.*, **380**, 7–17 (1963).

20. Dawson, S. B. and Stephens, W. E. Statistical classification of garnets from kimberlite and associated xenoliths. *J. of Geol.*, **83**, 589–607 (1975).

21. Dombrowski, T. and Marray, H. H. Thorium — a key element in differentiating Cretaceous and Tertiary kaolins in Georgia and South Carolina. Abstract presented for 27th IGC, Vol. VII, p. 255 (1984).

22. Drozdov, V. P., Komov, I. L. and Vorobyev, E. I. *Methods of Prospecting and Estimation of Deposits of Precious and Piezoraw Materials.* Nedra, Moscow (1986).

23. Dudkin, O. B. *Geochemistry and Regularities of Phosphorus Concentration in the Alkaline Massifs of the Kola Peninsula.* Nauka, Leningrad (1977).

24. Dyck, W. Helium methods of prospecting in Canada. In: *The Second Int. Symp. Methods of Applied Geoch., Irkutsk*, Vol. I, pp. 53–54 (1981).

25. Egin, V. I., Belousov, V. M., Boyarko, G. Y. and Leonov, A. O. Geochemical search of apatite ore within the Aldan Province. In: *The Second Int. Symp. Methods of Applied Geoch., Irkutsk*, Vol. II, pp. 50–51 (1981).

26. Faizullin, R. M., Kozlov, Ye. N. and Ablyamitov, P. O. Lithogeochemical survey along secondary dispersion haloes when prospecting for apatite-carbonatite ores. *Razved. Okh. Nedr.*, **6**, 22–25 (1977).

27. Fersman, A. E. *Geochemical and Mineralogical Methods of Exploration for Raw Materials.* USSR Academy of Sciences Publishing House, Moscow–Leningrad (1940).

28. Filimonova, L. G. Potentiality of geochemical prospecting for phlogopite deposits. *Geol. Rudnykh Mestorozhd.*, **6**, 116–122 (1970).

29. Fridman, A. I. In: *Natural Gases of Ore Deposits,* pp. 173–181. Nedra, Moscow (1970).

30. Frolov, A. A. Formation of complex deposits in carbonatites. In: *The Principles of Prediction and Evaluation of Mineral Deposits,* pp. 43–83, Moscow (1977).

31. Gamage, S. J. K., Rupasinghe, M. S. and Dissanayake, C. B. Application of Rb–Sr ratios to gem exploration in the granulite belt of Sri Lanka. *J. Geoch. Explor.*, **43** (3), 281–293 (1992).

32. Gapontsev, G. D., Nesterova, A. A. and Sarapulova, V. N. Primary dispersion haloes of the Kalangul fluorite deposit (Eastern Transbaikal Region). In: *Dispersion Haloes of Deposits in Eastern Siberia*, Vol. 6, pp. 75–82, Moscow (1971).

33. *Geology of Mica-bearing Areas and Mica Deposits in the Karel Kolsky Region and their Prospecting.* Karelian ASSR Publishers, Petrozavodsk (1975).

34. Gerasimovsky, V. I. Modern problems of alkaline rock mineralogy. *Zapiski All-Union Mineralog. Soc.*, **106**, 30–33 (1977).

35. Goetz, A. F. H. Imaging spectrometry for Earth observations. *Episodes*, **15** (1), 7–15 (1992).

36. Gordienko, V. V., Kryvovichev, V. G. and Syritso, L. F. *Metasomatites of Pegmatite Fields.* Publishing House of Leningrad Univ., Leningrad (1987).

37. Grigoryan, S. V. and Ovchinnikov, L. N. The common geochemical zonality of primary haloes of sulphide-bearing hydrothermal deposits and its practical significance. In: *The Second Int. Symp. Methods of Applied Geoch., Irkutsk*, Vol. I, pp. 28–29 (1981).

38. Gurney, J. J. and Moore, R. O. *Kimberlite, garnet, chromite and ilmenite compositions applications to exploration.* ICAM'91, CSIR, Pretoria, 2–4 September, Papers, Vol. 1, p. 21 (1991).

39. Hall, G. E. M. Inductively coupled plasma mass spectrometry in geoanalysis. *J. Geoch. Explor.*, **44**, 201–249 (1992).

40. Hawkes, H. E. and Webb, J. S. *Geochemistry in Mineral Exploration.* Harper and Row, New York, NY (1962).

41. Henley, K. J. Ore-dressing mineralogy — a review of techniques applications and recent developments. *Spec. Publ. Geol. S. Afr.*, **7**, 175–200 (1983).

42. Hishida, H., Minami, H. and Tsujimoto, T. Application of resistivity tomography to mineral exploration. In: *29th Int. Geol. Congress, Kyoto*, Abstracts 3 of 3, pp. 768–769 (1992).

43. Hodgson, A. A. *Fibrous Silicates.* Lecture series, No. 4. Royal Inst. of Chemistry, London (1965).

44. Hoffman, E. L. Instrumental neutron activation in geoanalysis. *J. Geoch. Explor.*, **44**, 297–319 (1992).

45. Ilupin, I. P., Sobolev, S. F., Zolotarev, B. P. and Lebedev-Zinoviev, A. A. Geochemical specialization of kimberlites in the different fields of Yakutia. *Geokhimia*, **4**, 499–513 (1974).

46. *Instruction on the Geochemical Methods of Prospecting for Ore Deposits.* Nedra, Moscow (1965).

47. Ivanov, V. G. Certain geochemical specific features of rock formation in lazurite deposits of Southern Baikal region. *Geokhimia*, **1**, 47–54 (1976).

48. Jarvis, I. and Jarvis, K. E. Inductively coupled plasma-atomic emission spectrometry in exploration geochemistry. *J. Geoch. Explor.*, **44**, 139–200 (1992).

49. Jiri, S. *Geochemistry of bentonites from Bohemia.* Abstract presented for 27th IGC, Vol. VII, pp. 306–307 (1984).

50. Jones, M. P. The characterization of ores and mineral products by automatic image analysis of mineralogical features. *Spec. Publ. Geol. Soc. S. Afr.*, **7**, 475–478 (1983).

51. Kabanova, Ye. S. and Plotnikova, L. Ya. Geochemistry of admixture elements in phosphorites. In: *Geochemistry, Mineralogy, Petrography*, Vol. 7, pp. 143–191 Moscow (1973).

52. Kahn, H. and Sant' Agostino, K. *The influence of mineral characteristics and distribution of mineral processing of the Anitapolis, Brazil, eluvial phosphate ore.* ICAM'91, CSIR, Pretoria, 2–4 September, Papers, Vol. 1, p. 27 (1991).

53. Kalinkin, M. M. and Anzel, V. V. Apatite-bearing Carbonatites in the Kola Peninsula. *Razved. Okh. Nedr.*, **6**, 13–15 (1977).

54. Kaminsky, F. V. and Potapov, S. V. The content of trace elements in kimberlites of the Ingeliysk region and their dispersion haloes. *Izv. Higher Educ. Establ. Geol. Razved.*, **6**, 38–40 (1967).

55. Kane, J. S. Reference samples for use in analytical geochemistry: their availability, preparation and appropriate use. *J. Geoch. Explor.*, **44**, 37–63 (1992).

56. Kanevsky, A. Ya. Petrochemical parameter M'/S as a criterion for subdividing serpentinites in the Bug river middle region according to bedrock nomenclature. *Geokhimia*, **8**, 995–998 (1968).

57. Kashaev, N. I. Geochemical indications of the emerald content in biotites. *Razved. Okh. Nedr.*, **3**, 24–29 (1973).

58. Kharkiv, A. D. *Mineralogical Fundamentals of Prospecting for Diamond Deposits.* Nedra, Moscow (1978).

59. Klyuev, O. S. Methodical peculiarites of prospecting of rare-metal deposits by primary haloes. In: *The Second Int. Symp. Methods of Applied Geoch., Irkutsk*, Vol. I, 208–209 (1981).

60. Komarov, P. V. Fluorometric method of geochemical prospecting. In: *Application of Geochemical Method in the Search and Prospecting for Ore Deposits*, pp. 5–42, Moscow (1975).

61. Komov, I. L. *Irradiation in Mineralogy.* Moscow (1982).

62. Kosolapova, M. N. and Kosolapov, A. I. Application of chemical methods in the search for kimberlite bodies. *Geol. Geofiz.*, **2**, 95–100 (1962).

63. Kovalevski, A. L. Mapping of volatiles and metals in rocks by plant. In: *29th Int. Geol. Congress, Kyoto*, Abstracts 3 of 3, p. 772 (1992).

64. Kresten, P., Fels, P. and Berggren, G. Kimberlitic zircons — a possible aid in prospecting for kimberlites. *Miner. Deposits*, **10**, 47–56 (1975).

65. Kukharenko, A. A., Il'insky, G. A., Ivanova, T. N., Galahov, A. A., Kozireva, L. V., Gelman, E. M., Borneman-Starinkevish, I. D., Stolarova, I. N., Skrijinskay, V. I. and Ridzova, R. I. Clarkes of the Khibini alkaline massif. *Zapiski All-Union Mineralog. Soc.*, **97**, 133–149 (1968).

66. Kuzvart, M. *Prospective, potential and substitute non-metallic mineral raw materials.* Abstract presented for 27th IGC, Vol. VII, pp. 276–277 (1984).

67. Laubenbakh, A. I., Levina, S. D., Skosyreva, L. N., Slavjagina, I. I., Stepanova, A. I. and Zitovish, B. V. Regularities of distribution of radioactive elements in phosphorites and their use in search for phosphatic raw materials in Siberia. *Trans. Inst. Geol. Geophys. Sib. Branch USSR Acad. Sci.*, **386**, 68–72 (1975).

68. Liddy, J. C. The role of geochemistry in mineral exploration. *Austral. Mining*, **65** (3), 60–65 (1973).

69. Lisitsyn, A. Ye. *Geological Fundamentals of Prospecting for Endogenous Boron Deposits.* Nedra, Moscow (1974).

70. Litsarev, M. A. *Physico-chemical environment of Precambrian phlogopite-bearing scarn formation.* Abstract presented for 27th IGC, Vol. VII, pp. 279–280 (1984).

71. Lobzova, R. V. and Litsarev, M. A. *Role of hydrothermal processes in graphite deposits forming.* Abstract presented for 27th IGC, Vol. VII, pp. 283–284 (1984).

72. Lukashev, V. K. and Lukashev, K. I. Some new ways of application of hydrogeochemical prospecting methods. In: *The Second Int. Symp. Methods of Applied Geoch., Irkutsk*, Vol. II, pp. 76–78 (1981).

73. Mamchur, G. I. Isotopic composition of carbon in vein graphites. *Dokl. Ukr. SSR Acad. Sci.*, **2**, 115–117 (1975).

74. Markov, K. A., Mikhailov, B. M., Predtechensky, N. N., Denisenko, V. K., Markova, K. A. and Rizk, J. E. *The Criteria for Prognostic Evaluation of a Territory for Useful Minerals.* Nedra, Leningrad (1978).

75. Matukhina, V. G. and Sukhoverkhova, M. V. Remarks on the use of metallometric sampling in the search for secondary phosphorites in the Altai–Sayany folded area. *Trans. Siber. Res. Inst. Geol., Geoph. and Miner. Raw Mater.*, **197**, 71–76 (1975).

76. Meyer, O. A. Genesis of diamond: a mantle saga. *Amer. Miner.*, **70** (3–4), 344–355 (1985).

77. Meyer, H. O. A. Mineralogy of the upper mantle: a review of the minerals in mantle xenoliths from kimberlite. *Earth-Science Reviews*, **13**, 251–281 (1977).

78. Moroz, I. I. and Lobanov, V. K. In: *Geochemical Zonality Primary Haloes of Emerald-bearing Biotitic Complexes. Gems and Precious Stones.* Moscow (1980).

79. *Muskovitic Pegmatites in the USSR.* Nedra, Leningrad (1975).

80. Myers, J. B. In: *Vermiculite. Industrial Minerals and Rocks,* 3rd edn, pp. 889–895. J. L. Gillson (Ed.). AIME, New York (1960).

81. Nasedkin, V. V. *The water-bearing glass: properties and deposits.* Abstract presented for 27th IGC, Vol. VII, p. 286 (1984).

82. Naumov, G. B. Areals of fluid inclusions in geochemical prospecting. In: *The Second Int. Symp. Methods of Applied Geoch., Irkutsk*, Vol. I, pp. 34–35 (1981).

83. Nesmelova, Z. N. and Travnikova, L. G. Radiogenic gases of old saline deposits. *Geokhimia*, **7**, 716–722 (1973).

84. Nichol, I., Garrett, R. G. and Webb, J. S. Automatic data plotting and mathematical and statistical interpretation of geochemical data. Paper. *Geol. Survey of Canada*, pp. 195–210 (1966).

85. Notholt, A. J. G. Phosphate exploration techniques. *Miner. Res. Devel.*, **32**, 214–228 (1968).

86. Oganesian, L. V. and Yazova, R. V. Procedure and result of exploring endogeneous dispersion haloes of hydrothermal crystal-bearing veins in Eastern Siberia. In: *Prediction, Search and Prospecting for Rock Crystal Deposits*, pp. 57–66, Moscow (1975).

87. Ozol, A. A. and Kovyazin, A. N. Volcanogenic-sedimentary borates in the Pamirs. *Sov. Geol.*, **12**, 142–143 (1975).

88. Parry, S. J. The role of neutron activation with radiochemistry in geoanalysis. *J. Geoch. Explor.*, **44**, 321–349 (1992).

89. Pavlenko, Yu. V. *Geology of fluorite deposits of the Solonechnaya group and directions of prospecting surveys*. Author's abstract, Cand. thesis. Irkutsk (1975).

90. Petkof, B. In: *Mica Minerals Yearbook*, pp. 783–791. U.S. Bureau of Mines (1973).

91. Petrichenko, O. I., Shaidetskaja, V. S. and Kovalevich, V. S. New geochemical criterion of prospecting for potash salts. *Doklady Ukrain. SSR. Sci.*, **6**, 512–515 (1975).

92. Petrova, N. S. Experimental study of rubidium distribution between crystals of carnallite and solution in the system NaCl–KCl–MgCl–H$_2$O at 25 °C. *Geokhimia*, **6**, 919–924 (1973).

93. Petrov, V. P. New kinds and new ways of utilization of non-metalliferous commercial minerals. In: *New Kinds of Non-metalliferous Commercial Minerals*, pp. 5–23, Moscow (1975).

94. Potts, P. J. and Webb, P. C. X-ray fluorescence spectrometry. *J. Geoch. Explor.*, **44**, 251–296 (1992).

95. Pridmore, D. F. Geological mapping using air-borne geophysics — new frontiers. In: *29th Int. Geol. Congress, Kyoto*, Abstracts 3 of 3, p. 777 (1992).

96. Romanovich, I. F. (Ed.) *Deposits of Anthophyllite-asbestos in the USSR*. Nedra, Moscow (1976).

97. Rudnik, V. A. and Velikoslavinsky, S. D. Prospecting and forecasting of deposits within the areas of Precambrian consolidation by geochemical data. In: *The Second Int. Symp. Methods of Applied Geoch.*, Irkutsk, Vol. II, pp. 53–54 (1981).

98. Saet, Y. E., Astrakhan, E. D. and Kargapolov, N. V. The epigenetic secondary dispersion haloes over deposits in Precambrian sequences. In: *The Second Int. Symp. Methods of Applied Geoch.*, Irkutsk, Vol. II, p. 31 (1981).

99. Saet, Yu. V., Igumnov, N. Ya. and Nesvizhskaya, N. I. *Geochemical Prospecting for Endogenous Boron Deposits*. Moscow, Nauka Publishers (1973).

100. Sarasadskikh, N. N. and Bagul'kina, V. A. Petrographic and petrogenetic differences of kimberlites from rocks similar by some indications. *Zapiski All-Union Mineralog. Soc.*, **98**, 415–421 (1969).

101. Schmetzer, K. and Bank, H. The colour of natural corundum. *Neues Jb. Miner. Mh.*, **2**, 59–68 (1981).

102. Seletsky, Yu. B. and Nikolayeva, K. N. Fluorine in natural waters in the region of the Khibini deposits. *Trans. of GIGKHS*, **10**, 305–308 (1965).

103. Shao Yue, and Liu Jimin A geochemical method for the exploration of kimberlite. *J. Geoch. Explor.*, **33**, 185–194 (1989).

104. Shmakin, B. M., Zagorsky, V. Ye. and Makagon, V. M. *Geochemical indicators from the exploration and assessment of granitic pegmatites*. ICAM'91, CSIR, Pretoria, 2–4 September, Papers, Vol. 2, p. 49 (1992).

105. Sinha, R. C. and Singh, R. C. Chemical evolution of mica-pegmatites of the Gaya-Hazaribag belt, Bihar (India). In: *The Second Int. Symp. Methods of Applied Geoch.*, Irkutsk, Vol. I, pp. 50–51 (1981).

106. Singhroy, V. H. Radar geology: techniques and results. *Episodes*, **15** (1), 15–21 (1992).

107. Smolin, P. P. *Brucite — the new high-magnesian industrial mineral*. Abstract presented for 27th IGC, Vol. VII, pp. 308–309 (1984).

108. Sobolev, N. D. On petrochemistry of ultrabasic rocks. *Geokhimia*, **8**, 679–695 (1959).

109. Sobolev, H. V. Deep inclusions in kimberlites and upper mantle. In: *Problems of the Earth's Crust and Upper Mantle Petrology*, pp. 103–111, Novosibirsk (1976).

110. Sochevanov, N. N. New methods of calculating and plotting the results of exploration of multicomponent geochemical haloes. In: *Geochemical Prospecting for Ore Deposits*, pp. 218–222, Moscow (1972).

111. Steinberg, D. S. and Chashchukhin, I. S. *Serpentinization of Ultrabasites*. Nauka, Moscow (1977).

112. Summerhayes, C. P., Hazelhoff-Roelfzema, B. H., Tooms, J. S. and Smith, D. B. Phosphorite prospecting using a submersible scintillation counter. *Econom. Geol.*, **65**, 718–723 (1970).

113. Suturin, A. N. and Zameletdinov, R. S. Geochemical aspects of nephrite prospecting in hyperbasite massifs. In: *The Second Int. Symp. Methods of Applied Geoch., Irkutsk*, Vol. I, pp. 197–198 (1981).

114. Sviridenko, A. F. and Smirnov A. A. Conditions of formation of rock-forming and jewelry jadeites. In: *Geology, Methods of Searching. Prospecting and Evaluating Deposits of Jewelry, Semiprecious and Decorative-facing Stones* (Abstracts of Papers of Seminar), pp. 22–24 Moscow, (1975).

115. Tatarinov, A. V. and Shmakin, B. M. Parameters of distribution of rare alkalis in potash feldspar of pegmatites of different origin (on the example of the Borshchevochny ridge). *Geokhimia*, **3**, 401–409 (1977).

116. Timchenko, V. A., Yagnyshev, B. S. and Strel'tsov, V. L. Potentiality of lithochemical sampling along secondary dispersion haloes when prospecting for kimberlite pipes. In: *News of Geology in Yakutia*, Issue 3, pp. 135–138 (1973).

117. Tokmakov, P. P. *Industrial Vermiculites in the USSR.* Abstracts presented for 27th IGC, Vol. VII, pp. 316–317 (1984).

118. Tole, M. P. Correlation between chemically and physically transported barium anomalies over the Vitengeni lead/barite mineral deposit: Kenya. In: *Int. Symp. of Geoch. Prosp., Prague*, p. 215 (1990).

119. Trueman, D. L. Exploration methods in the Tanco mine area of south-eastern Manitoba, Canada. *Energy*, **3**, 293–297 (1977).

120. Valls, R. A. and Nunez, F. Geochemical methods for talc prospection in the Escambray metamorphic massif. *Rev. Techn.*, **19** (1), 9–15 (1989).

121. Valyashko, M. G., Zherebtsova, I. K. and Sadykov, L. Z. *Geochemical Methods of Potash Salt Prospecting.* Moscow UN Press, Moscow (1966).

122. Varma, O. P. *Structural history and formational conditions of wollastonite deposits at Belka, Sirohi District, Rajasthan, India.* Abstract presented for 27th IGC, Vol. III, pp. 320–321 (1984)

123. Vasil'ev, Ye. N. Lithogeochemical prospecting for concealed beds of phosphorites. *Razved. Okh. Nedr.*, **2**, 22–25 (1978).

124. Viets, J. G. and O'Leary, R. M. The role of atomic absorption spectrometry in geochemical exploration. *J. Geoch. Explor.*, **44**, 107–138 (1982).

125. Vishnevsky, P. V., Pinagina, N. M., Zverev, N. N. and Gerasimova, V. V. *Search and Prospecting for Mining and Chemical Raw Materials by Geophysical Methods.* VIEMS Publishers, Moscow (1975).

126. Vorobyev, E. I. Methods of search for useful minerals by measurements on calcite. In: *The Second Int. Symp. Methods of Applied Geoch., Irkutsk*, Vol. I, pp. 226–227 (1981).

127. Xiarpei, C. Geochemistry of barite deposits in China. Abstracts presented for 27th IGC, Vol. VII, pp. 251–252 (1984).

128. Zarevish, I. P. and Kurman, I. L. *Geochemical Fundamentals of Prospecting for Boron Deposits.* Nedra, Moscow (1989).

129. Zhernakov, V. I. Geochemistry of chromium in emerald-bearing biotitic complexes. *Trans. Ural. Polytech. Inst.*, **131**, 134–140 (1976).

130. Znamensky, V. S. Genesis of native sulphur and related minerals in modern hydrothermal process. Abstract presented for 27th IGC, Vol. VII, pp. 335–336 (1984).

131. Boyd, F. R. Olivine megacrysts from the kimberlites of the Monastery and Frank Smith Mines, South Africa. *Carnegie Inst. Wash., Annu. Rep. Dir. Geophys. Lab., Yearb.*, **73**, 282–285 (1974).

132. Boyd, F. R. Ultramafic nodules from the Frank Smith kimberlite pipe, South Africa. *Carnegie Inst. Wash., Annu. Rep. Dir. Geophys. Lab., Yearb.*, **73**, 285–294 (1974).

133. Boyd, F. R. and Finger, L. W. Homogeneity of minerals in mantle rocks from Lesotho. *Carnegie Inst. Wash., Annu. Rep. Dir. Geophys. Lab., Yearb.*, **74**, 519–525 (1975).

134. Boyd, F. R. and Nixon, P. H. Origins of the ultramafic nodules from some kimberlites of Northern Lesotho and the Monastery Mine, South Africa. *Phys. Chem. Earth*, **9**, 431–454 (1975).

135. Danchin, R. V. and Boyd, F. R. Ultramafic nodules from the Premier Kimberlite pipe, South Africa. *Carnegie Inst. Wash., Annu. Rep. Dir. Geophys. Lab., Yearb.*, **75**, 531–538 (1976).

136. Ginzburg, I. I. *Principles of Geochemical Prospecting, Techniques of Prospecting for Nonferrous Ores and Rare Metals*. Pergamon Press (1960).

137. Ginzburg, A. I. Features of the concentration and dispersion of rare elements during endogene processes. In: *Chemistry of the Earth's Crust, Vol. 2; Israel Prog. Sci. Transl.*, pp. 202–210 (1967).

138. Hawkes, H. E. and Webb, J. S. *Geochemistry in Mineral Exploration*. Harper and Row (1962).

139. Hearn, B. C. and Boyd, F. R. Garnet peridotite xenoliths in a Montana, U.S.A., kimberlite. *Phys. Chem. Earth*, **9**, 247–256 (1975).

140. Jaques, A. L., O'Neill, H. St. C., Smith, C. B., Moon, J. and Chappell, B. W. Diamondiferous peridotite xenoliths from the Argyle (AK1) lamproite pipe, Western Australia. *Contributions to Mineralogy and Petrology*, **104**, 255–276 (1990).

141. Krauskopf, K. B. Sedimentary deposits of rare metals. In: *Econ. Geology 50th Ann. Vol.*, pp. 411–463 (1955).

142. Krauskopf, K. B. *Introduction to Geochemistry*. Mc-Graw-Hill (1967).

143. Lappin, M. A. and Dawson, J. B. Two Roberts Victor cumulate eclogites and their re-equilibration. *Phys. Chem. Earth*, **9**, 351–365 (1975).

144. Lucas, H., Muggeridge, M. T. and McConchie, D. M. Iron in kimberlitic ilmenites and chromian spinels: a survey of analytical techniques. In: *Kimberlites and Related Rocks, Vol. 1. Their Composition, Occurrence, Origin and Emplacement*, J. Ross (Ed.), pp. 311–312. Geological Society of Australia Special Publication No. 14, Blackwell, Melbourne (1989).

145. Lucas, H., Ramsay, R. R., Hall, A. E., Smith, C. B. and Sobolev, N. V. Garnets from Western Australia kimberlites and related rocks. In: *Kimberlites and Related Rocks, Vol. 2. Their Mantle-Crust Setting, Diamonds and Diamond Exploration*, J. Ross (Ed.), pp. 809–819. Geological Society of Australia Special Publication No. 14, Blackwell, Melbourne (1989).

146. Mason, B. *Principles of Geochemistry*. 2d edn. John Wiley and Sons, Inc. (1958).

147. Meyer, H. O. A. and Boyd, F. R. Composition and origin of crystalline inclusions in natural diamonds. *Geochim. Cosmochim. Acta*, **36**, 1255–1273 (1972).

148. Meyer, H. O. A. and Brookins, D. G. Eclogite xenoliths from Stockdale kimberlite. *Kansas. Contrib. Mineral. Petrol.*, **34**, 60–72 (1971).

149. Meyer, H. O. A. and Svisero, D. P. Mineral inclusions in Brazilian diamonds. *Phys. Chem. Earth*, **9**, 785–795 (1975).

150. Meyer, H. O. A. and Tsai, H. M. Mineral inclusions in natural diamond – their nature and significance: A review. *Miner. Sci. Eng.*, **8**, 242–261 (1976).

151. Meyer, H. O. A. and Tsai, H. M. Mineral inclusions in diamond: Temperature and pressure of equilibration. *Science*, **191**, 849–851 (1976).

152. Mitchell, R. H. Ultramafic xenoliths from the Elwin Bay kimberlite, the first Canadian paleo-geotherm. *Can. J. Earth. Sci.*, **14**, 1202–1210 (1977).

153. Nixon, P. H. and Boyd, F. R. Petrogenesis of the granular and sheared ultrabasic nodule suite in kimberlites. In: *Lesotho Kimberlites*, P. H. Nixon (Ed.), pp. 48–56. Lesotho Natl. Dev. Corp., Maseru (1973).

154. Nixon, P. H. and Boyd, F. R. The discrete nodule association in kimberlites from northern Lesotho. In: *Lesotho Kimberlites*, P. H. Nixon (Ed.), pp. 67–75. Lesotho Natl. Dev. Corp., Maseru (1973).

155. Moore, R. O. and Gurney, J. J. The development of advanced technology to distinguish between diamondiferous and barren diatremes. *Geol. Survey of Canada Open File Report*, **2124**, part i. 1–90 (1989).

156. Shacklette, H. T. Phytoecology of a greenstone habitat at Eagle, Alaska. *U.S. Geol. Survey Bull.*, 1198F (1966).

157. Smith, Chris. B., Lucas, H., Hall, A. E. and Ramsay, R. R. Diamond prospectivity and indicator mineral chemistry: a Western Australian perspective. In: *Proceedings of Fifth International Kimberlite Conference, Araxa, Brazil*, Extended abstracts, CPRM – Special Publication; 2/91, p. 584 (1991).

158. Sobolev, N. V., Botkunov, A. I., Laurent'yev, Yu. G. and Pospelova, L. N. Peculiarities of the composition of minerals coexisting with diamond from Mir pipe, Yakutia. *Zapiski All-Union Mineralog. Soc.*, **100**, 558–564 (1971).

159. Sobolev, N. V., Gnevushev, M. A., Mikhailovskaya, L. N., Futergendler, S. I., Shemania, E. I., Laurent'yev, Yu. G. and Pospelova, L. N. The composition of garnet and pyroxene inclusions from the diamonds of the Urals. *Dokl. Akad. Nauk SSSR*, **198**, 190–193 (1971).

160. Tsai, H. M. and Meyer, H. O. A. Coexisting Cr-rich and Cr-poor garnet in an enstatite xenocryst from Frank Smith kimberlite, South Africa. *Abstr. Progr., Geol. Soc. Am.*, **8**, 1146–1147 (1976).